D0794334

Pitt Series in Policy and Institutional Studies

Pesticides and Politics

The Life Cycle of a Public Issue

Christopher J. Bosso

University of Pittsburgh Press

Published by the University of Pittsburgh Press, Pittsburgh, Pa. 15260

Feffer and Simons, Inc., London
Manufactured in the United States of America

Library of Congress Cataloging-in-Publication Data

Bosso, Christopher John, 1956–
Pesticides and politics.

(Pitt series in policy and institutional studies)
Includes index.
1. Pesticides—Law and legislation—United States—History. 2. Pesticides—Government policy—United States—History. I. Title. II. Series.
KF 3959.B67 1987 344.73'04633 86-19245
ISBN 0-8229-3547-3 347.3044633

Contents

Tables

Figures

Glossary

I attempt to keep the use of acronyms to a minimum, but one can never avoid them entirely. Pesticide types are referred to in the text by their trade names, not their formal chemical signatures.

AFBF	American Farm Bureau Federation
ARS	Agricultural Research Service (USDA)
CPR	Campaign for Pesticides Reform
CSMA	Chemical Specialties Manufacturers Association
EDF	Environmental Defense Fund
EPA	Environmental Protection Agency
FDA	Food and Drug Administration
FCDA	Food, Drug, and Cosmetics Act (1938)
FCPC	Federal Committee on Pest Control
FEPCA	Federal Environmental Pesticides Control Act (1972)
FIA	Federal Insecticide Act (1910)
FIFRA	Federal Insecticide, Fungicide, and Rodenticide Act (1947)
FPCRB	Federal Pest Control Review Board
HEW	U.S. Department of Health, Education, and Welfare
NACA	National Agricultural Chemicals Association
NAS	National Academy of Sciences
NEPA	National Environmental Protection Act (1969)
NFU	National Farmers Union
NRDC	Natural Resources Defense Council
OPP	Office of Pesticides Programs (EPA)
PSAC	President's Science Advisory Committee
RPAR	Rebuttable Presumption Against Registration
USDA	U.S. Department of Agriculture
USDI	U.S. Department of the Interior

Preface

This study is about the birth and maturation of a policy, forty years of policymaking that, I hope, tell us something about how politics works. I am less interested in judging the virtues of federal pesticides regulation than in how politics affects policy, and vice versa—though process and outcome inevitably get intertwined. My main goal is to understand something about political change.

The study of a single issue area over time is a comparative undertaking, since any longitudinal case study in fact examines many snapshots of policymaking taken at different times. These snapshots are not left separate, however, but are edited together into a broader documentary on policy transformation. Such an approach is iterative; that is, I begin with the theoretical framework with which the case was selected, and then through induction and learning, I venture some generalizations about politics.

The post–World War II evolution of federal pesticides regulation represents a major type of postwar federal policy activity. The regulatory web in place today in fact results almost entirely from policies enacted during the past two decades, so the Federal Insecticide, Fungicide, and Rodenticide Act (FIFRA) is in good company. FIFRA in fact is one of the granddaddies of federal regulation, at least in terms of sheer longevity. Why FIFRA became law in 1947, long before the regulatory wave to come, should tell us a great deal about why some problems are acted upon by government and others are not.

A policy's longevity also helps us to study political change, since FIFRA's life spans a period that witnessed some of the most dramatic and fundamental transformations in the American social and political landscape. Studying this case over time may provide more subtle insights into the oscillations of political conflict than one might harvest from more aggregate (and nonlongitudinal) studies of public opinion, participation, or interest-group activity. Such studies tell us that the participatory community in the United States has expanded dramatically over the past four decades, but may not be as useful in explaining the dynamics of political conflict. What do we learn, for example, about the tensions that arise between those seeking change and those protecting the status quo? What do such studies tell us about the methods and strategies deployed by competing policy claimants over time, and why different actors use different tactics at any one moment? Finally, what do such approaches tell us about *politics*? Aggregate studies give us the broad sweep, which is essential to a knowledge of context, but deal less with the nuts and bolts of political conflict over a long period.

Time changes everything, natural or manmade; politics and policies are no exception. We can compare any policy to an organism: it is created, it matures, and (sometimes) it dies. Young policies, not unlike young lives, differ fundamentally from the old in looks, imperatives, and impact. Organisms are dynamic, so are policies. I study FIFRA over time to understand the dynamics of policy stability and change that ebb and flow through almost four decades of the act's life.

Pesticides regulation differs from most other environmental protection activities in one important respect. The chemicals that go into pest-control products, unlike other pollutants, are created *intentionally*. FIFRA provides a rare instance in which the government authorizes the manufacture, sale, and use of products carrying potential hazards to public health and the environment. Pesticides are not "externalities" of the market system, like air and water pollution, nor are they the unfortunate by-products of urban industrial society. We intentionally apply these chemicals for important agricultural and public health reasons. Pesticides are in many respects the key to America's agricultural abundance, and have proved instrumental in eradicating such diseases as typhoid and malaria. Pesticides, as we shall see, might be characterized simply as "good things that can cause harm," or, "bad things that can do good," depending on your perspective.

To extrapolate from a single case into more general statements about politics is to assume that the case accurately represents some broader problem into which this study might make some insights. The case of pesticides has certain unique qualities, to be sure, but it is not idiosyncratic, particularly when compared with other problems endemic to contemporary society.

Some problems seem relatively simple. Building a bridge, for example, involves well-established engineering and construction principles, with few "moral" problems. Other problems exhibit general agreement on ends, particularly when they are expressed in abstract terms, but there is little agreement on techniques. A good example of such a "technically complex" problem is economic growth. Virtually everyone agrees that growth is a good thing, but far fewer agree on which economic theory or practice will best achieve desired results—leaving aside for the moment the reality that "growth" is linked to other values.

Some problems are morally complex, since the dominant focus is on some moral dilemma. Capital punishment, for example, is primarily a moral question; any argument about the most "humane" way to execute somebody is subordinate to the more compelling debate over the morality of taking life in the first place. Abortion is another moral dilemma, though it too involves subordinate scientific questions about the nature of human life. Science takes a back seat to moral, ethical, religious, and democratic values during such debates, and one wonders whether the abortion question can be "solved" through scientific means.

Finally, problems like the use of nuclear power, disposal of toxic wastes, and regulation of pesticides are both morally *and* technologically complex—what I call "intractable" problems. There is, in practical terms, no clear line here between means and ends, nor much agreement on either. With respect to pesticides, it is almost impossible to find agreement on ends because the ends themselves often are clearly incompatible: environmentalists may seek to rid the earth of pesticides at any cost; chemical firms may seek to maximize profits; farmers want inexpensive and effective pesticides to maintain high crop yields at lower costs; public health officials want to eradicate disease. Consumers, for their part, are caught in a bind; they want cheap food, which might lead them to support the wide use of pesticides, but they also fear the possibly carcinogenic effects of pesticides residues in that food or in the environment.

Value conflict is accompanied by disputes over means and methods. Whose scientific data are more "correct"? Which analytical techniques do we accept as valid? Who decides? Values intrude mightily into every facet of science and technology during such debates, muddling even more our perspectives about just what is going on. I label such problems "intractable" because they constitute some of the most perplexing, frustrating, and closure-defying questions in our society. And they are becoming more common as our scientific and technological sophistication—and risk-taking—increases. Our knowledge becomes both our boon and our bane; we don't know enough to "solve" the problem to everyone's satisfaction, but more than enough to prevent the suspension of disbelief. The age of complexity dawns, both with its promises and its pitfalls, its shining accomplishments and its dark sides. Chemical pesticides as an issue typifies the age of enlightened uncertainty.

The centrality of pesticides to contemporary life poses important questions about competing values and tradeoffs. Unlike other pollutants, pesticides present no easily caricatured villian, no smokestack or hazardous waste dump to be singled out as a public menace. With pesticides we find it incredibly difficult to pose policy positions or solutions in stark "good-versus-evil" terms, since every "good" may have an "evil" side. We find instead multiple gray areas, and this issue involves perhaps some of the toughest choices confronting society. It is one thing to identify and clean up a toxic waste dump, a far different matter to ban the use of a specific pesticide when that product may eradicate known public health hazards or may, in the eyes of its supporters, provide for agricultural plenty and cheap food. Tough choices indeed.

Acknowledgments

Thanks of the most extravagant sort go to Bert A. Rockman of the University of Pittsburgh, who pressed upon me the virtues of clear thought and concise prose. Raymond E. Owen, Susan B. Hansen, and Paul Y. Hammond, all of the University of Pittsburgh, and George Hoberg of MIT read the entire manuscript and gave me immeasurable support and critical comment. My thanks to each of them.

I am also grateful to William J. Keefe and Morris S. Ogul, of the University of Pittsburgh, who taught me about Congress and its foibles, and to Charles O. Jones, now of the University of Virginia, who instilled in me a love of that institution and of public policy.

Much of my research took place while in residence as a guest scholar at the Brookings Institution, which graciously provided a haven for work, thought, and discussion. Thanks go to Martha Derthick (now at the University of Virginia) and Diane Hodges for making my stay so enjoyable, and to several newfound contemporaries—Gary Mucciaroni (Brown University), Mary Musca (Harvard University), Margaret Nugent (College of the Holy Cross), and Robert Katzman (Brookings)—who provided both companionship and intellectual stimulation to one seeking both.

Thanks also to the kind and expert folks at the Congressional Research Service, and especially to Roger H. Davidson, who unlocked so many doors and filled in so many gaps. Congress may be a massive institution, but people like Roger, John Blodgett, Jim Aidala, and Steve Hughes successfully reduced it to human scale, and taught me much. The good people at the USDA's National Agricultural Library give civil servants everywhere a good name. To them, and to all those who consented to interviews, I offer my thanks. I hope that I have misrepresented neither their words nor their perspectives.

Finally, to Marcia A. Goetsch, who puts up with me, and to my parents, who understood early on what possessed me to follow this path. This study is for them.

Pesticides and Politics

Chapter One

Images of Policymaking

Painting Politics: Policymaking as Art Metaphor

American political science is a museum of many canvases, each displaying some artist's concept of political reality. Some represent the "elitist" school: bold splashes of color arrayed horizontally to convey strata of power. Others emerge out of the "pluralist" school, with vectors coalescing almost randomly to produce new, but always temporary arrangements. Still others are jumbles of geometric shapes, each depicting some concise and more permanent network of political interaction; these canvases find their roots in the "subgovernment" school. And so on.

The problem, as the museum-goer quickly discovers, is to make the historical and conceptual links among these competing and partially contradictory depictions of the world. Many art museums have resident historians whose job it is to research and explain how art itself evolved, and how the paintings in a particular collection fit into the broader history of art. In doing so, the historians link together these various "schools" and blend them into a more illuminative tapestry, a documentary on the ebb and flow of art.

An activity that is the norm in the art world is not so in political science. The student finds in the discipline countless discrete images of policymaking, yet relatively few "art histories." Policymaking, like art, evolves; it is in a constant state of flux. Yet our images of policymaking have not been as fluid, and herein lies the problem with many of our conceptions of American political life. We have—to change our metaphor a bit—many snapshots of policymaking at various stages, but relatively few documentaries seeking to depict how these images relate to one another, or how they coalesce. We are not talking about mere storytelling, but about political analysis rooted firmly in theory.

I seek to study political change and its impact on public policy to make some analytical and conceptual sense out of this morass of competing images. I want to know what political change and policy transformation look like, what factors promote or retard change, and, most critically, how our competing images of policymaking fare over time. I want to edit a documentary about politics.

Such goals bend my focus inevitably toward interest groups and their policymaking roles. The simple reason is that all of our dominant concepts about policymaking in the United States are, in some sense, group theories, concerned with the interplay among interest groups and their connections to government. Students of American politics assume that groups do matter, though how or how much they matter is in dispute. We talk about winners and losers, about actors and interests enjoying variable success in their ability to influence—or even mandate—policy decisions. The question that daunts us, however, is why some succeed while others fail.

Not that we lack rival explanations. Elitist critiques of American democracy, rooted in sociology, find the answer in the inequitable distribution of economic and political power. For them, power accrues primarily by virtue of wealth, education, or social ties. This perspective allows Floyd Hunter to speak of a "unified power elite" when studying Atlanta, or C. Wright Mills to argue that America is dominated by a "power elite."[1] Pluralists, rooted primarily in political science, reject such depictions of power as overly reputational and static. If the elitist sees power as residing almost indefinitely within a narrow sector of society, the pluralist argues that we can neither understand nor measure power until we focus on a specific conflict. Power absent such a referent is of little use in the pluralist view, so the scholar must study actual conflict, must know whether a particular actor or interest prevails over opposition, vetoes initiatives promoted by others, or initiates policy without opposition, before calling somebody or something powerful or impotent.

It comes as no surprise that where elitists see concentration, pluralists uncover competition. Hunter sees a unified elite in the more stable distribution of social and economic power, while Robert Dahl finds multiple and partially overlapping elites, but none possessing comprehensive and enduring political power.[2] Pluralists agree that the "strongest" still prevail in a given conflict, but note that the victors in one instance may not even be involved in the next. Different issues in fact attract different interests, and on some matters the wealthy may lose out to better organized coalitions of the less affluent.

These two viewpoints also diverge on the role of democratic government. Elitist critiques of U.S. politics generally are social theories that either ignore government totally—power resides in the topmost social strata—or view it simply as the instrument for elite domination. Pluralists tend more often to portray government as a "referee": government provides the arena for conflict resolution within which social forces lock horns; government supplies the procedures for conflict resolution, sets the rules of the game, but *takes no sides* in the ensuing fight. Government is neutral, so any group, if it so wishes, can participate in policymaking. "A central and guiding thread to American constitutional development," says Dahl, "has been the evolution of a political system in which all *active and legitimate* groups in the population can make themselves heard at some crucial stage in the process of discussion."[3]

The pluralist state, simply put, is all process, and government's role is to provide relatively nondisciminatory procedures for conflict resolution, a structural "everything-else-being-equal" that gives the interested and the active their battleground. Government then gets out of the way, later to ratify the outcome in the name of the public good. Earl Latham offers a classically pluralist view of the legislative process:

> The legislature referees the group struggle, ratifies the victories of the successful coalitions, and records the terms of the surrenders, compromises, and conquests in the form of statutes. Every statute tends to represent compromises because the process of accommodating conflicts in group interests is one of deliberation and consent. The legislative vote in any issue tends to represent the composition of strength, i.e., the balance of power, among the contending votes at the moment of voting. What may be called public policy is the equilibrium reached in this struggle at any given moment.[4]

E. E. Schattschneider once observed that "the flaw in the pluralist heaven is that the heavenly chorus sings with a strong upper class accent."[5] Pluralist assumptions that the "interested and the active" will mobilize and actively defend collective interests are challenged by studies that find political awareness and participation to vary directly with socioeconomic status and education.[6] To be fair about it, pluralists have never denied that society is clearly unequal; rather, they deny that political outcomes invariably benefit the most affluent and educated. Their critics argue that this defense is flimsy, that pluralists gloss over the maldistribution of power in society for the sake of some procedural egalitarianism. Pluralists may contend that every group involved in the fight has a plausible chance of success, but their critics counter that the game itself is stacked against the lowermost strata. Not every interest is or will be represented, and no democratic facade can hide that reality.

Central to our concerns is the second assumption in the pluralist thesis—that government is the "neutral referee" over the group struggle. To critics like Schattschneider, government is anything *but* neutral. Latham's view of the legislative role is sadly misconceived, Schattschneider contends: "It is hard to imagine a more effective way of saying that Congress has no mind of its own or that Congress is unable to involve new forces that might alter the equation" among interests.[7] Not only do certain sectors of society go unrepresented, he continues, but those controlling important junctures in the political system also consciously structure the *opportunities* for access to decision making. Every form of political organization, be it a school board or a federal agency, is structurally biased in favor of some interests and against others. Government is not neutral; it is another political actor, with its own biases and agendas for action. Government takes sides through its structures, its rules and laws, and through a systematic consensus about what does or does not legitimately fall within its purview. Pluralists may describe the fight accurately, but tend to ignore the roles that such "predecision" factors as the "rules of the game" or system structure play in conflict. Outcomes may indeed hinge on the battle between interest groups, but the choice of battlefield and weapons often is predetermined, to the benefit of those able to make such choices. The handicap is not only sociological—some groups are more powerful than others—but also structural. Some interests, and some issues, are brought in by government. Others are kept out.

Such gaps in the pluralist response to elitism have generated a series of classic studies of policymaking in agricultural, water, and public works programs that constitute the core of the "subgovernment" approach.[8] This approach presents an image of sectoral politics, where policy dominance "is said to be vested in an informal but enduring series of 'iron triangles' linking executive bureaus, congressional committees, and interest group clienteles with a stake in particular programs."[9] Subgovernment theorists agree that issues define political action, but they also suggest that the arena for action is *not* open to all, and that policymaking is not so much a brawl as a private club.

Subgovernments find their foundations in the clientelism endemic to all narrowly defined participatory communities. Effective (as opposed to symbolic) access to decision making is open only to those with the most direct economic or political stakes in the policy. The issues settled within such configurations of policy actors tend to show low public visibility, and decisions typically emerge through mutual accommodation (or "logrolling"), where all involved tend to gain. Such enduring arrangements may simplify governance because they narrow the range of actors and choices attendant on any issue, and because they make policymaking less global than sectoral. Subgovernment policymaking also is inherently incremental, since great changes only attract unwanted attention. Whether it is "democratic" is another matter.

Clientelism obviously pervades a political system where ascent to national office depends powerfully on local constituencies, and there seems little doubt that the American system of fractionated political power nurtures subgovernment dynamics. The utility of this thesis as an overarching theory, however, is questionable. One problem with this image, like so many we use, is its rigidity, because it assumes (often explicitly) that subgovernments are enduring and therefore able to maintain indefinite hegemony over whole slices of public policy. This *may* be true, yet we know very little about the effects of time on such policy communities. In some ways the subgovernment thesis falls prey to the same failing of the elitist school; the distribution of power is constant. As Nelson Polsby observes, pluralists provide a powerful rebuttal: "Pluralists hold that power may be tied to issues, and issues can be fleeting or persistent, provoking coalitions among interested groups and citizens, ranging in their duration from the momentary to semipermanent. . . . To presume that the set of coalitions which exists in the community at any given time is a timelessly stable aspect of social structure is to introduce systematic inaccuracies into one's description of social reality."[10]

We rightfully should wonder if a thesis that evolved during the early 1960s still can explain decision making in today's political and economic climate. Indeed, we should wonder whether subgovernments, which thrive on logrolling and secrecy, can function well in a political atmosphere characterized by fiscal austerity and inquisitive news media. Perhaps this image, like the other two, is an incomplete analytic device. Perhaps no single image of policymaking *can* convey the rich complexity of policy roles and lines of influence evident in this political system. Perhaps, as one observer noted pessimistically, the modern welfare state is incomprehensible.[11]

Hugh Heclo suggests that our standard conceptions of power are ill-suited to the more amorphous and inchoate configurations of participation and influence that characterize American politics. "We tend to look for one group exerting dominance over another," Heclo states, "for subgovernments strongly insulated from other outside forces in the environment, for policies that get 'produced' by a few 'makers.'"[12] But, he argues, such notions of "iron triangles" or subgovernments inadequately convey the complexity of policymaking. In most initiatives taken over the past two decades, he asserts, "It is all but impossible to identify clearly who the dominant actors are. . . . Looking for the few who are powerful, we tend to overlook the many whose webs of influence provoke and guide the exercise of power."[13]

These "webs" of influence, which include elected officials, bureaucrats, the media, and policy experts of every stripe, comprise what Heclo calls "issue networks." "Subgovernments" imply relatively stable and exclusive sets of relationships among public and private-sector actors. Issue networks, by contrast, "comprise a larger number of participants with quite variable degrees of mutual commitment or of dependence on others in their environment; in fact, it is almost impossible to say where a network leaves off and its environment begins."[14]

This latter point poses tremendous analytical problems. If it is almost impossible to know where an issue network separates from its environment, how then are we to delimit a circle of actors and interests that we can identify as a specific network? We may as well renew the search for the Loch Ness Monster. The analytical murkiness of the issue network concept, however, may indicate a similar murkiness in the actual relationships among relevant policy actors, and hence between them and external political forces. If the "iron triangle" concept conjures up images of exclusive and autonomous decisional units, the issue network is far more evanescent; but it does *not* suggest pluralism-as-usual—or, for that matter, the lack of discernible connections among those who know each other through an issue.

Charles O. Jones, studying American energy policy in the 1970s, suggests an image of "sloppy large hexagons" that might help to clarify the issue network concept. If national energy policy (such as it was) prior to the OPEC oil embargo was dominated by "cozy triangles," these arrangements were scrambled completely during the "energy crisis" of the 1970s. Jones details what happened next:

> Demands by environmentalists and public interest groups to participate in decisionmaking, involvement by leadership at the highest levels in response to crisis, and the international aspects of recent energy problems have all dramatically expanded the energy policy population. Note that the expansion is up, out, and over—*up* in public and private institutional hierarchies, (e.g., the involvement of presidents of companies and countries, rather than just low-level bureaucrats, and of congressional party leaders rather than just subcommittees); *out* to groups that declared an interest in energy policies (e.g., environmentalists, public interest and consumer groups), and *over* to decisionmaking processes in other nations or groups of nations.[15]

The resulting expansion in the number of policy claimants produces uncertainty and conflict, greater complexity, and protracted delay, since the formerly stable sets of actors and procedures *no longer control the parameters of policymaking* in this issue area. Crisis politics transforms existing arrangements into new configurations, and those gaining access to policymaking during the upheaval are likely to remain, even after mass attention to the issue wanes, because these newly enfranchised insiders quickly develop their own stakes in policy outcomes. Crisis, it appears, breeds stakeholders. These new policymaking configurations, however, are likely to be less exclusive and more permeable than their predecessors. They are, in fact, probably rather inchoate, more susceptible to momentary political pressure. The issue network conceptually is much more than the subgovernment, but it is also much less than mass politics. We know intuitively what an issue network is not, but are not quite sure what it is.

The reader by now appreciates the dilemmas presented by the differences among these competing images of political life, which I portray more graphically in table 1. It is useful to picture these images as continua rather than as mutually exclusive cells, because issues themselves are never stable; they expand, they contract, the "move" from the narrow realm of technical discussion or logrolling among a limited number of actors to the more highly salient condition of mass conflict—and often back again. Yet this dynamic is precisely what we seem to lack in our analysis of the political world. What I seek here is to understand political change in a way that allows us to bring these conflicting images into play. My overriding supposition is that these images, as contradictory as they may seem, all compose parts of a broader spectrum of political change and policy transformation. Each image may in fact be partially valid in itself, but not as a wholistic device for explaining—or even describing—the dynamics we know exist. Each image may be but one facet, or one phase, of an ever changing condition in the body politic.

Table 1. Three Images of Policymaking

	Policymaking Dominated by		
	Pluralist Competition	*Subgovernments*	*Issue Networks*
Range of participation	Unlimited	Limited	Somewhat unlimited
Political alignments	Unstable	Stable	Somewhat unstable
Access to decision making	Unrestricted	Restricted	Somewhat unrestricted
Source of access	Group mobilization and power	Economic or political stake	Issue knowledge and expertise
Visibility of decision making	High	Low	Episodic
Range of decisional loci	Multiple	Few	Multiple
Exclusivity of decisional jurisdiction	Low	High	Moderate
Sensitivity to changes in partisan control	High	Low	Moderate
"Play" of influence	Disjointed (fluid)	"Cozy" (mutual accommodation)	"Loosely jointed" (messy)
Role of expertise	Peripheral	Moderate	Central
Nature of policy outcomes	Majority coalition-building	Clientelism and accommodation	Entrepreneurship and expertise
Function	Agenda setting, policy formation, "big" change	Policy stabilization, incremental change, defense of status quo	Implementation, evaluation, policy tinkering

Studying Political Change

Our politics are dynamic, but our analysis seems far less so: recall that our dominant images of policymaking emerged largely during the 1950s and early 1960s. And, despite major contributions to our knowledge made during the 1970s, it is striking how few studies in recent years have sought to reinterpet these competing images as political and social change redefined the landscape of American governance.[16] The problem, somewhat restructured, is this: if policymaking once was characterized as either pluralistic or dominated by subgovernments (or some combination thereof), and if contemporary views increasingly question the utility of these concepts, what *do* we have today, and—more important—how did we get from "then" to "now"? What does change look like?

Let me illustrate the problem more concretely. This study is about policy transformation concerning a single, well-defined issue: the Federal Insecticide, Fungicide, and Rodenticide Act of 1947 (FIFRA). I chose FIFRA because it represents the dominant type of policy activity of the past generation; it has a long and well-documented life history, alternating between periods of lengthy policy stability and then rapid change; it has involved, at one time or another, virtually every institution and process embedded into the American political system; it reflects major changes in the social, political, and organizational fabrics of the nation. In short, the politics of pesticides is a microcosm of the general trends evident in this nation since World War II. Table 2 presents a shorthand view of two major junctures in the development of federal regulatory policy.

There was no fundamental change in FIFRA between 1947 and 1972, despite concerted and escalating efforts to overhaul the statute during the interim. The 1947 act (to be discussed in chapter 3) essentially was the product of close cooperation among members of the House Committee on Agriculture, mid-level personnel within the U.S. Department of Agriculture (USDA), and those repre-

Table 2. The Federal Insecticide, Fungicide, and Rodenticide Act (FIFRA) in 1947 and 1972

	1947	*1972*
Problem	To assure product quality	To prevent harmful effects
Initiation	Industry, agriculture, and USDA	EPA, environmentalists
Participation in policymaking	"Narrow": —industry groups —farm groups —USDA —Agriculture Committee members in both houses	"Broad": —EPA —USDA —USDI —environmentalists —industry groups —farm groups —House and Senate actors —state governments —news media
Facts	Minimal visibility, little controversy; bill written by USDA, with industry help; little House committee action; no Senate hearings or debate; passed with little publicity	High visibility, much controversy; conflict within committees and between chambers; intense floor debate, floor amendments and substitutions offered; conference committee needed
Key decision arenas	House Agriculture Committee	House and Senate Agriculture Committees; Senate Commerce Committee; both floors; conference

senting the major agricultural pesticides makers—a classic "iron triangle." Pesticides policymaking in the 1970s, however, was far different, with protracted conflict among many and diverse actors, in multiple decision arenas (see chapter 7).

There is a certain irony to this. These events have all the characteristics of "subgovernment" behavior in the late 1940s, which was about the time that the pluralist image began to dominate American political science. This image no longer reigns so easily by the 1970s, yet actions regarding pesticides policy throughout the decade display all the characteristics of intense group conflict. One plausible answer to this apparent paradox is that pesticides policy *as an issue type* has been redefined at some point in the interim. Both pluralists and their critics recognize that types of issues or policies generate, and therefore are surrounded by, distinctive configurations of policy actors and relationships. Or, as Schattschneider argues cogently in his study of the Smoot-Hawley Tariff, "New policies create new politics."[17] Further, as Ripley and Franklin suggest, types of issues "in turn help to determine substantive, concrete outcomes when policy decisions emerge."[18] Subgovernment policymaking, they argue, is primarily associated with "distributive" issues, such matters as water projects, public works programs, or farm subsidies (the "porkbarrel"), where overall public awareness of the issues tends to be low and opportunities for logrolling high. "Regulatory" policymaking, by contrast, is more visible to the public. It attracts a wider array and greater number of policy claimants, more direct high-level executive involvement, and, as a result, far less stable relationships among those participating.[19]

According to this framework, as table 2 shows, pesticides policymaking rests comfortably in the regulatory policy niche, at least after 1970. But the political relationships surrounding the genesis of FIFRA, and the subsequent twenty-five-year hiatus in substantive policy change, point to subgovernment dominance, if not their autonomy. Perhaps the 1947 act was not regulatory at all; rather, it could be called more "self-regulatory," or even symbolic in nature. Chemical makers were required only to display credible labels and reliable instructions for use on their products—largely to keep unscrupulous operators off of the market—and the USDA had neither the mandate nor the inclination to check the veracity of the information or the safety of the pesticide. The 1972 act, by contrast, propelled government into the active regulation of product safety. Producers no longer regulated themselves, but instead faced scrutiny by a government intent on insuring against harmful effects. The nature of the issue thus was transformed, and its politics redefined.

We are still left to assess *why* this change occurred. What were the dynamics of policy transformation? There are several possibilities. We might find, for example, simply that there was no opposition to business-as-usual during the interim. Schattschneider early on recognized the variability of pressure-group activity, commenting, "Apathy and indolence are the rule and excitement is the exception."[20] Bruce Oppenheimer, in his analysis of the oil depletion allowance, found that the status quo resulted less from the industry's overpowering influence in Congress than from a longtime *absence* of effective opposition by other actors: "Clearly, a substantial part of the legend of the oil industry's power in American politics stems from its ability to maintain the depletion allowance. But with little non-governmental opposition, I question whether this was an accurate reason for ascribing power to the industry. Similarly, the improvement in the quality of work and the number of organized groups active in water pollution legislation is reflected in stronger legislation and enforcement in the oil pollution area."[21]

This perspective evolves in many ways out of Schattschneider's notion of the "scope of conflict." There is, he argued, an innate difference in the style and tenor of politics that results from whether the scope of conflict regarding a given issue is narrow or broad. Changes in the scope of issue conflict, by extension, correspond to shifts in policymaking outcomes: "Every change in the scope of conflict has a bias; it is partisan in its nature. That is, it must be assumed that every change in the number of participants is about something that the newcomers have sympathies or antipathies that make it possible to involve them. By definition, the intervening bystanders are not neutral. Thus, in political conflict, every change in scope changes the equation."[22]

Three factors powerfully influence the scope of issue conflict and subsequent political actions regarding any policy area: the degree of competition over the issue, the visibility of the issue to the public (the "audience") or to potential allies, and the role of government—actual or potential. A broader scope of conflict incorporates more competitors than does a narrower one, involves more highly visible issues, and includes government as a central player, with subsequent differences in policy results. It is not enough, however, merely to count the number of competitors on each side of a fight, nor to tot up the number of minutes devoted to an issue on the nightly news. It is not even enough to know that government is or is not involved. What is essential is to examine *how* competitors seek to enlist allies in the fight. We must understand how competitors seek to broaden or narrow the scope of conflict—what Schattschneider calls the tension between the "socialization" and "privatization" of conflict—the outcome of which profoundly affects issue visibility, recruitment of allies, and, eventually, the role of government.

To be able to control the scope of issue conflict, to determine the extent to which an issue is publicized or allies are recruited, is to have a powerful resource. In battle, all other things being equal, the larger army is likelier to win. But in politics, as in warfare, things rarely are equal. Just as the smaller army can prevail if it controls strategic mountain passes, a political minority in control of strategic junctures in the political system will be positioned to resist changes promoted by the majority, or pass initiatives in spite of them. Thus, we must know more than that credible opposition to the status quo finally evolved that, as a result, forced policy change. Oppenheimer, in his discussion of the oil industry's power in Congress, suggests that access to decisionmaking and policy influence followed, and was conditioned by, several structural and procedural factors—among them alterations in congressional rules and procedures: "Rules and procedures, in affecting the course of public policy and the actions of congressmen, have a simultaneous effect on interest groups concerned with various pieces of legislation. The way the process operates has a good deal to do with the success of an interest group's position and where interest groups will center their attention to influence those results."[23]

Rules and procedures are neither neutral nor mere reflections of the strengths of competing factions. The "rules of the game" powerfully affect who wins, who loses, or who even is allowed to play. Whoever decides what the game is all about lays down the preconditions for who plays, under what rules, and for which issues. "All forms of political organization," argues Schattschneider, "have a bias in favor of the exploitation of some kinds of conflict and the suppression of others because *organization is the mobilization of bias*. Some issues are organized into politics while others are organized out."[24] Government, after all, is that aggregation of rules and procedures designed to regulate human behavior, and, as Schattschneider states emphatically, "Procedures for the control of the expansive power of conflict determine the shape of the political system."[25] The rules we play by clearly matter.

We now return to the original question: how can we comprehend the dynamics of policy transformation? How, specifically, does federal pesticides policy change from an apparently "self-regulatory" activity dominated by a limited and relatively exclusive set of actors to one that is regulatory in intent and accompanied by broad issue conflict among multiple competitors? To understand this transformation, we need to know more than that certain actors or interests have gained access to decision making. We also need to know how and under what conditions this access was established.[26] To know these things is to understand more about the dynamics of issue conflict. Why do some interests dominate policymaking at any one moment? How does the passage of time affect such dominance? How do "outsiders," those actors or interests previously kept out of the game, gain access? Under what conditions? To what effect?

Conventional wisdom holds that U.S. history is one of inexorable, relentless democratization, of ever expanding popular participation in policymaking. This wisdom holds true at the aggregate level (for example, the franchise), but is it so at the level of issue conflict? And, even at the aggregate level, is it necessarily unilinear? Mass participation, at least the *opportunities* for mass participation, undoubtedly will continue to expand, barring the collapse of the system. Upon closer examination, however, it is likely that this broad tide of expansion contains subtle participatory cycles that oscillate between expansion and contraction. Policymaking seems to involve a dynamic tension between opposing forces—some seeking to expand conflict, others fighting to constrict it. Is it not possible, then, that such oscillations in the scope of participation exist, and that they produce shifts in policy direction? If so, we need to understand better the conditions, both social and governmental, that underlie these participatory swings.

Congress as Ordering Device

The assumptions supporting this study are that political change and policy transformation can be studied within the confines of a single issue over time, and that such an approach can tell us a great deal about the social, political, and structural requisites for change. The goals of this study compel me to focus on the process of policy formation, that process (or processes) wherein competing stakeholders promote divergent claims and seek to recruit others to their respective causes. To talk about policy formation, inevitably, is to focus on Congress. Other factors certainly affect policy, but Congress, as both institution and policy process, is perhaps the best place to start. If Congress is the "keystone" of the U.S. political system, it also is that single institution where all conditions and trends evident in the body politic inevitably coalesce. Congress, after all, has the constitutional mandate to make the law, and no other institution in government better reflects the warp and woof of the national mood—or should.

Congress, like any organization, has a structural bias, a framework of decisional forums and processes that provide differential access to outsiders and supplies its members with the tools necessary to absorb and channel external political and social pressures. Roger Davidson suggests that to view Congress as an organization is also to consider congressional *change* as a form of organizational innovation or adaptation. "Like all organizations," he argues, "Congress strives for self-preservation, protecting its autonomy and influence. To maximize its survival in such terms, it must adjust successfully to external demands while coping with internal pressures."[27] Congress hardly exists in isolation, nor are such pressures static, though how the institution over time copes with its external environment, and with what effects on policy, is not well understood.

We assume that changes in the structures and processes of congressional decision making subsequently should influence the structure of decisional bias, which, in turn, should produce a substantive redirection of policy. This is tricky to prove, but we can suggest that, at a minimum, new rules or procedures should have a profound impact on the array of access points and power resources within the legislative system. Exclusive committee jurisdiction over a subject area, the formal power of committee chairs, the visibility of congressional action, and the power of individuals at strategic procedural junctures—all powerfully influence who gains access to decision making, and on whose conditions. Such factors, in turn, ought to influence policy decisions.

To define change as a tension between competing efforts to privatize and socialize confict is to expand our focus outward, since Congress exists within a much broader political context. We need always to plant our study of that institution firmly in its surroundings, its social, political, and economic milieu, and, particularly, its issue agenda. After all, argues Amitai Etzioni, "There is no one effective strategy of decision-making in the abstract, apart from the societal context in which it is introduced and from the control capacities of the activists

Figure 1. A Framework for Case Analysis

Problem:
objective conditions
cause
undesirable effects.

Perceptions
of problem are mediated by
the visibility of the effects and our knowlege.

Interest aggregation and mobilization,
are conditioned by

Incentives (and disincentives) for action
and screened by the

Structure of "bias"
(the social values, laws, processes
that determine the legitimacy of
some interests and screens out others).

Demands
seek out

Decision arenas,
where they are (or are not)
aired and debated, and where a

Decision
is (or is not) made.

Implementation,
which depends on government

Support, resources, capabilities, and intentions,
may result in

(1) amelioration of the problem
and/or
(2) unintended consequences, which may result in
new conditions, new problems.

Return to beginning.

introducing it."[28] The framework presented in figure 1 is an attempt to display graphically the context within which the case evolves. Otherwise, states Hugh Heclo, case studies only "represent confused realism in search of an analytical framework."[29]

Any study of policy change should begin with the "problem" itself. The reason is simple, yet carries tremendous implications both for government and those deigning to study it. As Charles O. Jones explains, "Certain effects of events in a society are objectively verifiable; others are not. And there is no necessary coincidence between the *objective* (what is) and the *subjective* (what we think to be)."[30]

Some objective condition arguably can exist, and can be "bad," without society perceiving that it has undesirable consequences. If so, there is no "problem" to speak of. And, if this is so, we may need to ask why, just as we may ask how it is that some in society see a problem and others do not. Such dynamics may depend powerfully on our capacity to make the perceptual linkages between conditions and unpleasant results, which in turn can hinge on the state of our scientific knowledge. Broader knowledge, both in absolute and relative terms, logically should lead to greater perception of a problem. Or we may deny that a problem exists because to recognize it might challenge our core beliefs or values. Whatever the case, we often speak blithely about a "public problem" without ever tackling the intrinsic meaning of that problem. To do so seems to put the analytical cart before the horse.

Our observations of human nature tell us that people may not seek to ameliorate undesirable conditions at all times, which should lead us to wonder just at what point those affected do try to redress perceived problems. At what point do "common interests" organize, and under what types of dynamics? What incentives exist in society or in government to promote interest aggregation and mobilization in pursuit of change, or, conversely, to undermine such processes? What resources can those interested in some problem draw on to aid their cause? Such questions assume that problems have different impacts on potential publics—those who, logic suggests, should feel compelled to act. Indeed, such publics may not act even if a problem is perceived, which should force us to examine the structure of incentives and disincentives that affect the shape and direction of interest mobilization. Pluralists generally assume that economic self-interest is central to such dynamics, but so too are morality, a sense of equity, or some (perhaps ethereal) notion of "the public good." All political systems contain components that support or deter one's capacity or willingness to pursue goals, and not all incentives are equally legitimate in the eyes of society. The incentives or sanctions that dominate at any one time in history can tell us a great deal about which interests are mobilized which are not, and why.

The political system is one such resource. It determines who even gets to play in the game, and both the system of governance and prevailing social values combine into a powerful "structure of bias" that screens out values or claims not deemed legitimate or worth the government's time. Some demands are more legitimate than others, but we don't always understand why. Some claims may never be made because those who could make them know—instinctively or otherwise—that they never will be taken seriously. An issue may head the agenda of government less because of its innate quality than because society may be willing to grant it legitimacy. Setting the agenda for action, after all, is that step "whereby large number of problems government *could* address is reduced to the much smaller number of problems government *will* address."[31] Understanding why, and how, some problems reach the pinnacle of the national agenda may help to explain why other matters never get "solved," or even acknowledged.

Demands, thus filtered, proceed to some decision arena, though the specific avenue for resolving the problem is not predetermined. Government itself is a structure of variegated institutions and processes, each possessing peculiar internal biases and dynamics, and it is essential to understand why demands flow into one arena and not into others, or why policy claimants suddenly switch from one arena to another. What are the rules for access and decision making that make choosing the right venue so critical? For whom? Where a decision evolves can be critical to what decision emerges—a simple fact of political life often overlooked.

Finally, decisions must be implemented if policy change is to occur; policies rarely, if ever, implement themselves. Solutions to recognized problems "are difficult enough to achieve when the law is clear and the federal agency has organizational strength and political support," notes Jones, but, absent such conditions, the effort to match performance to promise becomes all the more problematic.[32] "Solutions" also may produce their own unintended consequences, which, of course, starts the whole cycle over again.

Conclusion: What Does Change Look Like?

"If we are to talk about politics in terms of conflict of interests," argues Schattschneider, "the least we might do is to stop talking about interests as if they were free and equal. We need to discover the hierarchies of unequal interests, of dominant and subordinate interests."[33] All theories of policymaking essentially concern such hierarchies: who gets included in policymaking, who doesn't, and why. This study asks such questions to examine the multiple images of policymaking and make general statements about dominant and subordinate interests.

I want to know how dominant interests seek to maintain policy hegemony, how the defenders of the status quo seek to manage conflict so that policy decisions time and again reflect their predispositions.[34] "There are an incredible number of devices for checking the development of conflict within a system," Schattschneider suggests. "Some issues are organized into politics while others are organized out."[35] This study seeks to assess how dominant policy actors and interests utilize the formal rules and structures of the political system, as well as its informal processes, to check issue conflict and ensure that their issues remain dominant. It also seeks to understand how subordinate interests utilize this same system to force action on their behalf. The U.S. system of diffuse and fractionated power exhibits multiple veto points, but this also suggests that the system exhibits mutiple points of potential access for those on the outside. It runs both ways, depending ultimately on whether at a specific juncture in the policy process certain actors block access by others or gain access themselves.

This study is chronological because it concerns political conflict and policy *change*—dynamics that may not surface through snapshots of policymaking at any moment. We begin in chapter 2 by examining the general parameters of policy genesis—the dominant issues, cleavages, and attitudes influencing the pesticides phenomenon—and then proceed inward, unwrapping layers of contextual, structural, and procedural factors until we focus directly on the formation of policy in chapter 3. Subsequent chapters examine issue dynamics and responses by policy actors as time passes, studying how dominant interests defend their prerogatives and how subordinate ones seek both access and influence, if only for a while.

This is more than mere storytelling. To understand what policymakers do, we must first know the limits to what they can do. To understand why certain policy outcomes occur demands that we know first which outcomes were deemed possible, which were not, and by whom. Behavior such as this does not exist in a vacuum; it is channeled by the political system's peculiar structures, rules, and procedures. Change the structures, alter the rules of the game, and we should find changes in the intrinsic nature of that game, not to mention the issues allowed to reach the public agenda for action. Such alterations in turn mightily affect the range and type of political behavior relevant to the issue, which inevitably affects policy outcomes. Different policy types create different politics, but it is equally plausible that different political games create different policies. How mechanisms for conflict management affect outcomes, and how time alters both such mechanisms and relevant interests, are central questions.

All this ultimately concerns the question of change. To even begin to depict change is to focus on three interrelated factors, and then to watch how each fares with time. First, how *visible* is the issue? We assume here that issue visibility affects who is attracted to the issue in the first place. Why some interests are attracted and others are not is worth pondering. Second, how *competitive* are the interests attracted to this issue? This question presumes that a "noncompetitive" decision-making condition—where one interest or set of interests clearly dominates—differs strikingly from a "competitive" one insofar as winners and losers are concerned. Third, what role does *government* play? Does it preempt interest-group activity? Does it acquiesce to interest-group entreaties? Is it neutral or, perhaps, immobilized because of sharp internal disputes? The answer may help us understand the worth of our competing images of policymaking. Posing these three questions to study change also presumes that each factor is central to the question of conflict, the scope of which seems to be a catalyst (or, maybe, a precondition) for policy outcomes. Change is chameleonlike; the surface status quo may in fact hide subtle shifts, while the more flashy pronouncements of change may prove more chimera than substance.

One a priori expectation is that the dynamics studied here gradually broaden the scope of effective decisional participation to the point that, by the 1980s, some sort of equilibrium has been reached. That is, by the late 1970s the structures and rules governing policy conflict in this issue area—and, perhaps, only in this issue area—have changed enough so that the system itself no longer discriminates as readily as before. Where once we talked of "iron triangles" or subgovernment dominance with some degree of certainty, today we may not be so ready to draw such conclusions. Subgovernments must control the parameters for policy debate and action to maintain issue hegemony, and those that lose control over the mechanisms for conflict resolution cede policy dominance. We could argue that what has resulted after a generation of social and structural "democratization" is a form of pluralism—twenty years after critics took that image to task.

But the pluralism I speak of is not so much social as it is structural. That is, the structure of policymaking in the United States has changed to the point that a far broader range of demands and views are included in the debate. "Institutionalized" pluralism of this sort does not entail the mobilization of temporary coalitions. Rather, it involves the more or less permanent "presence" of those who have previously gained access as the structure of governance became more permeable and far less exclusive. If, as I suggest, more interests than ever are dealt into the game on a more permanent basis, and the system itself does not discriminate among policy claimants as easily as before, then policymaking becomes all the more open, difficult, and—perhaps for those on the inside—more frustrating. The political market, in short, is no longer monopolized by the few. Says Schattschneider: "The pressure system makes sense only as the political instrument of a segment of the community. It gets results by being selective and biased; *if everybody got into the act the unique advantage of this form of organization would be destroyed, for it is possible that if all interests could be mobilized the result would be a stalemate.*"[36]

One does not need "everybody" to crowd the political market and, perhaps, cause a stalemate; nearly everybody will do nicely. And we need only look to the universe of policy "insiders" now included in the Washington game to understand this point. Everybody—from policy consultants and think tank experts to public interest law firms and media representatives—is in the act, occupying positions of policy leverage either astride or adjacent to the lines of power that grant them entree to policymaking. Access is based on a sense of "presence," on having a permanent stake in and knowledge of the issue at hand, not on temporary mobilization. Indeed, the pluralist thesis makes sense, at least in part, only if one takes into account the expansion of the permanent Washington establishment and the more open nature of the political marketplace.

"Presence politics" has implications for governance, if only because such conditions make it all the more difficult for governing institutions to forge consensus or resolve conflicts. The core of governance is the capacity to make decisions that stick, no matter how unpopular or hard they may be. Presence politics may in fact erode that capacity simply because there is no overwhelming disequilibrium among policy claimants to buttress or compel conflict resolution. If all interests are in the game, and if the structure of governance appears relatively permeable to all comers, then what mechanisms exist to construct policy? We will return to that question in chapter 10.

Chapter Two

The Pesticides Paradigm

Had you read the *New York Times* on June 26, 1947, you probably would have devoted your attention to the labor strikes sweeping the nation after Congress overrode Truman's veto of the Taft-Hartley Act. You might then have turned to the international news, which focused on the nascent United Nations, or to the sports, where rookie Jackie Robinson was packing Brooklyn Dodgers fans into Ebbetts Field. Trailing not too hard after the sports pages was the "News on Food" section, devoted primarily to recipes, articles on new products, and other household information. It wasn't much of a section, and easy to pass by, but the most dutiful of readers might have noticed a small Associated Press item tucked into the extreme right hand corner of page 26:

> New Law to Color Poisons
>
> A bill requiring color in some poisons to lessen the chance of housewives putting bug instead of baking powder into their biscuits became law today. President Truman signed the measure which tightens a 1910 insecticide control law, bringing rat and weed poisons under the act. It also requires coloring of any dangerous poisons that might be mistaken for flour, sugar, salt, and the like, registration of poisons before they go on the market, and warning labels.

Few probably took much note of this rather innocuous item—even, I imagine, the housewife to whom it evidently was directed. This "coloring law" in fact was the Federal Insecticide, Fungicide, and Rodenticide Act of 1947 (FIFRA), a major action aimed at a broad spectrum of chemicals then entering the postwar economy. That FIFRA received so little attention in the national news may, in retrospect, seem surprising, but it indicates how the issue of pesticides at that time was defined and how it ranked in the public eye.

This chapter examines the imperatives surrounding the genesis of postwar federal pesticides policy, the social, technological, economic, and political forces that impelled the direction and shape of federal action. Politics comes wrapped in layers, which we must first slice through before coming to study the act of policymaking itself.

Setting the Agenda: The Farm Problem

The number of public problems is so great as to be incalculable, particularly when defined in terms of how various people view the effects of events in society. Clearly, not all these get on the agenda of government. The process by which some issues get there and others do not is extremely important because the specific demands resulting from awareness of problems is the "stuff" of government in a democratic society.[1]

We begin at the "top." I mean this in two ways: we begin not only in the immediate post–World War II period, for reasons of policy genesis, but also by considering the issues dominating that period. We start by examining the social or "systemic" agenda. In the words of Roger Cobb and Charles Elder, these are the issues "that are commonly perceived by the members of the political community as meriting public attention and . . . *involving matters within the legitimate jurisdiction of existing governmental authority.*"[2]

"Policymakers are not faced with a given problem," Charles Lindblom argues; "they have to identify and formulate the problem."[3] Some issues or problems emerge out of the universe of potential problems to dominate the public psyche. Others dwell in the shadows, perhaps because they are not perceived by the public or by decision makers to merit attention. Some matters conceivably remain "nonissues" because they are screened out of the political arena by social norms, traditions, or by commonly held notions about government's role. National health care, to offer one example, simply was not a legitimate issue for much of this century because it apparently was so alien to traditional "American" values.

Some issues remain invisible because they challenge the power of dominant political institutions or elites, because important political forces act to keep such matters off of the public agenda. Pluralist preoccupation with decision-making processes tend to ignore *predecisional* dynamics—which issues become public matters in the first place, how, and why—and, critics charge, pluralists also appear to assume that "key" issues appear almost randomly. Bachrach and Baratz, for example, argue that "to measure relative influence solely in terms of the ability to initiate or veto proposals is to ignore the possible exercise of influence or power in limiting the scope of initiation."[4] There is, according to Schattschneider, an "unequal intensity of conflicts" shaping the political system, because politics deals ultimately with the domination and subordination of conflicts:

> For this reason it is probable that there exist a great number of potential conflicts in the community which cannot be developed because they are blotted out by stronger systems of antagonism. . . . Political conflict is not like an intercollegiate debate in which the opponents agree in advance on a definition of the issues. As a matter of fact, *the definition of the alternatives is the supreme instrument of power*. He who determines what politics is about runs the country, because the definition of alternatives is the choice of conflicts, and the choice of conflicts allocates power.[5]

There may be a bias in the sample of issues that emerge for consideration, so we must know which issues dominate, and why, before turning our thoughts to actual decision-making processes. Some issues do arise randomly—disasters and other unforeseen events obviously impel political response—but we nonetheless ought to consider why some issues dominated the agenda of government in the mid-1940s, and to what effect.

The two great issues facing postwar America, according to the *New Republic,* were food for Europe and the survival of domestic price controls administered by the Office of Price Administration (OPA).[6] The Gallup Poll in March 1946 revealed that 73 percent of respondents supported continued OPA controls to stave off inflation, while 67 percent were willing to consume less meat and flour so that the United States could increase food shipments to Europe.[7] Gallup in fact conducted at least twenty separate polls on food, food shortages, price controls, and related matters between 1946 and 1949—indicating some degree of national concern with this bundle of issues.

If a review of the *Congressional Record* and other government documents is indicative, four broad issues appeared to dominate presidential and congressional agendas during this period: redirecting a wartime economy without introducing a depression; absorbing millions of returning veterans into that economy; rebuilding the shattered economies of Europe; and, solving, for once and for all, the recurring dilemma of American agriculture. This last issue, as we shall find, is woven inextricably into the others, for the "farm problem," like most other types of problems, proves to be a collective entity. As such, it is divisible into clusters of interrelated components, encompassing demographic factors, agricultural economics, farm productivity, commodity prices, and personal income.

Agriculture, like all sectors of postware society, was undergoing revolutionary change. "Revolution" is a strong word, of course, but for agriculture the changes under way were just that momentous and irrevocable. Postwar agrarian America contrasts starkly with today's urban society, but it was already distinctly different from its prewar counterpart. The rural population in 1940 accounted for roughly 43 percent of the nation's total, and the farm population—those who actually worked the land—counted for almost a quarter of all Americans. Six years and a war later, the rural population was down to about 40 percent, and the farm population down to 17 percent—shifts presaging the massive migration patterns about to hit the farm belt. Rural demography was influenced mightily by the nonreturn of the farm boy at war's end. Veterans remembered all too well the travails of prewar rural life—a life synonymous with poverty and isolation—and went instead to the cities, to the rapidly growing industrial belts of the Middle West, and to the newly emerging Levittowns springing up on the urban periphery. Urban, and suburban, life was the future; rural life was the past. Tables 3 and 4 indicate postwar population trends.

The U.S. Department of Agriculture (USDA) viewed this decline as paradoxically regrettable and welcome. Fewer rural dwellers meant fewer USDA constituents, but also reduced labor surpluses—a trend the department hoped might forestall any repeat of the devastating prewar farm depressions. To others, and especially to those serving in the U.S. House of Representatives, the declining rural population boded only ill, since many in that chamber publicly equated rural emigration with lower farm production, crop shortages, and, equally important, eroded rural political power. The USDA proved far more prescient in this regard, however. Wartime farm production already had skyrocketed by some 33

Table 3. Urban and Rural Populations, 1940–1980

(in millions)

	Urban		*Rural*	
	No.	*%*	*No.*	*%*
1940	74.4	56.7	57.2	43.3
1950	96.8	64.0	54.5	36.0
1960	126.3	69.9	54.1	30.1
1970	149.3	73.5	53.9	25.6
1980	167.1	73.7	59.6	26.3

Source: Statistical Abstract of the United States, 1982 (Washington, D.C.: GPO, 1982), pp. 21, 26.

Table 4. Farm Population, 1930–1984

(in millions)

	Total Farm Population	*As % of U.S. Population*
1930	30.5	24.9
1940	30.5	23.2
1950	23.1	15.3
1960	15.7	8.7
1970	9.7	4.8
1980	6.1	2.7
1984	5.7	2.4

Source: Statistical Abstract of the United States, 1986, p. 633.

percent since 1940, and department officials saw no reason why that trend should change; they in fact worried more about product surpluses.[8]

The apparent contradiction in the relationship between farm population patterns and production trends is explained by the massive changes in the nature of farming already under way. Wartime needs prompted the relaxation of many land conservation, crop rotation, and farm management policies enacted under the New Deal—when overproduction and the Dust bowl were dominant issues—and farmers who were guaranteed markets for their crops planted fencepost to fencepost. But these measures in many ways were reimposed after the war, without noticeably harming production. The roots of the postwar farm boom lay, instead, in the almost overnight technological transformation in farming that would eventually generate massive dilemmas for those dedicated to rural life.

The American farm in 1946 in this respect already differed markedly from its prewar counterpart. There were, as table 5 indicates, fewer but larger farms—the trend away from the small family farm had begun. Farms were increasingly mechanized, hence the need for fewer farm hands and the resulting slack in the farm labor force. About 1.5 million tractors were in use on American farms in 1940—a figure that almost *doubled* by 1946 and would double yet again by 1960. Widespread use of hybrid strains dramatically increased yields per acre; corn yields, for example, rose from 23 bushels per acre in 1933 to about 62 per acre by 1964.[9]

Table 5. Farms, by Number and Average Acreage, 1930–1985

	No. (in thousands)	*Average Acreage*
1930	6,546	151
1935	6,814	155
1940	6,350	167
1945	5,967	191
1950	5,648	213
1955	4,798	251
1960	3,963	297
1965	3,356	340
1970	2,949	374
1975	2,767	391
1975[a]	2,521	420
1980	2,433	427
1985[b]	2,285	445

Source: Statistical Abstract of the United States, 1986, p. 631.
Note: Hawaii and Alaska excluded.
a. Figures from 1975 on are based on a 1974 redefinition of categories.
b. Estimated.

But farm production boomed most of all because of chemicals. Use of lime to reduce soil acidity and to enhance fertilizers tripled during the war, while application of phosphates and nitrogen almost doubled. The future, at least in this respect, was bright: the new era of farming in America promised greater technological sophistication as the products and techniques developed in war were shifted to the civilian sector. We typically think of innovations like nuclear power, the Jeep, and maybe Spam, when discussing the postwar conversion to a peacetime economy, yet farmers probably benefited most from the knowledge gleaned through wartime research. This technological revolution in farming was to have a tremendous social impact on farming and on the "social meaning" of being a farmer. "Science," noted one observer, "made agriculture a business . . . not a relatively low-paying way of life. . . . Science saved the American farmer from agrarian peasantry."[10]

At the same time, however, technology and its impacts on farm life were to have perhaps ironic consequences for the agricultural economy. New technology made farming more efficient, but in the aggregate also generated huge crop surpluses that drove down commodity prices and threatened another depression in the farm belt. Only the postwar European famine and the coming of the Korean War ate up those surpluses.[11] The fact remained that technology, while it made farming easier and more cost-efficient, did not make it significantly more profitable. The farm problem was (and is to this day) the problem of raising farmers' incomes while simultaneously ensuring optimal total production, and the American farm economy tends to work in highly inelastic ways. That is, if each farmer maximizes production to protect personal income, the inevitable aggregate outcome is a glut. The ensuing surplus pounds down commodity prices, but, for most crops, does *not* expand consumption appreciably. The farmer thus produces more but earns less, particularly if the initial planting is capitalized through bank loans. The individual farmer will not or cannot lower production unilaterally to reduce the surplus, because such a decision would be economically irrational. A third party, government, thus is recruited to manage the relation between aggregate production and farm income, to create an equilibrium where the market apparently cannot.

So central was the farm problem to the nation prior to the war that the New Deal attempted to remold agriculture perhaps more than any other sector in the national economy, creating a web of policies and programs that exist in some form to this day. The Commodity Credit Corporation floated low-interest loans to capital-starved farmers. Rural Electrification Act projects laid the foundations for farm modernization. Commodity price supports—fixed by the Agricultural Adjustment Act—provided farmers with equitable returns for their crops. Direct farm subsidies, intended as temporary Depression-era measures, became, by the mid-1940s, rural entitlement programs defended both by farmers and their congressional patrons. Farmers, who as individuals typically espoused Jeffersonian ideals of self-sufficiency and limited government, came to expect such payments, if only because they saw their plight as an aberration in the market system. By 1946, in fact, federal subsidies for eighteen different commodities totaled almost $1.6 billion, and agriculture was fast on its way to becoming the economy's first "collectivized" sector.[12]

So the farm problem was, in many respects, *the* domestic issue of the immediate postwar period. Many feared that surpluses, which had caused the collapse of the farm economy after World War I, would return. "By autumn 1946," notes Matusow, "in spite of growing evidence that Europe needed food, [Secretary of Agriculture] Andresen was considering a call for reduced output, and he foresaw food surpluses and price supports rather than shortages and price controls as his real problems."[13] That fear, real or not, would dominate debate both on the farms and within government, and the "politics of abundance" would have a tremendous impact both on overall farm policy and on government's response to the new age of chemical pesticides.

Defining the Issue: The Pesticides Revolution

Many people are beginning to realize that it is no longer necessary to put up with fleas, mosquitos, and other pests that annoy man and carry diseases, or to continue suffering the heavy economic losses caused by a multitude of insect pests that attack growing crops, forests, livestock, and all agricultural products.[14]

Issue definition is at best an inexact science, yet is critical to understanding why government does what it does. Issues surely divide people, but they just as easily can unify the like-minded, particularly when they are defined in a way that gathers together a broad array and number of adherents. How an issue is defined, in what terms the debate is set, comes before any sort of "decision making." Issue definition influences our choice of conflicts to be entered into—which, in turn, allocates power.

Defining an issue itself is a form of conflict, because how an issue is presented structures the number and range of participants and policy alternatives. Define an issue broadly—"national defense" always is a good example—and you open up the floor to all views, no matter how crackpot or out of favor. Define the issue narrowly—the "MX missile" instead of "national defense"—and you likewise narrow the range and number of interested or legitimate parties to the debate. After all, more people are interested in and able to speak to the broader and more philosophic question of defense than can argue capably about a single piece of military equipment. "The number of people involved in any conflict determines what happens," Schattschneider argues, because "every change in the number of participants, every increase or reduction in the number of participants, affects the result."[15]

How, then, was the pesticides issue defined immediately following World War II? Two terms—"agriculture" and "progress"—come to mind, terms with broad intrinsic meanings that staked out the boundaries for policy debate. Pesticides in the 1940s were equated predominantly with agricultural progress, while most other concerns, such as "the environment" (a term not yet in use) or "safety," did not intrude so mightily into the debate. In fact, the Adams Act of 1907, which established federal entomological research, stipulated that such monies "shall be applied *only* to paying the necessary expenses for conducting original research or experiments on the agriculture industry of the United States."[16] All other matters, if recognized at all, clearly were subordinate to the farm problem.

Few in the urban 1980s comprehend the importance that farmers in the 1940s attached to the newly emergent pesticides technology. Most Americans today are several steps removed from the farm, and food is something to be purchased in prepackaged form in supermarkets, not something to be wrested from nature. Few urbanites can picture the devastation wrought on wheat crops by stem rust, on corn by the corn borer, on cotton by the boll weevil, or on livestock by scabies or cholera. To the farmer, however, these problems are both direct and compelling, since such maladies have debilitating impacts both on crops and personal income. According to the USDA in 1945, the *average* annual loss in farm income from pests and crop diseases was about $360 million.[17] Even if this figure is debatable—estimating the market value for crops never harvested is a tricky business—crop pests undoubtedly were costly banes.

Major trends in agricultural practices threatened to exacerbate the problem. Economic imperatives increasingly forced farmers to turn to *monoculture,* the practice of growing a single crop each year. Farmers prior to the twentieth century regularly rotated their crops because, as the South discovered with cotton, consistently planting the same crop in the same place eventually destroyed soil nutrients. To avoid this fate, the farmer might plant half the acreage with the cash crop, for example, corn, and the other half with a soil enricher such as alfalfa, and then switch acreage a year later. Crop rotation made tremendous scientific sense, but its economics sometimes were another matter. The American farm always has been a "factory in the field," typically capital-intensive and focused clearly on cost effectiveness. This orientation, rooted in the nineteenth-century upper Midwest, impelled American farmers to rely heavily on a single, high-yield cash crop. Widespread introduction of chemical fertilizers in the early 1900s made crop rotation unnecessary, since one now could enrich the soil chemically. Crop rotation remained common in dairy areas, where forage crops served a double purpose, but the major cotton, corn, and wheat regions of the Midwest and South turned increasingly to single-crop farming. "Monocultural mass production," argues James Whorton, "appeared as the surest path to profits, and if few farmers could afford the huge 'bonanza farms' of the northwestern wheat regions, many could at least imitate them on a modest scale."[18]

But widespread monoculture brought with it unforeseen perils. Multicrop farms are complex ecological systems where insects nurtured by one crop feed on insects threatening the others. Clover, for example, harbors ladybugs, which feed on aphids and other pests. Monoculture, by contrast, is a simple system, far more delicate and amenable to major pest blights since the single crop no longer is guarded by "friendly" parasites. Lacking such "natural" predators, farmers turned increasingly to chemical technologies.

Federal agricultural policy apparently also provided imperatives for wider pesticides use. "The suddenly availability of pesticides after World War II coincided with a farm policy that was a pesticides salesman's dream," argues Harrison Wellford, long a critic of federal pesticides policy. "Controls were placed on acreage, not production. The farmer poured technology into his remaining land to increase his yields."[19] Wellford's point is valid, particulary when juxtaposed with the trends toward monoculture, because federal acreage restrictions provided an unforeseen incentive to farmers dependent on a single crop. Farmers were paid a flat rate per acre left fallow, but were allowed unrestricted production on remaining land. Each farmer's rational response could be but one thing—take the cash, and then maximize profits by producing as much as possible on the unrestricted acreage.

Acreage restrictions thus paradoxically generated more surpluses, and may have forced farmers onto an endless pesticides treadmill. Greater use of pesticides helped to produce greater crop yields, which drove down prices. Low commodity prices hurt farmers burdened with high capital investments—which meant most American farmers—compelling them to increase yields per acre even more to recoup losses. This dynamic in turn required even greater use of pesticides so as to eke out the last possible stalk or ear, producing yet another round of surpluses. And so on. That such surpluses undercut price levels and propelled this cycle (or massive federal subsidies to prop up farm incomes) suggests a measure of the frustration farmers must have felt.

Farmers saw technology as the answer to what probably were far deeper problems, and they not surprisingly regarded the emerging pest-control technology as the panacea in their fight against pests. The prewar farmer possessed a limited pest-control arsenal, since most available products were rather inefficient, comparably expensive arsenic and lead-based compounds. Such widely used inorganic compounds as copper arsenite and lead arsenate were rather brute instruments, short-lived in their efficacy, and requiring repeated applications by hand. They were acutely toxic to their users, and had disastrous impacts on farm animals and such "beneficial" insects as bees. And, more critical, such products came under increasingly sharp debate within government and professional circles just prior to the war. Farmers constantly caught in the crossfire over arsenicals were only too happy to give them up for something more effective and less overtly toxic. Formulations distilled from plants, such as pyrethrims, were far less toxic, but also short-lived, inordinately expensive for wide use, and distilled largely from plants imported from Asia, a supply cut off by the war.

World War II was the catalyst for real research breakthroughs on pesticides, just as it was for countless other scientific areas. Research was spurred on by dual goals of increasing food supplies and for saving soldiers from the debilitating effects of malaria and typhus. Protecting soldiers in fact was the primary goal, since many more died from such diseases than from enemy bullets. The stimulus to research was victory, a goal no different from that guiding the Manhattan project, and the entomological community both in the USDA and the land-grant colleges wholeheartedly joined in the battle.

What resulted was no less revolutionary than the atom bomb, for the discoveries made in war were to transform the way Americans farmed, the abundance of their food, and the quality and price of what they bought. Wartime research produced a wide array of uses for DDT, a synthetic organochlorine developed by the Swiss in the 1930s. DDT was, compared to then available compounds, a dramatic answer to a spectrum of health and agricultural problems. It was "persistent" where earlier formulations were not; that is, once applied, DDT remained lethal to pests far longer, requiring fewer applications. It was more lethal in concentration, so it could be more diluted and still achieve desired effects. It was regarded essentially as nontoxic to human health. In fact, as wartime showed, it was central to eradicating typhus and malaria, and scenes of soldiers "dusting" refugees with DDT powder became commonplace. It was relatively inexpensive, far less costly than were the organics and inorganics preceding it. And it had multiple uses: vector-borne diseases, crop pests, common household pests—the list went on and on, as those in the USDA unlocked additional uses for this wonder chemical. In short, DDT was *the* panacea, and widely hailed as such. "The possibilities of DDT are sufficient to stir the most sluggish imagination," stated Brigadier General James Simmons in 1946. "In my opinion it is the war's greatest contribution to the future health of the world."[20]

DDT was but the most famous of a wide array of new technologies developed during the war that would dominate the national pest control market through the 1970s: persistent organochlorines like chlordane, heptachlor, and dieldrin; herbicides like 2,4-D and 2,4,5-T; organophosphates like parathion and malathion. It was a chemical revolution, one forged in war. New technologies augmented new products. The USDA refined aerial spraying techniques during the war, which brought a revolution of its own because farmers no longer needed to walk along the furrows, laboriously applying arsenicals or leads by hand. One had to do it this way, else the toxins would kill everything besides the pests. Aerial spraying proved ideal for the new compounds, which could be less concentrated but more effective than the old, and the sight of cropdusters dodging power lines and laying trails of white powder over croplands became commonplace.

Wartime priorities shunted most of these new compounds to the military, so agriculture continued to rely on available products. But farmers knew that revolutionary new technologies existed, and chemical companies in turn recognized a pent-up demand for these products that would explode at war's end. Indeed, chemical makers, sensing a vast new market in sythetic pesticides, pumped some $3.8 billion into production expansion between 1947 and 1949. Their expectations were met: DDT sales in 1944 totaled roughly $10 million, mostly for military use, but sales of DDT *alone* in 1951 surpassed $110 million, the bulk going into agriculture.[21] New formulations were snapped up as soon as they became available, with those both in the agricultural community and entomological circles shedding nonchemical methods without a backward glance. John Perkins observes,

> Not only were the older chemicals replaced by newer ones, but the new inventions were sufficiently cheap and effective to warrant adoption for control of insects not previously considered to be subject to chemical control. . . . Not only were biological methods of control replaced or disrupted by use of chemicals, but entomologists who persisted in attempting to develop control procedures using both biological control and chemicals were . . . ridiculed by the dominating chemical control proponents as a lunatic fringe of economic entomologists.[22]

The dramatic success of DDT and its kin in fact prompted the USDA to virtually abandon all nonchemical control research and practices. This reversal of long-time departmental efforts no doubt was due in part to the sudden infatuation with the new technology, but the fact that an equally enthusiastic Congress gutted biological control programs after the war may have been even more influential. The chemical age had begun.

There were, to be sure, cautious voices raised, many by USDA researchers. DDT soon was found to be lethal to honeybees (which are essential to crop pollination), and ineffective on certain crops or pests. Researchers also were concerned about the chemical's "astounding" rate of accumulation in body fats and organs, and recommended that it not be used on forage crops because it might contaminate dairy products. DDT was released for civilian use in 1945, but stocks of the chemical were scarce until late in the decade, a situation that some in the USDA saw as fortunate because it gave the department more time for testing.[23]

Such cautions, however, essentially were swept aside in the mad rush to acquire and use the new products. "The theoretical dangers of DDT," observes Thomas Dunlap, "were so distant, and its superiority over lead arsenate and its effectiveness in so many situations so obvious, that it quickly became used in agricultural and public health insect control programs throughout the country."[24] Overall opinion within the agricultural community was that, with "proper use," the new formulations were safer and more effective than anything known previously. Farmers, supported by the chemical industry and by their own economic needs, wasted little time, and DDT soon was used more widely and for more purposes than the USDA itself recommended.[25]

The Pesticides Paradigm

"The triumph of chemical pesticides," writes Thomas Dunlap, "was not due just to the visible results they gave, but to their acceptance by a public and a farming community that valued, above all else, convenience, simplicity, and immediate applicability."[26] Pesticides may have provided farmers with a new technological edge in their battle against pests, but the economic imperatives made chemicals absolutely necessary. The new products were relatively cheap, and promised to the farmer increased yields at lower costs. Such economic arguments were, if not totally compelling, highly persuasive.

The new synthetics also proved particularly suited to farmers' psychological predispositions. To succeed, biological pest control typically require both close cooperation among all farmers in an area and an active government role in the program, since even one uncooperative farmer's pests could ruin matters for the rest. Chemicals, by contrast, suited farmers' individualist ethos—each in control of his own destiny—and provided relatively inexpensive, effective, and quick technological fixes. Chemicals also required far less labor, and farmers easily gathered sufficient capital to buy the new products. Chemicals, in short, provided to each farmer individual and *individualized* mastery over his particular chunk of nature, a mastery requiring little, if any, government meddling.

The pesticides panacea of the postwar era thus was propelled on by technological, economic, and, later, political imperatives that powerfully sculpted the agricultural community's attitudes toward both pest control and government regulation. The "pesticides paradigm," as some scholars came to call this set of orientations, would prove a powerful orthodoxy. In Dunlap's words, "It was this frame of reference, a set of unspoken and almost unconscious assumptions about the need for, and the uses, disadvantages, and possible dangers of chemical pesticides, that accounts for the enthusiastic and almost uncritical acceptance of DDT found in 1945, and that helps explain both the wide use and passionate defense conducted later."[27]

This set of assumptions shared among those in the agricultural community, the USDA, and the congressional farm bloc, would dominate policy debate and subsequent government action for a generation. The imperatives *for* use clearly outweighed any signaling caution. James Whorton observes, "Chemical pesticides were developed to meet a pressing legitimate need; they were developed at a time when the danger of epidemic chronic intoxication from environmental contamination could not be fully appreciated. By the time this danger was appreciated, chemical pesticides were a *fait accompli;* predictably, agriculturalists, convinced that insecticides, as commonly used, were both necessary and safe, reacted strongly to the charges that they were neither."[28]

The pesticides paradigm thus narrowed the range of permissible public policy options. Pesticides were among the keys to postwar agricultural abundance, the road to greater production at lower costs that would lift farmers up from their status as second-class citizens. No longer would agriculture be an economic lag sector; it would take the lead. Technology would make food cheap and abundant, and technology would keep it that way, creating in the process a set of consumer expectations that, in turn, would fuel even greater technological dependencies. Agriculture's key economic role would extend soon to foreign policy as America became the world's breadbasket, with farm commodities eventually topping the nation's list of exports. The pesticides paradigm figured heavily in agriculture's new centrality, and those standing to benefit most directly from widespread pesticides use would lead in formulating appropriate federal policy. It is to these interests, and to the contexts within which they operated, that we now turn.

Pesticides and the Structure of Policy Bias

One interesting way to understand the politics of a generation is to study its standard textbooks. Texts are, in the words of Anthony King, works that "convey the common understanding of the time in which they were written."[29] Political textbooks can be, but seldom are, controversial. They are if anything the mainstream lore of political reality.

Robert K. Carr, et al., writing in 1961, considered what they thought to be the "spectrum of pressure groups": "business, labor, agriculture, the professions such as law, medicine, and education, regional, racial, and nationality groups, war veterans—almost every type of group interest or allegiance known to man."[30] Nowhere, notice, did the authors mention environmental, consumer, or any other type of "public interest" organization, but this was no oversight. Such groups simply did not exist, or were not very important, when the text was published.

This minor illustration highlights a problem underlying many group theories to emerge curing the 1950s and early 1960s. The aforementioned list may seem lengthy, yet, as Schattschneider argued, "The range of organized, identifiable, known groups is amazingly narrow; there is nothing remotely universal about it."[31] The range of known groups relevant to the pesticides issue during the 1940s was equally, if not more, constrained—with critical consequences for policy outcomes.

Having set the broad social, economic, and technological imperatives that set the parameters for the debate over pesticides policy, we now examine the distribution of interests that, in retrospect at least, were *potentially* linked to the issue area. Pluralist theorists often seem content to compare relative group strengths, agendas for action, and policymaking strategies in studying conflict. Such factors are important, but before considering any of these we need to know which interests were attached to this issue in the first place. We also need to examine their relative ease of access into government before we even start to talk about group power.

In this respect we need to remember that the pesticides issue in the 1940s was defined by the pesticides paradigm, buttressed by the centrality of agriculture to postwar America. This reality ensured that the pesticides issue, when it arose, was debated almost entirely in agricultural terms. This general consensus about pesticides meant two things. First, few nonagricultural policy actors were attracted to the issue. The major "conservation" groups of the day displayed little if any interest in pesticides, and instead devoted their limited resources and energies to land use or wildlife management. The Audubon Society at this time was a relatively modest collection of affluent bird-watchers who campaigned tirelessly against the use of plumage in apparel or for decoration; its "old ladies in tennis shoes" reputation was indicative. The Sierra Club was a California-centered organization dedicated to land use projects, while the National Wildlife Federation worked most to create and maintain wildlife preserves. Even if such organizations had worried about the new technologies, few probably understood the stakes involved in the widespread use of pesticides. The debate would arise almost entirely within the agricultural community—a configuration of interests already bound together by adherence to the pesticides paradigm.

Second, definitional consensus among a constrained set of policy actors translated into low overall public attention to the issue; no fire, no smoke to attract bystanders. The "audience," those potentially drawn to this issue, either did not recognize that the issue existed or undervalued its importance. FIFRA was a "nonissue" to most Americans, and in fact was not even the top agricultural issue of the time; the newspaper headlines focused instead on the acrimonious and highly public battle over the first postwar omnibus agricultural act.

Two interrelated factors—lack of issue conflict and low issue visibility—produced a representation of interest skewed entirely toward the agricultural community. If pluralist theory hinges on some notion of participation by "active" and "legitimate" groups, there is little question that the definition of the pesticides issue in postwar America produced a range of policy claimants heavily constrained by and skewed toward the agricultural perspective, and that policymaking would include only those with direct and largely economic policy stakes in the issue. This condition, in all fairness, is no surprise, given what we know about incentives for group mobilization, but it does raise questions about a central tenet of the pluralist image.

In 1951 David Truman lay the foundation for postwar group theory with the argument that "disturbances" prompt those with similar interests to emerge, organize, and act on behalf of collective needs.[32] Disturbances no doubt do spur on group mobilization, but this notion implies that an issue is highly controversial and visible—at least, the latter. Both conflict *and* visibility were absent from the postwar pesticides debate. This is significant because critics of pluralism long have argued that this image of policymaking overemphasizes "big" issues, that it relies too heavily on controversy, with its attendant group mobilization and conflict. If competition and visibility are keys to the scope of issue conflict, then the scope of conflict over postwar pesticides policy was so limited as to be almost nonexistent. The pluralist thesis as a result suffers in this case, because it cannot explain policymaking in "routine" or "small" issue debates. The emergence of the new pesticides technology certainly was disturbance enough for a set range of policy claimants, but no groups were mobilized that did not already have direct stakes in the widespread use of chemicals.

The disturbance theory also falters with respect to the questions of group resources and their access to decision-making power. Jack Walker argues accurately that "no matter what propensities exist, large amounts of capital are needed to form and maintain most groups."[33] Disturbances may indeed catalyze group mobilization, but these still nascent interests need both money and an entree into government to coalesce into political forces. Group influence hinges less on "storming government" as if it were the Bastille than it does on exploiting existing connections or cultivating potential ones. Interest groups, to succeed, need patrons—be they providers of capital or opportunities for access into government. Every policy claimant needs someone on the inside willing to bend a sympathetic ear or lend a helping hand; few groups can twist arms and get away for it very long. It is this notion of patronage, insofar as it applies to the farm community, that we now turn. It will also recur in later chapters.

USDA and Its Clients

Wesley McCune concludes his study of the agricultural community by arguing,

> More than any other Cabinet department, Agriculture is the protagonist, the pleader for its constituents. It teaches them to be self-sufficient, to use government credit, and to live inexpensively. Like the Department of Labor and the Department of Commerce, it grinds its ax *officially* for one part of the national economy—only more so. That is not to say that it should not do so; it is only to say that the farm bloc is not made of thin air. It is nurtured, consciously or unconsciously, twenty four hours a day. It is a facet of the democratic way of lending a helping hand to one chunk of society.[34]

Hyperbole aside, McCune is correct: the agricultural subsystem was created and is to this day nurtured by the U.S. Department of Agriculture, the farm community's earth mother and government nanny. The USDA, in conjunction with the congressional committees on agriculture, were so crucial to the organization of that subsystem and its primary interests that clearly the two major farm organizations of the 1940s did not emerge "spontaneously" at all; they instead were generated by government and its policies.

The American Farm Bureau Federation (AFBF), which spoke for roughly 1.2 million farm families in 1947, emerged *following* the passage of the 1914 Smith-

Lever Act that established the USDA's Agricultural Extension Service. This service relied on its network of county agents to bring modern information and techniques to the farm, and to organize farmers on their own behalf. The county agents in fact nurtured the development of local farm bureaus "to help them establish good relations with the farmers that their task so obviously required, and to overcome the hostility that might have attached itself to agents of the federal government."[35] The Smith-Lever Act made county agents the conduits for federal agricultural aid, and in this sense was as much a charter for the farm bureau system as it was for the Extension Service.[36]

The AFBF was the umbrella for the local bureaus, and its size and local ties often led it to regard itself as *the* voice of agriculture. Whether agriculture ever was so united in actuality is debatable, but there never was any argument that the federation's breadth and grass-roots linkages made it a powerful force in American agriculture during much of this century. V. O. Key's comments are illustrative:

> From the circumstances of its origin in local groups of farmers organized by county, it developed and continues to maintain close relationships with the land-grant colleges and their state extension services. In a sense the Federation speaks for a coalition of cotton, corn, and agricultural services associated with the land-grant colleges. Its great legislative success in the 1930s resulted in part from the fact that it brought the South and the Midwest into alliance, a problem that had plagued agrarian politicians of the 1920s.[37]

Critics of the federation and its allies argued that it had been "conceived by businessmen and county agents, born in a chamber of commerce, nurtured on funds from industry, and has never completely left its home and its parents."[38] The farm bureau system was always open to such attacks largely because its members consistently emphasized the farmer's role as a producer for the market. With monoculture came large-scale farming, and farmers were exhorted to think like businessmen and transform their operations into commercial enterprises. The county agents, for their part, helped farmers to organize and generate financial support for their interests because they wanted agriculture to be as organized as business and labor.[39]

Tied into the farm bureaus and local business communities are the farmers' cooperatives, collective enterprises—exempt from federal antitrust laws—that provide members with a wide array of services, products, and marketing outlets. The National Council of Farmers' Cooperatives claimed some 4,500 affiliates in 1940, with heavy concentrations in the wheat and corn belts of the midwest.[40] if the local bureau gave the farmer information and technical expertise, the cooperative provided purchasing and marketing power—a role to become increasingly large-scale and profitable as agriculture evolved from the family farm to the corporate farm.

Federation policy in both instances flowed upward from the grassroots, with constituencies well organized at the local level so as to include business, local colleges, and relevant political subdivisions. This structure has historically proved effective in Washington, since each federation gives the USDA and congressional farm bloc members a formidable voice both in collective terms—the federation speaking as one—and purely in terms of each local bureau or co-op. That both federations are organized along lines paralleling the decentralization and local ties embedded into the American constitutional system is no coincidence; the structure of the political system itself powerfully conditions that of the universe of corresponding interest groups.

Other farm groups, though smaller and more regional in character, also provided important sources of support for their government patrons.[41] The National Farmers' Union (NFU), as liberal as the AFBF was conservative, evolved in the early 1900s out of the radical populism of the upper Great Plains. The NFU found its institutional home in the USDA with the passage of the Agricultural Adjustment Act of 1933, and what the Extension Service was to the Farm Bureau Federation the Adjustment Administration became to the NFU during the New Deal—a patron, and sure source of continued access. As Allen Matusow describes them: "Though not too effective as reapers of votes, the farm organizations have always been of considerable political importance. In the 1940s, the Farm Bureau was exercising power over farm legislation rivaling that of the Department of Agriculture, while the Farmers' Union and the AAA had great influence with certain members of Congress in the Middle West."[42]

Size and structure may have served these organizations well in Washington, but their popularity with farmers stemmed not so much from enthusiasm as from the benefits they provided to the farm community. The problem of incentives never was too significant for the cooperatives—which are commercial enterprises anyway—but the Farm Bureau, in seeking a range of indivisible benefits for the farm community, traditionally has relied on a range of particularistic benefits to bolster membership. Federation members enjoy special insurance rates, discount merchandise, and other attractions to augment the organization's central services. Such benefits also ease the task of mobilizing members because farmers are better able to perceive that they have a direct stake in organizational goals, and receive tangible benefits. While the AFBF's sheer size might reduce incentives for individual participation—the problem of "collective goods"—the economic nature of group goals accounts for higher overall participation than one might otherwise expect.

This notion of group incentives is key to understanding why some groups dominate policymaking and others do not. E. E. Schattschneider, for example, takes pains to distinguish between "private" and "public" interests. Private interests such as the AFBF seek "exclusive" benefits, which accrue solely to group members or to some defined sector of society. "Public" interests, by contrast, seek "inclusive" benefits that all members of the community enjoy regardless of organizational membership or social status. Wheat subsidies, using this distinction, are "private" goods, since the benefits accrue solely to wheat growers. "Clean air," on the other hand, is far less divisible, being theoretically beneficial to all. Groups seeking private goods prove better able to mobilize their memberships than those seeking public goods, since the latter are burdened more heavily by the "free rider" problem. Indeed, Mancur Olson observes, private interests are more likely to mobilize members successfully because the benefits sought are more direct and tangible.[43] The cumulative impact is that the incentive system favors private groups, which in turn dominate politics as they seek to defend and expand their entitlements. That the USDA nurtured this bias is of no small consequence.

Bias in the Legislature

Incentives, however, are not always enough to explain why some groups dominate an issue area. If incentives alone had been the key, the AFBF, despite its services and size, might not have succeeded. The farm groups succeeded because they enjoyed access to key government actors. Jack Walker argues that "the number of major interests, the mixture of group types, and the level and direction of political mobilization in the United States at any point in the country's history will largely be determined by the composition and accessibility of the system's major patrons of political action."[44] Mobilization is one thing, but political influence may hinge more on how open the political system is to group demands, or in what way the system nurtures, accommodates, or rebuffs those demands. We thus are compelled to turn away from incentives and relative group strength to the question of access, or, more correctly, the opportunities for access and influence open to various claimants.

That the farm groups enjoyed unique access to the USDA is no surprise, but the structure of the entire political system in the 1940s in many ways discriminated markedly in favor of those seeking greater use of pesticides. William Jump, upon retirement from the USDA, nicely summed up the webs of influence and interaction characterizing the agricultural subsystem and how it dealt with common needs:

> The great connectionalism (I know about the word "nexus," but it's too simple a word to do justice to the agricultural organization in the United States) in which the Department is involved with the land grant colleges and other State agricultural departments and institutions, the farm organizations, the farmer committee systems, the cooperatives, the agricultural credit committees, the industry and trade groups, the congressional committees, and so on, has developed a philosophy of collaboration and cooperation that has made an imprint on the Department and its methods of thought and action that is unique in Government.[45]

Jump's "great connectionalism" depended ultimately on the centripetal force generated by the congressional committees on agriculture—forces superior to any provided by the USDA. Congress as a whole has traditionally represented rural and small-town America well, sometimes far in excess of actual rural population. This bias was always a structural one in the Senate, since each state is represented equally. In the House it was based essentially on demographics—the rural nature of our population during much of our history—though it also was structural, as the apportionment battles of the 1960s would show.

House members are particularly vulnerable to populations shifts, and postwar migration patterns cautioned rural representatives that their political power was eroding inexorably from beneath their feet.[46] However, such trends were only beginning in the 1940s; approximately 140 members of the House comprised its powerful farm bloc, and rural states far outnumbered urban ones in the Senate. More important, agriculture, especially southern Democratic agriculture—that is, cotton—dominated congressional committee chairs, its members comfortable in one-party states and safe seats that ensured them long and influential lives in the legislature. Farm policy has long been synonymous in political science with subgovernment politics, so there is little need here to describe in detail the congressional role in farm policymaking.[47] In 1946, suffice it to say, agriculture still was king, particularly in the congressional committee system.

But Congress in 1946 was on the edge of sharp institutional change. The reforms embodied in the Legislative Reorganization Act were intended, among other things, to streamline the postwar Congress and "modernize" its committee system. Members increasingly argued that the institution's plethora of standing committees—forty-eight House and thirty-three Senate in 1946—hamstrung congressional efficiency, fragmented policymaking, weakened the body's overall organizational capacity to deal with postwar problems, and, therefore, eroded the legislature's ability to counter the growing power of the executive branch. Such motivations for reform would recur almost thirty years later, but the changes at this time were hailed as a "legislative miracle . . . the most sweeping changes in the machinery and facilities of Congress ever adopted in a single passage."[48] Congress slashed the number of standing committees, prescribed explicit committee jurisdictions, and formalized committee procedures and record-keeping—actions that were sought cumulatively to alleviate internal fragmentation and to give congressional leaders greater leverage.

Yet, upon reflection, the changes only enhanced the power of the committee chairs and further entrenched already powerful policy biases. First, the reorganization broadened the jurisdictions of the remaining committees, consolidating greater legislative domination in fewer decisional loci. Second, rationalization of committee jurisdictions, while not totally eliminating "turf wars," reinforced committee hegemony over substantive policy areas, thus eroding the ability of the leadership to avoid assigning legislation to hostile work groups.[49] Thomas

Jefferson, in his rules of Congress, once argued, "The child is not to be put to a nurse that cares not for it." The reorganization of 1946 in many ways reinforced the probability that pesticides policy, for one, indeed would go to "nurses" known to oppose strict regulation.

The reorganization also left the subcommittee system intact, which may have been a price extracted to guarantee passage of the bill. There were, prior to 1946, about 180 subcommittees, a number to grow to about 250 by 1960. Subcommittees as loci of congressional policy leverage are even further removed from control by the leadership, and in areas like agriculture become "miniature legislatures" whose specialization and even more constricted access enhance the power of the members and, specifically, that of the chairs. Charles O. Jones points to the agricultural commodity subcommittees as classic examples of policy devolution in Congress, which was particularly favorable to those interests attached to the subcommittee.[50] "Given an active subcommittee chairman working in a specialized field with a staff of his own," commented one congressional aide, "the parent committee can do no more than change the grammar of the subcommittee report."[51] The subcommittee as a cozy realm of policy dominance came to characterize federal pesticides regulation throughout the postwar era, offering to the faithful preferential status in the policy community. The "heretics," those not wholeheartedly supporting the pesticides paradigm, would find the subcommittees tough bastions to crack.

The Tax Code and Preferential Pluralism

Critics of pluralism tend to focus on socioeconomic status and its impact on political behavior, or they point to the power of subgovernments, to prove that the pluralist thesis contains woeful flaws. Few scholars, however, focus on the tax system, which channels political behavior in subtle yet systematic ways. Fewer still observe how tax systems affect participation; the emphasis instead is on how groups generate tax breaks. Yet tax laws are part and parcel of the structure of bias in America, and levy inequalities as a matter of public choice, congressional directive, or bureaucratic whim. Robert Holbert argues that "the unequal treatment of political actors under tax laws is a result of biased value system which dissuades certain actors from entering the political arena by creating penalty disincentives against political activity."[52] It is plausible, then, that the tax system may inhibit those seeking policy influence even if they *were* able to mobilize on behalf of some good. Taxes are coercive, creating both incentives and disincentives for behavior through deductions or penalties. When the Internal Revenue Service, through the authority granted to it by Congress and various court decisions, determines what constitutes deductible and nondeductible expenses, it is in fact determining who can and who cannot gain public subsidies for some activity. A group with tax-exempt status for itself and tax deductions for its contributors can offer to potential members clear and tangible benefits through tax writeoffs. To lose this credit is to jeopardize both membership and contributions, since many members may decide that the expenditure outweighs any nonmonetary benefits received.

The tax code in the 1940s (as today) largely dealt with business deductions far differently from "nonprofit" organizations. Section 162 of the tax code, a provision practically coterminous with the Sixteenth Amendment (1914), allows a business or business organization to deduct from its taxes all expenses deemed as "ordinary and necessary" for business affairs. Outright lobbying expenses are not deductible—though the line between "lobbying" and "legitimate" activity is rather blurry—but business can deduct the costs of advertising, certain expenses related to legislative activity (such as travel or lodging), and any expenses incurred as part of membership in a business association. Organizations like the Sierra Club or the National Audubon Society, however, faced stricter limits on their activity. As section 501(c)(3) organizations—those established for "religious, charitable, scientific, or education purposes"—such groups were tax-exempt, and members could deduct their dues and contributions so long as the groups did not engage in "substantial" legislative activity. The interpretation of "substantial" was left to the IRS (a reality that would haunt environmental groups some years later), but the law in 1946 on its face forbade any such group from engaging in overt attempts to influence public policy beyond relatively passive "education" programs.[53]

One doubts that the tax laws had any impact on the pesticides policy debate of the late 1940s, given the narrow definition of the issue and the lack of general knowledge about the potential health and environmental hazards of pesticides. Hence the debate would remain among those fully behind pesticides use. However, the tax code and the IRS would be major issues in themselves some twenty years later. If organizations like the Sierra Club had wanted to participate in 1947, extant tax laws might have hampered their range of activities with the threat of losing their tax-exempt, tax-deductible status—a potential loss sure to give any group pause. One should also note that Congress in 1946 passed the Lobby Registration Act, a provision requiring all "lobbyists" to register annually, as part of its reforms. Whatever the law's deficiencies, and there were many, one fact remained: registration as a lobbyist by any 501(c)(3) group was sure grounds for the revocation of tax-exempt status. The major conservation groups in fact confined their activities largely to polite advice (when solicited) or rather genteel "public education" campaigns—activities in marked contrast to those employed by business and agriculture. There may have been little real choice.

The Structure of Bias: Access to the Courts

"The complete independence of the courts of Justice is peculiarly essential in a limited constitution," wrote Alexander Hamilton in *Federalist* 78, adding that the "courts were designed to be an intermediate body between the people and the legislature in order, among other things, to keep the latter within the limits assigned to their authority." Democratic theory assures us that access to a court of law is a fundamental right, constituting an alternative avenue into the "black box" of government for citizen access and interest articulation. Yet access to the courts—known as "standing"—like so many rights, is limited by the political system and by prevailing judicial norms or traditions. The courts are potential arenas for group conflict, but only when the courts themselves decide to get involved, and only on their terms.

Essential to our discussion is the notion of standing that existed in the 1940s. American law, evolving as it did out of English common law traditions, long has held that a citizen's right to a court of law is limited to cases where the plaintiff can show a *direct* and *adverse* impact resulting from the actions or planned actions of the defendant. A citizen who sues a corporation because its factory spews noxious smoke must show direct injury resulting from that smoke. If the alleged harm affects only a narrow sector of society, the plaintiffs must show that they are part of that group, while those alleging harm to *all* of society can sue only as individuals—not as representatives of some class (the notion of a "class action" suit did not yet exist). Whatever the case, the compelling factor was proof of direct and adverse impact.

The right to sue also hinged on the magnitude of the alleged harm. In *Frothingham v. Mellon* (1923), the U.S. Supreme Court ruled that a taxpayer was without standing to challenge the constitutionality of a federal statute because that taxpayer's particular interest was "comparatively minute," and because the citizen failed to show direct injury.[54] The practical impact of *Frothingham* was that, by the 1940s, citizen access to the courts on environmental or any other "public interest" matters was limited solely to negligence, liability, or nuisance suits—all of which required the plaintiff to prove damage. A farmer could gain standing with relative ease by alleging that a certain pesticide damaged his crops—a liability suit—but a citizen lacking such direct consequences arising out of private or governmental actions was denied standing.

This strict interpretation of standing, which remained in force through the late 1960s, made it incredibly difficult both in theory and in practice to sue on behalf of some "public interest." Even if a conservation group gained access, which would be unlikely, the judicial norms of the time required the plaintiff to establish clear blame, to show cause-and-effect proof of negligence or intent. Establishing clear damage or culpability in such matters is more problematic than we might like to admit, so these legal requirements provided considerable obstacles to the prosecution of alleged hazards.

The courts in the 1940s, and for two decades after, played virtually no role in environmental matters. Even if some group of citizens did perceive that pesticides composed a threat, the judicial norms of the time disallowed access to the courts and effectively took the Third Branch out of the game.

Summary: The Structure of Bias and Its Impacts

James Q. Wilson argues, "One way to understand how an issue affects the distribution of political power is to examine what appears to be the costs and benefits of the proposed policy."[55] It matters less to Wilson whether policies are "distributive," "regulatory," or whatever, than how relevant policy claimants *perceive* direct costs or benefits resulting from policy initiation. "The perceived distribution of costs and benefits shapes *the way politics is carried on,*" he continues; "in particular, it determines the kinds of political coalitions that will form."[56] The question of pesticides, in this respect, was not salient to most Americans in the 1940s, but was of tremendous direct importance to the agricultural community and its allies. Clearly, the incentive system alone offered direct and tangible benefits to that subsystem: higher profits for farmers and chemical companies, political support for congressional farm bloc members, program and budgetary support for those in the USDA. The costs of using pesticides, in economic terms at least, seemed both negligible and widely distributed insofar as the mass public was concerned. Greater use of pesticides in fact brought the promise of cheaper, cleaner, more abundant food, as well as improved public health, so few consumers probably would have opposed the trend even if they had bothered to think about it. And, at this time, there was both little knowledge of, or attention paid to, any potential environmental or health hazards posed by the new chemicals, at least by the nonprofessional sectors of society. The incentive system plainly supported widespread pesticides use.

The structural or governmental side of the equation also favored the increased use of pesticides. "All policy systems have a bias," states Charles O. Jones. "Simply maintaining an open process favors the strong; doing anything else creates other biases."[57] Given an unbiased institutional and procedural structure—an assumption clearly not made here—the incentive system would favor farmers and their allies, those facing unambiguous, direct, and econimic costs and benefits. This condition was buttressed by a structural bias in government toward what Schattschneider termed "private interests." The system of governance in the United States—its array of laws, structures, and procedures—at this time discriminated markedly against "public interests," those without such clear economic stakes in the issue. In the history of American liberal democracy, in fact, "private interests" possess deeply rooted property rights that are often more sacred than any evanscent notion of the "public good." Such conditions allow private interests to dominate politically.

Given all this, then, perhaps it should come as no surprise that pesticides policy in the 1940s displayed all the characteristics and dynamics of subgovernment politics, even though it was not a distributive policy. As Wilson argues, "Whether these measures are regarded as distributive or regulatory is less important than the fact that they represent major redefinitions of the proper scope of government."[58] What will be made clear in the next chapter is that these dominant perceptions about the use of pesticides, and, more important, the dominant perceptions about the proper federal role toward such use, will structure the alternatives of possible federal action.

Chapter Three

The Politics of Clientelism

The Issue Arises: Imperatives for Policy Change

The technological breakthrough in chemical control of pests after World War II burst onto the agricultural scene without a professional technocracy to guide and control its use. Literally overnight, some of the most toxic and ecologically disruptive chemicals known to science were placed in the hands of farmers, agribusiness salesmen, and federal farm advisors almost totally unaware of their genetic and ecological implications.[1]

"The publicity given DDT," one food company executive remarked in 1950, "might well be envied by any Hollywood movie star."[2] Such was the case with pesticides generally during the immediate postwar years. Popular magazines breathlessly reported each new pest-control breakthrough, while a chemical industry excited by the enormous commercial potential for these products extolled the virtues of DDT and its kin to a society suddenly aware of the scientific revolution under way. "DDT is good for me!" sang a pert housewife (accompanied by a cow, apple, and carrot) in a Penn Salt Chemical ad in *Time*. The advertisement exulted, "The great expectations held for DDT have been realized . . . Beef grows meatier . . . bigger apples, juicier fruits . . . more milk . . . more comfortable homes . . . protects your family."[3] The *Science News Letter* informed its readers, "For those who like to experiment, Science Service has assembled a kit containing five samples of the insecticide that is doing such a job of controlling disease and pests. Two other insecticides included. Timely. Fun to do."[4] The nation clearly was enthralled by the wave of scientific and technological marvels spun out of wartime research, of which DDT was but one of the most ballyhooed.

Cautious voices nonetheless arose amid the hyperbole. U.S. Public Health Service (PHS) researchers as early as 1944 warned, "The toxicity of DDT combined with its cumulative actions and absorbability through the skin places a definite health hazard in its use."[5] A special committee of the American Association of Economic Entomologists cautioned that DDT had "difficulties" needing further investigation, even as the president of that association declared at the 1945 annual meeting that pest extermination, not simply control, was the association's new goal.[6] And John Terres, in *The New Republic,* revealed that USDA test sprayings of DDT on northeastern gypsy moth populations resulted in widespread destruction of "beneficial" insects and other wildlife. Said Terres: "DDT's greatest effect for use out-of-doors is in its non-selective killing power. Even in experienced hands, its use may be likened to firing a broadside at a throng of people in which we have both friends and enemies."[7]

Such warnings were largely confined to scientific publications or professional journals, though several of the more popular magazines carried an occasional short article about possible problems with the new chemicals. These cautions were drowned out by the more prevalent sentiments favoring their use, which is not surprising if one considers the technologies applied previously. In all fairness, too little was known about the new chemicals to warrant alarm. All available testing procedures suggested that DDT and the rest were nonhazardous when used properly, and, if not entirely safe, certainly were less acutely toxic to human beings than were the once-dominant arsenics and leads. Even those raising cautions admitted that the new synthetics, when mixed and applied properly, were preferable to the older chemicals.

Important sectors within the agricultural community did, however, worry about possible problems arising out of the new chemical age. Farmers, for example, worried about the quality of their purchases, a concern warranted by a postwar inundation of new products that allowed many an intrepid entrepreneur to peddle worthless or even dangerous goods. The postwar demand for DDT, commented one chemist, "created opportunties for every shyster and marginal manufacturer to produce and market DDT."[8] Farmers' concerns were shared by the major agricultural chemical firms, though for different reasons, and by the farm community's government patrons.

Such concerns in fact date back to earlier federal policy. Federal pesticides regulation time and again has been made necessary because technological, scientific, social, economic, or cultural change has outdistanced extant law. Indeed, one could argue that federal regulation in almost any area of national life is today's governmental response to yesterday's conditions. Such unplanned policy obsolescence creates gaps that become painfully apparent when yesterday's policy no longer addresses today's realities. This observation applies particulary to any policy area of great scientific or technological complexity. Such a gap increasingly characterized the Federal Insecticide Act of 1910 (FIA), a sense of policy inadequacy perceived particularly by those with the clearest and most tangible stakes in the post–World War II generation of pesticides. We turn next to the roots of those perceptions.

Pesticides Regulation Before World War II

Government's concern with chemical pesticides has traditionally moved along parallel paths, one directed toward consumers, the other toward the agricultural community. This parallelism, and the timely intersection of the two paths, is key both to the historic controversies surrounding pesticides and critical to almost every instance of federal policy formation and implementation.

The first path in the federal dyad concerns the quality and cleanliness of our food—a concern as old as urban society's division between producers and consumers. Yet protecting the purity of food as a public problem—one seen widely as meriting government attention—bursts onto the national scene largely with the publication of Upton Sinclair's exposé of the meat-packing industry in *The Jungle*. The objective condition of food adulteration no doubt existed long before 1906, but lack of coalesced public concern, coupled with the dominance of entrenched producer interests in the Congress (especially in the pre-Seventeenth Amendment Senate), undermined substantive government action. Proposals for pure food legislation increased with the onset of the twentieth century, but, as one leading proponent of federal involvement observed, "There seemed to be an understanding between the two Houses that when one passed a bill for the repression of food adulteration, the other would see that it suffered a lingering death."[9]

Sinclair and other muckrakers exposed conditions that shocked many and mobilized critical sectors of the society in support of statutory change. Producer interests in Congress had resisted such legislation almost a hundred times prior to 1906, but found this time that they faced a stronger, if temporary, coalition dedicated to redressing perceived outrages. W. H. Wiley writes,

> The Pure Food and Drug Act, as it was popularly known, was a victory of aroused public opinion over Congressional inertia, a victory engineered by a coalition of forces: by scandal-mongering journalists whose magazine exposures alarmed the public about the poisons in food and drugs; by womens' clubs that pressured legislatures to insure a wholesome food supply for their constituents' families; by state food and drug officials who argued for the necessity of regulation on a national scale; and by many of the food manufacturers themselves, who hoped to purge their business of its less scrupulous operators.[10]

Pure food advocates for the first time were part of a coalition more encompassing than that resisting policy change, but ultimate success apparently hinged most on Theodore Roosevelt's vigorous intervention on their behalf. Roosevelt actively pushed the legislation through a recalcitrant Congress—a precedent for executive action on behalf of the "public interest" that would prove critical in later regulatory controversies.

The 1906 Pure Food and Drug Act (PFDA) forbade the manufacture, sale, or transport of poisonous, adulterated, or misbranded foods, drugs, liquors, or medicines. The act authorized federal authorities to set standards, enforce them through fines, and seize products deemed contaminated or a public health hazard. The duty to protect consumers was placed in the hands of the USDA's Bureau of Chemistry—a responsibility soon to be juxtaposed with its role of patron to the farmer by the 1910 Federal Insecticide Act. That law, unlike the PFDA, passed through Congress with little public fanfare or debate, largely because it was drawn up within a farm community increasingly concerned about a new chemical age. John Perkins comments, "The commercialization of the insecticide industry was accompanied by substantial fraud, including adulterating legitimate products and making extravagant claims for absolutely worthless junk. Indeed, some of the insecticides sold in the latter part of the nineteenth century were decidedly destructive to the plants upon which they were used."[11]

Farmers feared increasingly that their purchases might be ineffectual or outright dangerous, while chemical makers worried about "unbridled competition" and less scrupulous competitors. The FIA required chemical makers to list product ingredients and to guarantee farmers that the poison inside the package did what the label on the outside claimed. Chemical makers were assured, at least in theory, that all competitors played by the same rules and operated honestly, thus making all producers equal in the marketplace. The law did not actually require that the product be guaranteed as safe, but stated simply that everyone should know what was inside the package. For farmers the FIA assured quality and safety; for producers it was the gatekeeper for an increasingly wild marketplace.

Responsibility for FIA enforcement went to an Insecticide and Fungicide Board within the Bureau of Chemistry; few at the time thought such a duality of roles contradictory. By the mid-1920s, however, the department's role as protector of farmers *and* consumers came under sharper attack as the problem of pesticides residues on produce became more widely recognized. Critics focused not on the FIA, which essentially was little more than a "truth in packaging" law, but on a Pure Food and Drug Act which had palpable regulatory intent. The Bureau of Chemistry was accused of allowing overly generous "tolerances" for spray residues (the amounts deemed safe for human consumption), and, worse, lax prosecution of even those standards. The 1906 law contained a number of problems, as reformers later recognized, but, according to one student of the law, the chief problem was "the resistance to vigorous enforcement within the Department of Agriculture itself. . . . The Department of Agriculture, after all, has had as its *raison d'etre* the protection and advancement of the prosperity of America's farmers, the same farmers who, as the ultimate producers of the country's food, were necessarily affected by the enactment and enforcement of pure food legislation."[12]

Bureau of Chemistry officials believed that quiet persuasion and education, not publicity and punishment, were the most effective means for preventing product adulteration. "It is the Bureau's theory," stated assistant chief Paul Dunbar, "that more is to be accomplished by acting in an advisory capacity under such conditions as will insure legal products than by accumulating a record of successful prosecutions with attending fines."[13] This view was not accepted widely by nonagricultural policymakers, and the bureau's determination to avoid conflict reaped considerable headaches for the USDA.

Critics of the department also protested the secrecy that usually surrounded its actions. The USDA in 1925, for example, campaigned to warn apple growers about British sanctions against excessive arsenic residues on American fruit (a perennial Anglo-American trade problem), but this information dissemination process avoided the general press. The Apple Growers Association later crowed, "All of this work has been conducted on the personal basis and with the strongest possible emphasis against newspaper, magazine, radio, or any other similar publicity. In fact the matter has been kept out of even the small sheets at country points because it is only a step from the four corners to the columns of the metropolitan dailies."[14] Such secrecy aimed to prevent widespread public concern about the quality and purity of agricultural goods; "negative" publicity could have a "chilling effect" on sales. So great was this paranoia that the department refused to divulge its "informal" (that is, hortatory) tolerance for arsenic on apples, even to many growers!

Criticism about USDA activities came largely from within government or professional circles, not from the general press or citizens' groups, and the debate rarely caught the public eye. This intramural debate nonetheless prompted USDA officials in 1927 to dissolve the Bureau of Chemistry and consolidate the PFDA's enforcement duties into a new Food, Drug, and Insecticide Administration, later renamed the Food and Drug Administration (FDA). Department officials hoped that this bureau-level reorganization would mute criticism and promote better enforcement, but the new FDA had no jurisdiction over the Federal Insecticide Act. "The reorganization," notes John Blodgett, "marked the separation of regulatory activities directed at protecting the farmer from those concerned with the consumer."[15] That separation, as we will see, created both problems and opportunities for those concerned about the health and safety effects of pesticides; it would complicate coordinated policy activity, but also would allow for policy change through indirection.

Jurisdiction over both laws remained in the USDA until the late 1930s, which created a situation increasingly irksome to department critics as concern about pesticide residues mounted. Finally—and largely through the efforts of a few New Deal activists—the Food and Drug Administration in 1940 was moved out of the USDA and into the new Federal Security Agency as part of Franklin Roosevelt's general consolidation of federal health and safety activities. Enforcement of the FIA, because it directly concerned farmers and not consumers, remained in the USDA's Agricultural Marketing Service. But, in the eyes of critics of the USDA, the FDA at least was free of the clutches of the bureaucratic patrons of agriculture.

Almost. The FDA now was administratively separate, but it still had to make the annual pilgrimage to the congressional committees on appropriations—work groups dominated by senior farm bloc legislators. Worse, for the FDA at least, its appropriations remained for years under the jurisdiction of the subcommittees on agriculture, who would be likely to oppose stronger pesticides regulation. And, if one came before the House Subcommittee on Agricultural Appropriations, one dealt inevitably with Clarence Cannon.

Rep. Clarence Cannon (D, Mo.) chaired that subcommittee, earning in the process a reputation both as a staunch patron of agriculture and a fiscal conservative on other matters. Such a duality recurs constantly on the subcommittee in the postwar era. Cannon, an apple grower, believed that "lead arsenate on apples never harmed a man, woman, or child."[16] He also argued consistently that the FDA attempt to enforce pesticide residue tolerances "has reached the point of absurdity and is without foundation of any fact or scientific reason whatever."[17] Cannon was no scientist, but he was a farmer, and, like most rural legislators, shared a view about pesticides that boiled essentially down to "If it don't kill you immediately it isn't dangerous." Such a view may seem naive today, but one had to appreciate the rough empiricism in this outlook. Cumulative effects and the dangers of chronic exposure were both too little known and seemed too indirect to bother those who worried most about crops and farm income.

One 1935 incident sheds light on Cannon's attitudes and actions towards pesticides. Cannon, during one FDA appropriations hearing, questioned agency officials about spray residue dangers and received answers that strongly contradicted his own views. When the public record of the hearings was published, however, all such interchanges had mysteriously vanished, while other passages were doctored in ways that buttressed Cannon's arguments. The chairman's role in the incident never was proven, but the committee was impelled by agency complaints to assure that all future hearings were published verbatim.[18]

A more critical incident occurred in 1937. FDA personnel discovered to their dismay that the USDA appropriations bill expressly forbade the use of funds for "laboratory investigations to determine the possibly harmful effects on human beings of spray residues on fruit and vegetables."[19] Agency officials were stunned to find that Cannon instead had granted total research jurisdiction to the Public Heath Service, which took a widely different analytical approach to the pesticides problem. The FDA relied primarily on laboratory experiments to assess the effects of residues, using results obtained from animals to extrapolate possible chronic (or long-term) effects on human beings. The PHS, by contrast, relied exclusively on more traditional field surveys, in which it questioned farmers and field hands about their health. This approach might reveal systemic toxic (or immediate) effects, but was notoriously inadequate (to the FDA, at least) for uncovering chronic effects that might not surface until years after exposure. Subsequent PHS findings that pesticides residues had toxic but not chronic effects thus came as no surprise to the FDA, but agency officials were dismayed nonetheless. Such results only supported Cannon's contentions—as Cannon knew they would. This aspect of the pesticides debate—whose testing protocols and data to accept—recurs with the DDT furor of the 1960s.

Such squabbles were important but subordinate to the simultaneous battle raging over revisions of the Pure Food and Drug Act. Assistant Secretary of Agriculture Rexford Tugwell, one of Roosevelt's original "brain trusters," long had criticized the 1906 law as woefully inadequate in the face of rapid changes in the drug and cosmetics industries. Tugwell's advocacy of stricter FDA enforcement, taking a stance markedly different from the USDA's normal passivity, soon earned him the enmity of many farm bloc members, particularly Cannon, and his presence at USDA partly motivated Cannon's assault on the FDA. Tugwell's nickname—"Rex the Red"—no doubt passed the lips of many a rural legislator and food company executive after Tugwell proposed sweeping changes in FDA regulatory authority.

Tugwell's proposals for stricter regulation languished in various legislative crannies until a new wave of scandals hit the nation. The new band of muckrakers attacked the drug and cosmetic industries' use of dangerous chemicals in their products and, more critical, federal laws that apparently failed to protect the public. The elixir sulfanilimide scandal of 1937, concerning a patent medicine alleged to have killed over 100 persons, proved particularly important in raising awareness of the issue. Subsequent outrage focused less on the deaths than on the light fines levied against the manufacturer—penalities limited by existing law. The 1906 law did not prohibit "toxic" ingredients in medicinal preparations, and the manufacturer in question was fined solely for a misbranded product label, which was a violation. It seemed so long as one stated honestly the contents of one's product, no matter how toxic, no punishment could be prescribed under existing law.

Despite the temptation to portray passage of the 1938 Food, Drug, and Cosmetic Act (FDCA) as a congressional reaction to widespread public outrage, such apparently was not the case. The proposed revision of FDA regulatory authority had been bottled up within the House Committee on Agriculture (the FDA still was part of the USDA), and the administration appeared reluctant to push hard on a farm sector that was so critical to the New Deal coalition. The scandals, however, prompted presidential intervention in the name of the public good, as had occurred in 1906. "Perhaps the most striking characteristic of the history of the FDCA," Thomas Dunlap comments, "is the fact that the measure . . . never became the object of widespread public attention, much less of informed public support."[20] One instead finds intervention by leaders concerned about public health, led by Tugwell and backed strongly by Roosevelt.

The FDCA required drug manufacturers to file test results for new products with the Food and Drug Administration—a modification that it hoped would improve the government's ability to keep dangerous products off of the market. Farm bloc congressmen were willing, if reluctantly, to support such "consumer-oriented" provisions, but proposals to strengthen the FDA's enforcement power over pesticide residues on fruits and vegetables provoked strenuous opposition. Farmers and their congressional patrons feared that the FDA would turn a deaf ear to agricultural needs in its zeal to protect public health, particularly because the bill gave the agency the sole authority to set legal tolerances for spray residues. The distinction between *legal* tolerances, which could stand up in court, as opposed to the *administrative* standards of the time, was central to this dispute. Administrative tolerances carried no inherent legal standing, but were seen instead as informal standards, possibly arbitrary thresholds of toxicity whose validity had to be proved by the FDA in court. Such proof was difficult to attain, particularly in the defendant's home federal district court, and prosecuting alleged violations during the decade consumed most of the FDA's time and budget. The authority to set legal tolerances would be key to the agency's regulatory power. Notes James Whorton: "Armed with official tolerances, the FDA would not have to waste its energies trying to persuade a jury that one-hundredth of a grain of arsenic might be harmful, but could concentrate its courtroom activity on the simple demonstration that the seized product carried residues in excess of established tolerances."[21]

The agricultural community, particularly fruit growers, opposed this "dictatorial" and "destructive" attack on individual rights. The House Agriculture Committee apparently agreed, amending Tugwell's proposal so that it allowed "aggrieved parties" the right to file suit against any newly issued tolerance in *any* of the eighty federal district courts, regardless of the plaintiff's residence. The court, for its part, could collect new evidence and judge the validity of the regulation. Judicial review of this sort answered growers' fears of arbitrary administrative fiat in principle, but in practical terms it meant that enforcement could be delayed indefinitely by a succession of appeals.[22] The FDA regarded this provision, despite its outward trappings of equity, as a wily attempt to scuttle the entire enforcement process. (That someone might oppose a tolerance because it was too lenient apparently had not occurred to anyone, though that realization would emerge decades later.) Whatever the case, this provision proved to be the primary point of dispute between the chambers, with Senate conferees refusing to agree with what many saw as a blatant erosion in FDA authority. Roosevelt soon announced his intent to veto any bill containing the House provision, and conferees agreed to replace it with one granting to all parties "grievously affected" by a regulatory action the right to appeal in the party's home federal court of appeals. Growers still had their judicial review, while the FDA gained greater flexibility over pesticide residues.

It was but a partial victory. The FDA now could set legal tolerances, but these standards were still based on research performed by the Public Health Service, thus continuing the inconsistency between testing methods so bothersome to FDA officials. Nor could the FDA set tolerances for spray residues until after the product in question had been applied in the field. Maintaining a product's innocence until proven otherwise seemed consistent with traditional regulatory philosophy, but it also prevented the FDA from testing and clearing a product before it hit the market. This inability, combined with standard-setting and enforcement processes known to drag on interminably, meant that the FDA was ill-positioned to prevent widespread use of potentially dangerous substances until perhaps too late.

Additionally, the legal burden of proof concerning dangerous products lay squarely on the FDA, which plagued enforcement for years. Dunlap, in his analysis of the DDT experience, argues that FDA officials as early as 1944 expressed quiet reservations about using the new chemical before testing was complete, but "there was no way for the FDA to hold DDT off of the market. The 1938 law only gave the agency the power to set binding tolerance levels after a long series of hearings. It could set interim levels at once, but it could hardly forbid the use of DDT without some good reason that would command public backing, and in the immediate postwar period there was no sentiment for such a ban or even for elementary precautions."[23]

Finally, the FDA's new authority ironically came at a time when use of the substances under attack—the leads, the flourines, and the arsenics—began to decline, largely because they increasingly failed to do the job. The agency's partial victory, obtained at great political expense, in fact proved almost moot with the dawn of the new pesticides age. The tools provided in 1938 were barely adequate for such blunt substances as arsenic; they would be obsolete less than a decade later.

The Cozy Triangle at Work: FIFRA of 1947

Thus, by 1945, federal pesticides policy, with its parallel paths, lay in a regulatory twilight zone. Federal law governing spray residues on fruits and vegetables—the "consumer" side of the regulatory ledger—enabled the FDA to protect public health only through rather cumbersome and expensive legal machinations. Federal regulation of the sale and use of pesticides, however, was even more problematic; the 1910 law applied solely to a narrow range of now-obsolete substances. Requiring valid product labels for leads or arsenics may well have been adequate, since the law basically told farmers that the product contained a poison that all knew was extremely toxic. The new formulations, however, contained chemicals whose toxic *and* chronic effects were so ill-understood as to be totally alien to all in the agricultural community. The gap between technology and policy yawned anew.

The postwar dilemma for farmers was their inability to assess adequately new product safety or efficacy. Arsenic is one thing, but DDT a far different matter.

The farmer hardly could be expected to interpret complex chemical formulas alone—and there were a multitude of new products inundating the market. Such concerns were echoed by the USDA officials, who now worried openly about their capacity to protect farm health and safety—not to mention investments—from potentially harmful or worthless products. The FIA offered to the department only ex post facto prosecution of mislabeled or fraudulent goods, forcing it to make random field checks of farms and manufacturing plants to assess label veracity. This system barely proved adequate with the old chemicals, and department officials realized that they soon would be sorely pressed to monitor technologies whose sophistication clearly outmatched available knowledge. Farmers and their bureaucratic patrons agreed that the USDA now needed the capacity to approve new products systematically *before* they hit the market, to register a product's label and proposed uses before allowing its sale. Consumers needed to be protected—the consumers in question being the farmers.

Chemical companies, for their part, anticipated tremendous profits in the new pesticides market, and, in essence, wanted the federal government to act as gatekeeper and screen out those fly-by-night operators who might sully the industry's reputation. A national pesticides policy also was desirable because chemical makers wanted to avoid the patchwork of state laws that might otherwise result. State-by-state policy variation would boost the costs of tracking regulations and make it harder for companies to plan market strategies, while a single national policy would create but one set of regulators to worry about. Pesticides makers at most wanted a federal stamp of approval for their products to assure farmers, and were willing to go along with premarket clearance so long as such provisions did not obstruct sales markedly.

More than anything else, then, those directly tied to the future of pesticides wanted to bring the federal government to bear on the uncertainties and potentially destabilizing impacts of significant technological change. All involved shared assumptions about the unlimited potentials of the new products, and all wanted to create regulatory and market mechanisms conducive to the orderly realization of those potentials. The calls for policy refurbishing came not from "outsiders," for none were involved, but from those who made up a rather insular policy community.

The arena for policy change was the House Committee on Agriculture. Its chair, John W. Flannagan, Jr. (D, Va.), in early 1946 submitted a bill drafted primarily by committee members after consultations with and technical advice from USDA officials and industry representatives. Hearings on H.R. 4851 began in early February and involved four days of testimony by trade association representatives, state departments of agriculture, and USDA officials. Not one witness came from outside the agricultural subsystem, not even from the FDA. The lead witness, S. R. Newell, assistant director of the USDA's Production and Marketing Division, summarized the bill's purpose: "The broad objectives of this bill is to protect the users of economic poisons by requiring that full and accurate information be provided as to the contents and directions for use and, in the case of poisons toxic to man, a statement of antidote for the poisons contained therein. It is also designed to protect the reputable manufacturer or distributor from those few opportunists who would discredit the industry by attempting to capitalize on situations by false claims for useless or dangerous products."[24]

The committee bill required manufacturers to state clearly all ingredients and directions for use of a new product, and to submit the label to the USDA prior to marketing. Newell emphasized that, except for premarket clearance, the proposed changes to the 1910 law were insignificant. Indeed, most of the traditional USDA procedures respecting agricultural pesticides were left unchanged. Said Newell, "It is not a drastic law. Its enforcement is through the courts, and adequate provision is made for hearings and for the use of warnings instead of prosecution when it is felt that such means will fulfill the requirements and purposes of the Act."[25] Newell also emphasized the unanimity of support for the bill, though he added, "It represents a compromise to meet objections raised during extended discussions with some of the groups represented here."[26] USDA and industry representatives disagreed somewhat over setting minimum standards on rodenticides, the industry preferring and getting voluntary thresholds, but overall no controversy of consequence surfaced during debate. Nascent disputes in fact were extinguished rather quickly by Chairman Flannagan, who consistently urged those involved to resolve their differences in private; the hearing room was no place for public dispute.[27]

The very idea of premarket clearance proved to be the major point of departure between the USDA and the industry. Some industry representatives saw it as rather unnecessary and wasteful, arguing that the USDA should make clearance voluntary and should focus instead on unscrupulous operators. Newell quickly sought to allay such fears, explaining that premarket clearance was better than the old system of field checks: "Experience over the years indicates that many manufacturers would welcome the opportunity to check their products and the claims made upon them with the administrative body. Recent experience in the examination of labels applying to DDT amply demonstrates this fact. It would be the one who is apt to evade the law or the one who most needs the assistance who would be most likely to fail to submit labels on a voluntary basis."[28]

Registration did not seem particularly onerous on its face, since manufacturers were required simply to submit the product's name, a copy of the label, and a statement about all product claims. Nowhere did the USDA insist on submission of test data or chemical formulas, which were seen commonly as industrial trade secrets. Premarket clearance would enable the USDA to track more easily what was being sold and would allow it to warn producers whose labels or claims appeared to lack completeness or veracity. Overall, however, the department vigorously argued that this provision would be of little consequence to the honest operator, and that everything possible would be done to make registration easy and inexpensive. "Registration need not be burdensome to the manufacturer,"

Newell assured industry representatives, "because every effort would be made to keep it as simple as possible.[29]

Just in case the USDA did reject a product, and if the chemical maker disagreed with the department's opinion, the product could be registered "under protest." This classification signified that the department was negotiating with the producer about the problem but had not yet initiated legal action against the product; the USDA could ban a product only through the courts—a provision retained from the 1910 law. Department officials quietly opposed this "escape clause" because it essentially derailed the USDA's right to deny an application, but went along with the industry's arguments that the mechanism allowed for challenges to potentially arbitrary administrative decisions. Industry representatives argued that the USDA could prosecute anyone marketing a product so registered, though all knew that to prove in court that a label was misleading or that a product was fraudulent was difficult at best. Protest registration essentially made FIFRA a label law with very little regulatory teeth, but all involved agreed that it was better than its predecessor. Besides, the honest chemical maker had nothing to worry about.

The Agriculture Committee favorably reported H.R. 4851 to the House floor in early April, where it was placed on the consent calendar. However, for reasons not discernible from the record, three Republicans objected to the bill and no further action was taken that year. Whatever the reason, Flannagan reintroduced the bill (now H.R. 1237) in January 1947. Flannagan no longer was committee chair, having been replaced by Rep. Clifford Hope (R, Kans.) following the Republican sweep in the 1946 elections, but partisan change in committee control had no apparent impact on the bill. Attitudes toward the pesticides issue in the 1940s did not follow party lines, and the committee was already known for its bipartisan stance on most agricultural proposals. In fact, the issue—as discussed within the agricultural subsystem—generated no fatal cleavages whatsoever.

Representative August Andresen (R, Mich.) convened his subcommittee in April to review the bill, but the hearing lasted only a day. The reason, as Andresen explained, was that little new ground was explored in the interim: "Extensive hearings were held on similar legislation in the 79th Congress, and in view of the fact that the industry and distributors, and others, are pretty much in accord on this legislation, this hearing will be comparatively brief, in order to hear the views of those witnesses who have indicated that they wanted to be heard, and to receive suggestions for the record on this bill."[30]

Agricultural community amity this time proved even more pronounced. E. L. Griffin, of the USDA's Insecticide Division, reiterated the department's support for the bill, adding that he had no desire to burden industry with unnecessary regulation nor intrusive government action. For example, Griffin noted, the USDA believed that FIFRA should apply only to labels on the original package, not on those repackaged for sale to consumers. Otherwise, he added, the provision might restrain distributors' right to repackage pesticides as they saw fit, which the department had no desire to do.[31]

Despite some minor arguments over such matters as a nominal registration fee (which the department lost) and an expiration date for each registration (which it won), the tone of debate revealed a general consensus on the need for premarket clearance and a national pesticides policy. J. M. George, representing the Interstate Manufacturers' Association, praised the bill as ahead of its time and pressed for its passage to a committee needing little pressing: "I want to restate that industry is together on this. Industry has agreed, probably reluctantly in some respects, to some of the provisions of this bill, but it is a step forward that I think should be passed the way it is written. In a few years, if there is anything wrong—we don't have to wait for 36 years to change it—we can change it. When I say 'we' I mean Congress, and with the assistance and advice they get from the department and from industry."[32] George, noting that the bill was in the public interest, worried that "it might not pass if it becomes controversial."[33] He need not have worried, because FIFRA avoided controversy. The Committee on Agriculture favorably reported H.R. 1237 to the floor in late April, and it came up for debate in May, with Andresen warning his colleagues that FIFRA was both a "highly technical bill" and that "the insecticide industry, the Department of Agriculture, the distributors, and the organizations representing the [farmers] of this country are all in accord on the need for this bill. Some of the manufacturers do not like the registration proposition, but it was deemed advisable that they should be required to register their products with the Department of Agriculture so that the public could be protected."[34]

Only minor debate ensued, yet the sole interchange of note exposed the protective attitudes held by farm bloc members towards the USDA and its jurisdiction over pesticides. Rep. Frank B. Keefe (R, Wis.) wondered why the FDA was not to implement the new law, since the agency already administered the Food, Drug, and Cosmetic Act. Andresen replied that the FDA "deals with food, and this deals with insecticides, and under the act of 1910 the Department of Agriculture has handled it."[35] This line of reasoning apparently did not impress Keefe, who then asked why the federal government should maintain two testing facilities when "we are trying to simplify government and do away with overlapping bureaus and that sort of thing."[36] He again asked why the FDA, which already operated such a lab, couldn't carry out the testing.

Keefe's question was valid, particularly for Republicans currently on a budget-cutting campaign. For Agricultural Committee members to promote new USDA activity, when another agency already had facilities and experience, would look like a clear defense of parochial interests. Flannagan came to Andresen's rescue. "This Act," he responded, "at present is being administered by the Department of Agriculture. It primarily affects the farmers. . . . It only amends the old act. It is not new legislation."[37] Proper consideration to the question of administrative consolidation could be given at a later date, Flannagan assured a still skeptical Keefe, while Andresen added that Keefe's concerns should be the subject of separate legislation. Administration duplication could be eliminated, Andresen continued, though it "might require the transfer of the enforcement of the insecticide law by the Pure Food and Drug Administration, or the transfer of the Pure Food and Drug Administration *over to the Department of Agriculture*."[38]

Keefe, a member of the appropriations subcommittee with jurisdiction over the FDA, had argued for transfer of USDA pesticides jurisdiction *to* the FDA, which already enforced spray residue standards under the FDCA. Flannagan and Andresen, however, were equally adamant about maintaining USDA jurisdiction and even proposed returning the FDA to Agriculture. Compromise on this matter was unlikely, but Keefe recognized a full-court press by committee members when he saw one and announced that he would not oppose the bill. Besides, nobody else appeared to want a battle over such a technical and uncontroversial bill, particularly one supported by powerful senior House members. There was no further debate, and H.R. 1237 passed by voice vote.

Senate treatment of H.R. 1237 proved even more perfunctory. The Committee on Agriculture and Forestry held no public hearings, and the committee report consisted only of its favorable recommendation and a copy of the House committee report. The bill came to the floor in mid-June, was explained briefly by Sen. Allan Ellender (D, La.), and passed by voice vote with neither comment nor debate. Neither chamber deigned to limit FIFRA's statutory authorization, thus giving the law an indefinite lease on life and the guarantee that it would not reenter the legislative arena unless called upon. President Truman signed the measure without comment on June 26, which the *New York Times* duly noted in the next day's food section.

Summary: The Politics of Clientelism

FIFRA simply required manufacturers to register new products and to promise that the product in question was both safe for use and effective as claimed. The fact that the USDA bore the onus of proof in court on regulatory actions, and the various loopholes peppered throughout the measure, made FIFRA a label law and little more. With hindsight, argues John Blodgett, we can see that FIFRA contained a number of serious flaws: "Fundamentally it was intended to serve the same basic function of the 1910 Act—to protect the farmer from adulterated, ineffectual, or unsafe products. The Act was not intended to provide direction and authority for dealing with health and environmental problems; nor did it recognize the need to protect farmers from damages to beneficial insects or the potential of insects to develop resistance."[39] Also, we should note in fairness that these "deficiencies" resulted primarily from ignorance about potential problems, though unalloyed faith in the pesticides paradigm no doubt closed adherents' eyes to other possibilities. The shared assumptions about pesticides as a panacea, and the imperatives motivating their widespread use, provided the parameters for policy debate: the new technologies would be used, and used widely, and those involved in formulating the law simply wanted some mechanism for monitoring new products. None wanted to impede the farmers' ability to obtain new products nor the companies' ability to sell them.

The politics of FIFRA commenced under conditions particularly conducive to small-group, or subgovernment, dynamics, in sharp contrast to the earlier debate over the FDCA. The Food and Drug Administration won a partial victory in 1938 largely because of controversy about the issue—which expanded the population of concerned congressional actors—and, more critical, because of active presidential intervention. Otherwise, the agricultural bloc might well have prevailed in opposition. The FDCA debate in many ways approximates what Ripley and Franklin call the politics of "protective regulatory policy": considerable issue visibility, shifting alliances, executive intervention, and sharp issue conflict.[40] The primary relationships under such conditions are those among congressional leaders, the president and the centralized bureaucracy, and the regulated interests. Conflict escalates from the subcommittee room to the congressional cloakrooms and the Oval Office, with a temporary coalition of interests favoring regulation outflanking (at last partially) entrenched opposition.

What, then, do we conclude about the politics of FIFRA? At first glance, a policy that is meant to be regulatory nonetheless displays all the dynamics of classic clientele politics, which we would expect to find largely under "distributive" policy conditions. The pesticides issue was not salient to any but those directly benefiting from pesticides, and the scope of debate was severely limited to those most intimately involved. Furthermore, relationships among these players were clearly accommodative in tone. The configuration of central players also approximates a classic "iron triangle," with decision making among congressional committee members, mid-level USDA bureaucrats, and those representing a few, well-organized interests—all sharing a relatively common perspective on the issue. No apparent presidential intervention, little noncommittee policy interest among legislators, and no interest-group "free-for-all" marked this issue area. Steven Kelman, in his comparison of Swedish and American decisional styles, argues that, regardless of cultural differences, it is under such conditions of small-group decision making that cozy relationships evolve.[41] The atmosphere surrounding pesticides policy formation proved similarly conducive.

Perhaps FIFRA wasn't regulatory at all. One could argue that the law was more "self-regulatory" than "regulatory," that government allowed pesticides makers to operate freely so long as they honestly labeled their products. This is conceptual hairsplitting, however, for it forces us to analyze substantive policy outcomes more than the politics that influenced them. The ex post facto typing of policies may in fact neglect the very incentives and perceived policy stakes that attract interests to an issue in the first place. It neglects interests that existed prior to decision making. Nor does it account for how perceptions of stakes influence the composition of policymaking communities. Pesticides as an issue was defined wholly by those perceiving direct and largely economic policy stakes in the use of pesticides, and the policymaking structure in place at the time reinforced dominant attitudes about possible choices.

There is, then, a perfect mesh between the interest-mobilizing system and the structure of political power relevant to this issue. This close weave promotes clientele politics, which prevails so long as the pesticides issue is defined in agricultural terms and so long as "outsiders" fail to perceive any stake in what goes on. The dominant assumptions about pesticides begin to fragment in the 1950s, however, and "outsiders" begin to press dissonant claims on the policymaking system. Challenges to orthodoxy thus mount, leading one to expect subsequent policy change. But, as we shall see, this was not necessarily what happened.

Chapter Four

The Apotheosis of Pesticides

Consensus, Stability, and Political Theory

These are the days in which even the mildly critical individual is likely to seem like a lion in contrast with the general mood. These are the days when men of all social disciplines and all political faiths seek the comfortable and the accepted; when the man of controversy is looked upon as a disturbing influence; when originality is taken to be a mark of instability; and when, in minor modification of the scriptural parable, the bland lead the bland.[1]

Thus John Kenneth Galbraith pungently summarized an entire decade, in a sardonic encapsulation of the 1950s that no doubt irked many a scholar and citizen. To some it seemed an idyllic period of social harmony, national purpose, and economic prosperity; to others, it was a time of stultifying conformity, virulent anticommunism, and economic inequality. Whatever the case—and personal ideology of course affects one's view—the 1950s inarguably were years of remarkable social calm and professed consensus on "traditional values." To those who recall the era fondly, it was characterized (as Dwight Eisenhower intended it to be) by "an atmosphere of greater serenity and mutual confidence"[2] than had been true in previous decades. Most Americans, at least on the surface, agreed with their Ike.

This atmosphere, not surprisingly, led many American political scientists to speak optimistically of "an end to ideology." The political turbulence of the previous decades had declined, and by the mid-1950s, the popular mood had slipped into general identification with and satisfaction about the political system and its institutions. The American system delivered economic prosperity and social peace—or so it seemed. And, as scholars noted with wonderment, such phenomena were not confined to home. Postwar developments in Western Europe and Japan produced unparalleled economic growth, stimulated social class mobility, and softened ideological polarization—evidence that the liberal democratic model indeed worked. Consensus politics reigned, at least at the elite level.

Samuel Huntington, examining those years when American institutions apparently upheld dominant social values, argues that the dominant image of politics during the decade emphasized social consensus and organizational pluralism. Americans in the 1950s, Huntington comments, worried little about the concentration of social and political power, comfortable in their belief that what was good for General Motors indeed was good for America—and vice versa: "Consensus existed and was good. Power existed in a plurality of large organizations—veto groups—that at times made governmental action difficult but also made the serious abuse of governmental power extremely unlikely. What is more, in a brief but notable inversion of the traditional American approach, thinkers in the 1950s became intrigued with the idea that large organization in itself could be good."[3]

In the 1950s, anyone who studied the politics of chemical pesticides probably would have been struck by the number of large farm and business organizations directly attached to this issue. One also might have observed a notable consensus about the issue that transcended cleavages of party, region, or political ideology. That consensus centered on the belief that federal regulation was a necessary but minimal evil, designed solely to provide for stable but relatively free pesticides markets. If, as I have argued earlier, all policy systems have a bias, in this case the bias tilted overwhelming toward the almost uncritical faith in the pesticides paradigm. Government policymakers nurtured the bias, but what is more important here, they also responded to "private interests," those organized and active groups motivated by direct and tangible stakes in the use of pesticides. Amorphous and unorganized interests—assuming one could identify them—by contrast fared badly. Bauer, Poole, and Dexter argue that "the structure both of an issue and the arena in which it is discussed shapes the way in which it is discussed."[4] It is clear that how the pesticides issue was defined and the clientelistic style of policymaking that prevailed in 1947, facilitated a federal policy strongly promoting the chemical option.

But, says the pluralist thesis, this is as it should be. After all, those perceiving the most direct and tangible stakes in using pesticides mobilized themselves to pursue their interests. The dynamics of interest mobilization and representation worked splendidly, since policy outcomes reflected those views most intensely promoted; that the "public good" was also protected would be a bonus. Yet there is little doubt that the pluralist concept of policymaking can be used to justify inequitable outcomes. The pluralist's reliance on some inherently benign "procedural democracy" as a be-all-and-end-all, combined with the view of the passive state or government as the neutral referee, appears to rationalize severe imbalances in social representation and influence. Pluralist faith in an "open" system clearly places the burden of success or failure squarely on social actors; the system itself is neutral, and cannot be faulted if some sectors in society go unrepresented. Losers, it appears, have only themselves to blame.

A faith in procedural democracy fails to hide the reality of power. Interest groups, E. E. Schattschneider argued in 1935, "are formidable and overwhelming only when they have become unbalanced and one-sided."[5] Such an imbalance characterizes the genesis of FIFRA, since the ledger of social interests involved clearly favored the agricultural subsystem. While there may have been some rough equilibrium within that narrow spectrum of farm and industry groups, there was a severe disquilibrium between the subsystem and outsiders. Nowhere do we find a true plurality of competing interests; an imbalance was also evident in, and woven into, the relevant decision arenas of government.

Little conflict over the pesticides issue marked FIFRA's passage. But conflict is fundamental to the pluralist concept, for it is only through a dynamic tension among competing views that "good" public policy emerges. According to pluralism, conflict is a gauntlet testing the merits of and support for alternative policy directions, with the surviving option becoming government policy. The logical corollary to this is that policy produced in the absence of conflict is less enlightened, but how the pluralist thesis fares without observable conflict is unclear. We can approach this question by examining attempts among those dominating the shape and direction of pesticides policy to maintain their hegemony as consensus breaks down and as challenges to orthodoxy emerge. The pluralist view would suggest that issue conflict eventually should produce policy change, but at what point this occurs is not obvious. We therefore need to understand which factors promoted the status quo, and which generated challenges, as the pesticides paradigm began to encounter opposition.

The Golden Age of Pesticides

> Crop pest control in this country rests fundamentally on a system of free enterprise in which the grower must be our primary objective. Other secondary objectives involving benefit to our society accrue indirectly as reduced costs to the consumer or through improved quality of the environment.[6]

Pesticides reigned supreme at the dawn of the new decade. Total production, below 100 million pounds in 1945, jumped to approximately 300 million pounds by 1950, to double again by the decade's end.[7] Farm production boomed, responding to the demands of rapid postwar population growth, the Korean War, and the requisites of a burgeoning export market. U.S. farmers responded not only to demand, however, but also to the peculiar and often perverse imperatives of the agricultural economy: produce-or-perish and produce-*and*-perish were but twin products of the farm problem. Federal agricultural policy, whether Democratic or Republican in origin, in many ways unwittingly promoted heavier pesticides use. To farm without chemicals was an option few dared to consider, and fewer still as yet recognized the interconnection among seemingly separate issues. Whatever the root causes, the pesticides paradigm wove itself, perhaps irrevocably, into the fabric of American agriculture during the fifties.

The transformation in farming begun after World War II accelerated dramatically. Farm migration hemorrhaged as almost 6 million rural dwellers moved to the cities. Some 18 percent of all farms folded, leaving behind only the larger and better capitalized operations. Farm society, and the social meaning of farming, underwent a similar transformation. Whereas subsistence-level family farms had been the rule, by the late 1950s average farm income reached middle-class heights. Farmers increasingly became managers of agricultural enterprises, though the public image of the simple family farm would endure much longer than the fact. This discontinuity between image and reality, shielded in no small way by those in agriculture, would have a powerful impact on national farm policy. Programs meant to alleviate farm poverty increasingly subsidized the rural business class.

Verticalization of agriculture—whereby giant corporations owned farms, processing plants, and even retail stores—emerged powerfully in the 1950s. This development had important implications both for farmers and consumers. Corporate farms are highly capital-intensive, given the imperatives built into food retailing, and also tend to lead the way in using new technologies to maximize profits. Accelerating population growth and urbanization promoted the "rationalization" of food production, since the vast urban market promised fantastic opportunities to the freshest, the cleanest, and the cheapest. Profits lay in picture-perfect fruits and vegetables, and pesticides helped to give Americans the food they wanted. Chemical companies, not surprisingly, aggressively promoted a wide array of new products to farmers eager for more effective and increasingly specialized technologies. The scope of the explosive pesticides market is indicated only in part by the fact that the USDA registered almost 10,000 separate products during FIFRA's first five years.[8]

It was also during this time that government charged into the war against insects and diseases. Vast stretches of forest, rangeland, and urban expanses fell under chemical treatment as public authorities sought to eradicate an increasing number of insect threats, real or feared, on both public and private land. Federal and state expenditures for pest control skyrocketed, with cumulative federal spending alone reaching into the hundreds of millions of dollars by 1960. Such outlays paled by comparison with those in agriculture, but it would be the more visible federal actions that later generated public attention to the use of pesticides.

It was, in this sense, the "golden age" of pesticides. Every golden age has its patrons—those who promote reigning policy and fight to maintain its dominance, and their own political power. The Medici of the pesticides paradigm resided in Congress, and not only in the two committees on agriculture. These committees were central to the formation of policy in 1947, but would be far less so as policy formation gave way to the routines of implementation. FIFRA's annual status, and its overall relation to overall federal pesticides activity, shifted during the 1950s to the committees on appropriations—the congressional watchdogs of the federal purse. More precisely, power over pesticides policy shifted to the House Subcommittee on Agricultural Appropriations, one of the renowned congressional fiefdoms of the postwar era and the fulcrum of the pesticides subsystem during its glory days.

Ties That Bind: Whitten, the Subcommittee, and the USDA

"Everyone," noted Representative Clem Miller in 1962, "trembles before the Appropriations Committee."[9] This assertion might seem hyperbolic today, but congressional appropriations committees in the immediate postwar era differed mightily from their contemporary brethren, with the primary distinction lying in the power wielded by committees and their chairs. The 1946 reorganization had created, in the words of Charles O. Jones, "committee suzerainty" by reducing the number of standing committees and by "rationalizing" committee jurisdictions.[10] The "barons" of Congress retained almost autonomous authority over their estates, authority buttressed by the institution's then-hallowed norm of seniority. The late Speaker John McCormack (D, Mass.) advised freshmen congressmen: "Whenever you pass a committee chairman in the House, you bow from the waist. I do."[11]

Seniority overwhelmingly favored southern Democrats, whose regional party supremacy, magnified by discriminatory electoral practices and apportionment schemes, ensured them long tenure. Southern Democrats in the mid-1950s chaired fifteen out of nineteen House committees and eight out of fifteen in the Senate—numbers quite disproportionate to their overall strength in Congress.[12] Control over key positions and stability in office gave southern Democrats considerable policy leverage, but equally crucial was their almost uniform conservatism, a traditional populist distrust of federal action (particularly of the social welfare type) that magnified their power. The "conservation coalition," that superficially curious amalgam of southern Democrats and midwestern Republicans, would dominate Congress long after liberal northern Democrats became more numerous. This combination of position, seniority, and ideology powerfully structured the legislative agenda, particularly between the end of the New Deal and the onset of the Great Society.

Southerners in each chamber controlled the appropriations and agriculture committees, work groups also with traditions of strong support for the conservative coalition.[13] Harold Cooley of North Carolina inherited the reins of the House Agriculture Committee in 1949, and, save for a brief interregnum during the early 1950s, held them until 1967. His Senate counterpart, Allen Ellender of

Louisiana, sat in the chair almost continuously from 1951 to 1970. Equally, if not more important, were the committees on appropriations, particularly in the House. The reason was obvious: "Of all legislative prerogatives," wrote Missouri Representative Clarence Cannon, "the power to appropriate is the most vital. Not a wheel of government can turn without motivation of an appropriation. Basic authorizing statutes for the most part are broad in scope and grant to the executive rather wide latitude. The one continuing and recurring procedure for congressional control over governmental activities within such statutes is the annual appropriations review and the legislative provision of funds."[14]

Cannon, the longtime scourge of the FDA, spoke from personal experience. Chairman of the House Committee on Appropriations from 1947 to 1964, Cannon became one of the most feared legislators and procedural tacticians in House history, a chairman whom neither Speakers Rayburn nor McCormack ever quite controlled. And Cannon's committee, dominated by like-minded southern and midwestern conservatives, wielded quiet but pervasive power throughout the federal establishment.

Richard Fenno, in his unparalleled analysis of congressional appropriations between 1947 and 1964, notes that the House committee's unique assets gave it overwhelming leverage within the chamber. Appropriations bills *alone* avoided the House Rules Committee, site of innumerable liberal Waterloos, and instead glided to the floor under privileged rules—provisions that, Fenno argues, were key to committee success:

> The more quickly the Committee can move its bill to the floor, the fewer contextual changes it is likely to confront. And the fewer the contextual changes, the less will the Committee control be weakened. Most important, a speedy trip to the floor inhibits the dissemination of information. The more time that elapses between the publication of reports and floor action, the smaller becomes the Committee's information advantage and the more resources become available with which to organize an opposition force in the chamber. For most of the period under study, the Committee utilized its prerogative to the maximum in keeping the elapsed time close to the minimum.[15]

The occasional floor challenge usually encountered remarkably unanimous support from committee members, many of whom held other chairs and thus were well positioned to foresee uprisings. During the 1947–1962 period, Fenno notes, about 94 percent of all appropriations bills went to the floor without public dissent, much less a minority report.[16] Public unanimity complemented private secrecy, as most committee deliberations went on behind closed doors, revealing a proclivity for avoiding publicity that irritated those outside.[17] It was equally difficult even to find out what the committee was doing. Congressional committees, according to the 1946 Reorganization Act, were required to report in the *Congressional Record Daily Digest* the time, place, and purpose of any meeting.

The House Appropriations Committee honored this provision in the breach; during 1957, for example, the committee held 369 meetings, reporting not one.[18]

Fenno paints a picture of a committee with extraordinary autonomy and power. And, to make matters more byzantine, the committee's internal structure was considerably compartmentalized, each subcommittee working in virtual isolation from the others. Chairman Cannon unilaterally allocated subcommittee seats and resources, but granted his chairs almost free rein over their respective fiefdoms. "We tend to be more an aggregation of autonomous subcommittees than a cohesive Appropriations Committee," observed Cannon's successor, George Mahon of Texas.[19] Another committee member, however, characterized the power of the subcommittee chairs in far less sanguine terms: "They are lords of their fiefs and duchys—each with power over his area of appropriations. There's a power elite in this Committee. And these subcommittee chairmen are as powerful as the legislative chairmen."[20]

Perhaps the most powerful and autonomous of these "feudal lords" was—and is to this day—Jamie Whitten of Mississippi, the seemingly permanent chair of the subcommittee on agricultural appropriations. Whitten's tenure as chair began in 1947 and continues to the present, even after he inherited full committee leadership from Mahon in the late 1970s. Whitten hails from Mississippi's First District, a poor, largely rural area with an economy based on the traditional southern cash crop, cotton. The chairman's populist conservatism and distrust of federal activism is tempered only by his equally traditional southern excellence in constituency service—that is, the "porkbarrel"—and it is in this respect that Whitten's unparalleled hold over the USDA and its appropriations gains true meaning. Whitten's use of his position to benefit his district is something about which he makes no bones: "My membership on the Committee which controls the purse strings means that I am in the middle of the show and have a real opportunity for service. Certainly, my membership on this Committee has contributed in my own state to the Boll Weevil Laboratory, the Poultry Laboratory, the Soils Laboratory, . . . and to many others which I have been able to promote."[21]

Whitten's title as the "Permanent Secretary of Agriculture" is well earned and documented.[22] His attention to fiscal and program details is legendary, and, aided by a large and decidedly hyperactive subcommittee staff, so is his tight rein over annual USDA appropriations. Whitten's hearings read like extended General Accounting Office reports, only greater in detail, and he carefully screens every expenditure, be it as large as the multibillion-dollar Commercial Credit Corporation or as minute as a new automobile for the Pawtuxet, Maryland, research center. And, as Whitten often warned USDA officials, he expects notice of *any* changes in expenditures made during the fiscal year. One thing he would not brook was "going around the committee and the Congress and doing it on the side, when you should have met the issue head on."[23]

Department officials, for their part, consistently went out of their way to maintain cordial relations with the chairman and his colleagues. "Agricultural officials," Fenno notes, "especially stressed services to the Committee members as a natural way to maintain informal communication. . . . More than any other department, the heads of agricultural agencies take it upon themselves to mend their committee fences."[24] The clientelism that pervaded committee-USDA relations matched those evident between the department and farm interests, producing working relations marked by mutual deference, avoidance of conflict, and accommodation. One should not, however, attribute USDA success solely to grass-roots pressure on Congress, though it clearly existed, because subcommittee members traditionally shared the values of their bureaucratic clients and home-town constituents. As Whitten said on more than one occasion, "I know the value of the various agricultural programs at the first hand."[25] Nobody ever doubted that claim.

Subcommittee members were as dogmatic as their Agriculture Committee brethren about the pesticides paradigm, and, in fact, over the next two decades would be even more militant than most other farm bloc representatives in promoting chemical controls. Whitten and his colleagues believed wholeheartedly that farmers should expect safe and effective pesticides, but they also displayed throughout this period an emphatic intent to rid the nation of as many insect pests—real or perceived—as was humanly and fiscally possible. And, with his hand firmly on the federal wallet, Whitten was singularly well positioned to carry out this objective.

The particular zeal with which Whitten and his colleagues pursued total pest eradication is explained in part by the peculiarities of southern agriculture and by their desire to free the region of its postbellum legacy of poverty and backwardness. Southern agriculture in the 1950s, while diversifying, still relied heavily on cotton, which had a history of pest problems dating back to the first boll weevil infestations of the late 1800s. Traditional cotton monoculture, combined with a large rural population, small farms, a subtropical climate, and a pronounced lag in the use of farm machinery, produced in the region a disproportionate dependence on chemical technologies to enrich the soil, eradicate pests, and overcome deep-seated structural problems and practices. More than any other region, the South saw agrichemicals as the key to alleviating rural poverty, and, as history would show, southern legislators adroitly manipulated the federal treasury to achieve their goals.

Whitten, like Clarence Cannon before him, professed an optimistic faith in scientific progress and its direct benefits to agriculture, but it was an enthusiasm tempered by a deeply rooted belief that the folks paying the bill and receiving the benefits should get the maximum return for their tax dollars. Dual themes thus recur throughout the 1950s: subcommittee members expected fiscally prudent but successful pest control. Terms like "effectiveness" and "efficiency" punctuated debate, though members took pains to emphasize that eradication at any cost superseded other norms. Rep. Walt Horan (R, Wash.), himself an apple grower, perhaps best stated subcommittee orientations: "We expect the Department of

Agriculture, in cooperation with the land grant colleges and experiment stations and in cooperation with insecticides producers and the chemical industry, to develop pesticides that will control what is left of the immune insects that attack what we produce and present to the American consumer."[26]

The Whitten Subcommittee in early 1951 ordered the USDA to convene a panel to study federal pest control programs and recommend future directions. This panel, composed entirely of those "having direct association with the program in one way or another"—that is, farm groups, federal and state agricultural department officials, and the chemical industry—gathered testimony from over 200 witnesses in forty-two states.[27] Its final report is perhaps the most succinct declaration of the goals and motivations guiding federal pest control through the decade and into the next. Efforts to suppress or eradicate pests and plant diseases should be intensified, the panel emphasized. Moreover, the report added, "It is the firm belief of the study group that there has been a general tendency to underestimate both the extent of damage and the costs incurred due to insects and plant diseases in producing and marketing agricultural commodities. This nation cannot afford existing and potential damage from plant pests in view of the present need for sustained high production. Added to this is the possibility that foreign powers might attack us by introduction or dissemination of injurious insects and plant diseases to cut our production of food and fiber."[28]

The pest threat in fact often was seen in almost apocalyptic terms, pest control seen as a "never ending struggle between man and nature."[29] Worse, a threat to agriculture was a threat to national security, and pesticides increasingly were defined as yet more weapons in the cold war. An "evil force is loose in the world," warned Whitten's Senate counterpart, Allen Ellender, in 1953, and food was essential to prevent the spread of dissatisfaction that would come from hunger. And, Ellender concluded darkly, "I do not need to tell you that dissatisfaction breeds communism."[30] Clearly, according to this geopolitical view of agriculture, it was far better to be fed than red.

Such motivations powerfully justified massive federal and state pest eradication efforts, expenditures for which rose relentlessly through the decade. Most subcommittee members may have shared a philosophy of limited government, but pest control was the exception. The proper federal role, Whitten argued repeatedly, was to intervene rapidly to prevent pest outbreaks like those already ravaging southern cotton. Local initiative was valuable, but state and local officials sometimes were unable or unwilling to respond quickly. "I can point out to you many hundreds of millions of dollars we have spent because at the outset somebody did not act."[31]

Subgovernment theorists assert that "small, specialized governmental structures, such as subcommittees, may be coopted by the very groups they were supposed to oversee or, at the very least, may provide differential access to government for certain groups, often at the expense of others."[32] One is hard pressed to prove that the Whitten Subcommittee *actively* discriminated against nonagricultural actors during the 1950s; nonetheless, only USDA officials and farm group representatives appeared before the body throughout most of the decade. With respect to co-optation, Fenno argues that the USDA owed its appropriations success during this period to the "support activity of influential grass-roots clientele groups."[33] However, at least in this case, any such pressure probably only reinforced prevailing subcommittee faith in the pesticides paradigm. This consensus required little, if any, "co-optation" of the subcommittee by social groups. One can in fact argue that, if anything, the relationship ran in the other direction, since Whitten and the others acted as patrons to the USDA and its allies. One might even question whether "patronage" is precise enough to describe the practices to be developed during the decade, since subcommittee members displayed a zealous sense of "trusteeship" on pest control questions, promoting programs for regions and communities whose residents may not have perceived a threat. Who, after all, knew better?

FIFRA, meanwhile, was a forgotten stepchild amid this overwhelming predilection for aggressive chemical control, the stepchild that should be seen but not heard. The law was little discussed in subcommittee hearings throughout most of the 1950s, and sometimes not at all.[34] Pesticides registration was, at least for those on Capitol Hill, a noncontroversial and routine administrative matter, subject to incremental funding hikes but of little overall consequence so long as the USDA carried out its functions quietly and efficiently. Fighting the pest menace was the overriding objective, and FIFRA's role in the grand scheme was to provide for rapid product clearance so that farmers could deploy the new technologies with a minimum of bureaucratic interference. The USDA wanted a record of all products on the market, but only in extreme cases would it obstruct a manufacturer's ability to produce and sell new compounds. "The proper administration of this act," commented Production and Marketing Administration head Ralph Trigg in 1951, "is important dollarwise to farmers and in terms of food and fiber for national defense."[35] Legislators could pretty much ignore FIFRA so long as Trigg's people performed their duties quickly and cheaply, and evidence throughout the decade indicates that bureau personnel performed well.

In sum, then, federal pest-control efforts mounted during the 1950s, an escalating war against pests made possible by new technologies and made imperative by common assumptions about national security and abundance. Whitten and his colleagues coordinated federal action through the appropriations process, giving shape and direction to broad statutory authority in ways too often overlooked by students of public policy. "The Committee," Fenno notes, "acts as a restraining force on the level of federal expenditures and on the aspirations of the federal bureaucracy. But it also has taken a bold lead in selected policy areas and pushed the executive branch to expand its horizons."[36] The Whitten Subcommittee's role in promoting the chemical option was critical, and during the next two decades often would be more adamant about pesticides than even farmers or agricultural bureaucrats. In fact, during the next twenty years, the subcommittee uncritically expanded federal eradication efforts beyond USDA requests, calling on the carpet any Agriculture official who dared utter a public word of doubt. If anyone was co-opted, it was the department itself. The experts and the politicians eventually diverged in the intensity of their support for control programs, but it would be the latter, in control of the levers of public power, who prevailed.

However, as discussed earlier, federal pesticides regulation has always had two heads. The one, tilting toward agriculture, dominated policy debate. The other, concerned primarily with protecting those purchasing food, had faded back into the shadows since the passage of the FDCA. But it was about to re-emerge, via a nonagricultural avenue, with notable ramifications for the pesticides status quo.

The Delaney Committee

The public has a right . . . when they buy a package, to be absolutely sure that it contains nothing harmful or injurious to public health.[37]

If the pesticides paradigm was gospel in the farm community during the 1950s, other actors who were less wholeheartedly concerned about agriculture began to doubt the reigning orthodoxy. Warnings about DDT and its kin had been raised early on, only to be dismissed by agricultural interests as only so much unscientific hysteria. Those outside agriculture, however, began to take the warnings seriously, and it would be from these outsiders, those without vested economic or political stakes in agricultural pesticides, that the critical challenges to the pesticides paradigm first emerged.

Years of wrangling over arsenic and lead residues on farm produce had sensitized the FDA and the medical community to the potential dangers of pesticides poisoning. Such was this early concern that both the agency and the Public Health Service had tested the original shipments of DDT for acute and chronic toxicity, and the tests indicated that the chemical accumulated in body fats, persisted in milk, and generally posed unprecedented, though ill-understood, long-term effects. But the FDA was unable to translate its concern into action because of the overwhelming support for DDT and because existing law simply did not allow the agency to prohibit its use. Drug manufacturers were required to clear new products with the FDA, but the 1938 Food, Drug, and Cosmetic Act pertained neither to food additives nor to agricultural pesticides. The agency could act on a suspect chemical only when it was submitted voluntarily for clearance or when harm occurred, and the FDA bore the legal burden of proof that a product was hazardous.

The postwar avalanche of new chemicals—ranging from Vitamin B to growth hormones and preservatives—made this situation even more troublesome, since the FDA had few tools to meet the challenge. In early 1950, *Consumer*

Reports noted that since 1938 the agency had issued but one minor regulation governing pesticides residues in food (a tolerance on fluorine allowed on pears and apples), and that it "was so clumsily drafted that it was invalidated on appeal by the courts."[38] A special American Public Health Association panel on pesticides in 1950 reported, "The scientific literature show clearly that most if not all of the useful insecticides also have toxic properties for man."[39] The new chemicals may not pose immediate dangers, the panel warned, but the possibly chronic effects merited greater scrutiny. That report prompted the association to call for changes in the FDCA, arguing that one should not "demand impressive statistics from the death records in order to justify this undertaking."[40] Unfortunately for the association, proponents of the pesticides paradigm would point time and again to just this lack of "impressive statistics" as proof that the new chemicals posed little threat. Critics would be hard pressed to convince legislators like Clarence Cannon, to whom laboratory tests proved little, of any danger when field studies indicated no systematic adverse effects.

Nonetheless, by the early 1950s a competing, if minority, voice began to challenge the prevailing orthodoxy. The view began to be advanced among many in the scientific and medical professions that stricter federal controls on pesticide residues were essential to protecting public health. There were, as time would show, widely divergent fears about the extent of the potential danger, but those in the FDA, the American Medical Association, and in the public health field, as cited in *Consumer Reports,* agreed in principle that "intentional chemical additives must not be permitted until the complete harmlessness of these agents can be demonstrated beyond reasonable doubt."[41] Such terms as "complete harmlessness" and "reasonable doubt" later would prove divisive, but there was, at a minimum, an emerging consensus that the chemical age brought with it legitimate health concerns.

One key advocate of stricter federal regulation was Rep. Frank Keefe, the only House member to question USDA jurisdiction over FIFRA in 1947. Keefe bore a public reputation as both a proponent of pure food legislation and a patron of the FDA through his position on the House Appropriations Committee. The Wisconsin Republican by this time was growing increasingly worried about the consequences of wider chemical use. Keefe fully understood the power of the agricultural bloc to derail any solo attack on the pesticides paradigm, so he turned to Rep. James V. Delaney (D, N.Y.) for help. Delaney, an urban legislator who sat on the strategic Rules Committee, recalled, "[Keefe] told me that he had spoken to several congressmen, but couldn't get them interested. Finally, he came to me because we were old friends, and also because he probably thought I knew something about pesticides. That's not a problem in the Bronx."[42]

Problem in the Bronx or not, Delaney apparently agreed that the FDCA required strengthening, so he approached Rules Committee Chairman Adolph Sabath (D, Ill.), who in turn spoke to Speaker Sam Rayburn about appointing a special committee to study the problem. Rayburn knew that rural legislators would oppose a stronger FDA, but he also sympathized with those concerned about food additives—not to mention important urban Democratic constituencies—so he agreed to an ad hoc committee that balanced the major interests involved. The House approved the proposal in May 1950, and Rayburn subsequently appointed Delaney to head the panel. Thus, through the intervention by the speaker and a few well-placed members of the Rules Committee, a new forum in the debate over chemicals emerged, one not to be dominated so easily by agricultural interests.

The Delaney Committee, as the panel came to be known, was granted a broad authority to investigate the use of chemicals in food production, their effects on health, and their centrality to the stability and well-being of the farm economy. Such authority was both sweeping and delicately balanced (as was committee membership), designed to weigh the valid interests of consumers against those of agriculture. It was an attempt to join together those traditionally at odds over the pesticides question, since all had a stake in the investigation to come; however, that union would prove to be ephemeral.

The difference between the Delaney Committee hearings of 1950–1951 and those held by the Agriculture Committee on FIFRA could not have been more striking. Delaney from the outset stressed that *his* hearings were open to all, including the press, and he took the panel on the road to several cities to bring the hearings "closer to the public," and, no doubt, to generate more publicity.[43] Food and chemical industry officials in fact found Delaney's penchant for publicity rather disconcerting, even threatening. Lea S. Hitschner, secretary of the National Agricultural Chemicals Association (NACA), lambasted the committee and many of its witnesses for "careless and unsubstantiated criticisms" that threatened to harm both farmers and chemical makers.[44] Such criticisms seemed to many as overdone, since the Delaney Committee did sympathize outwardly with industry needs. Nonetheless, industry critics were correct, insofar as *any* publicity given to the hearings probably would highlight potential chemical hazards rather than the more mundane litany of benefits created by science and technology. The history of food and drug regulation, dating back to the 1906 law, reveals an endemic dread within industry and agriculture about the impacts of "bad publicity" on sales—a fear fed by a deep and abiding suspicion that the public might react negatively out of "misguided" fear or sheer ignorance. Such publicity as existed at this time was confined largely to the print media—television was not yet a factor—but any publicity not generated by industry's own advertising campaigns was likely to create public consternation. And, always strongly image-conscious, industry wanted above all to avoid muddling the minds of consumers.

Such fears also proved partially valid because Delaney from the outset made no mistake about his panel's mission. Most food processors and related industries did a "commendable job" in testing and using new chemicals, Delaney noted, but it was the committee's intent to assess possible dangers to public health and decide whether the threat warranted federal regulatory reform. The chairman came not to praise the chemical revolution—there were enough on Capitol Hill already doing that—but to judge its consequences. When paired with the publicity surrounding the hearings, the committee's judgmental tone caused food and chemical industry representatives to avoid appearing before the panel if at all possible. "There has been a general reluctance on the part of the food industry to testify," commented committee counsel Vincent Kleinfeld to the press. "They have avoided us like the plague."[45] Such was industry's avoidance that Delaney threatened to use his subpoena powers should cooperation not be forthcoming.

Food and chemical industry representatives eventually appeared, but reluctantly, especially after the reception enjoyed by NACA's Hitchner. Pesticides, Hitchner acknowledged, were "within themselves often poisonous or deleterious," but argued that industry has protected consumer health and safety. No new legislation was necessary, since "pesticides are not sold without legislative control." In fact, Hitchner concluded, "I know no other industry which has to comply with more laws and legislation in order to sell their products."[46]

Committee members for the most part were unimpressed by such arguments. Noting that he himself as a citizen had to comply with thousands of laws, Rep. Keefe wondered aloud why the industry opposed federal legislation requiring premarket testing of pesticides and other food additives.[47] Hitchner replied that the industry was not opposed to regulation in principle; however, he added, "We want to be sure that we will not unduly restrict the wide range of chemicals which can be used during the next couple of years."[48] The industry, he observed, would also have to deal with two federal bureaucracies—the USDA and the FDA—which would cause it problems: "We have no quarrel with control. . . . But if that proposal went out, we would not only have to go to Agriculture and say 'Here, gentlemen, here is a product,' but we would have to go to Food and Drug, and Food and Drug would also determine whether it is necessary, whether it is toxic, and we might possibly get back two opinions, one from one agency and one from another."[49]

Keefe wondered, however, just how much testing the USDA actually performed under FIFRA, commenting that "there could not be any real testing of 22,000 newly made up insecticides or pesticides."[50] Hitchner could only agree, though he continued to argue that better enforcement of existing law was the real problem. Committee members, however, greeted such claims with skepticism, with Keefe and Delaney in particular criticizing the industry's arguments that new regulatory authority would only retard new chemical development and use. Few disputed the importance of pesticides to food production and farm income, but, unlike those on the Agriculture Committee, Delaney and his colleagues placed stronger emphasis on public health.

USDA officials encountered similar skepticism. Dr. Elmer Griffin of the Production and Marketing Administration admitted that FIFRA, though an improvement over the 1910 law, contained some key weaknesses. The law did not require premarket testing, but only registration; nor was it entirely clear about the problems of pesticide residues on raw agricultural commodities. And, Griffin added under questioning, the government still bore the burden of proof on product safety.

Committee Counsel Kleinfeld noted that FIFRA lacked any reference to the "consumer," and wondered whether the consumer really knew what was contained in pesticides containers or which pesticides were used on foodstuffs.[51] Griffin admitted that FIFRA's provisions applied solely to original containers—not to those repackaged for retail sale—and that consumer health depended almost entirely on whether farmers used pesticides judiciously. This was a complaint leveled traditionally by both the FDA and public health authorities. Implicit in this view was that the USDA, by virtue of its role as patron to agriculture, was ill-suited to protect public health and safety. USDA officials admitted that misuse existed, but argued consistently that the department depended heavily on other federal agencies for the basic information on which registration decisions were made. FDA officials agreed with this assessment, but added that they had neither the statutory mechanism nor the funds to generate such data early enough to prevent chemicals from finding their way to the consumer.

The hearings, which ran forty-six days and produced 2,700 pages of testimony, established two major points in the minds of those seeking legislative reform: (1) food and chemical industries did not have consumer health as their primary orientation, and (2) the FDA had no mechanism for knowing beforehand which chemicals reached the consumer, and with what effects. The hearings ranged widely from cosmetics to fertilizers, but the major focus was on pesticide residues on food. And, as Delaney observed upon ending the investigation in March 1952, the prevailing disposition among committee members was to require premarket testing of food additives and provide for stronger FDA regulatory muscle.[52]

Industry representatives and those in the USDA fought such directions, while FDA officials pushed hard to forbid use of any new chemical until it was proven "utterly and completely" without harm. The committee, torn between the exigencies of chemistry's role in contemporary society and its desire to protect consumer health, tended toward a middle ground of "reasonable testing" for toxic and chronic effects. Industry reaction was mixed, with cosmetics makers most vociferous in opposition, while the food industry, though initially apprehensive about stronger FDA authority, eventually accepted the reasonable testing possibility with relative equanimity. Baby food processors in particular proved sympathetic to stricter regulation, since their customers were most sensitive to the harmful potential of food additives. One Beech-Nut official testified that his company expended considerable resources to keep residues out of its products. In fact, they did so to such an extent that the company was accused by chemical makers of being "hysterical about the problem." But, he noted, evidence did show that newborn infants already displayed traces of DDT in body fat—residues that needed to be eliminated.[53]

The Delaney Committee proposed no new legislation in its final report, but the hearings did spawn several suggested reforms. Delaney in late 1952 proposed a strict pesticides regulation bill that proved too extreme for many of his own committee colleagues, and it never surfaced. Rep. A. L. Miller (D, Neb.), another committee member, in March 1953 proposed to require that the FDA consider "all chemical additives as being harmful and not safe until conclusive evidence has been furnished otherwise."[54] More important, H.R. 4271 required manufacturers introducing new chemicals, or new uses for existing formulations, to prove their safety—a distinct shift in the direction of the 1938 law. The FDA, upon petition by the chemical maker, would have ninety days to (1) establish a tolerance for the product's use on any raw agricultural commodity; (2) exempt the additive from any tolerance because it was deemed safe for any use; or (3) refuse to grant any tolerance because it deemed the product unsafe for any use and in any amount. Miller also proposed to create a federal advisory committee of "distinguished experts" to advise the FDA on scientific and technical questions. This proposal was designed both to placate industry fears about "dictatorial" FDA power and to provide the agency with "objective" expertise.[55]

Miller's bill did not emerge out of the House Committee on Interstate and Foreign Commerce during 1953, primarily because of objections by farm bloc legislators and industry lobbyists that the changes by-passed the USDA. Farmers feared exacerbating already bifurcated federal authority on pesticides; that the FDA might seize commodities bearing residues of chemicals previously registered by the USDA was a fear dating back to the acrimonious prewar disputes over arsenic. The chemical industry, worried that successive regulatory disputes might destabilize a pesticides market only recently structured under FIFRA, shared such concerns. Above all, the agricultural community and its government patrons sought to protect the USDA's dominance over pesticides registration, and wanted little part of legislation that gave the FDA—the primary advocate of consumer health—veto power over agriculture.

Farm bloc legislators were careful, however, neither to scuttle the Miller bill outright nor to remove FDA authority to establish spray residue tolerances on commodities. The reason, Thomas Dunlap suggests, was to avoid unwarranted publicity: "Because of the previous food-poisoning scares and the public overreaction, the pesticide industry believed that publicity would be more dangerous than useful, and the agencies, remembering their experience in the prewar period, were reluctant to draw the criticism of the [pure food] lobbyists and their congressional allies."[56] Indeed, as Representative Miller later exclaimed, "There was little ballyhoo about it. The radio, press, and television gave it little play."[57] The fight instead went on within the rather closed circle of affected industries, agencies, and health professionals, but the debate lacked the drama of a bona fide scandal. Miller's Pesticide Control Amendment to the FDCA passed the House by voice vote, without much discussion, in April 1954. The Senate Committee on Labor and Public Works held no separate hearings, and the upper chamber approved the bill in early July. The public, if it perceived anything at all, saw the dispute as a largely arcane debate over technologies already widely accepted as beneficial to society. Faith in the pesticides panacea clearly extended beyond the agricultural community.

The amendment forbade USDA registration of any pesticide that left residues on fresh fruits or vegetables until and unless the FDA first issued a tolerance that indicated "safe" residue levels. The FDA now could discriminate among "safe" and "unsafe" uses of the same formulation, so the USDA could register a chemical only for prescribed uses. Further, the FDA's judgment that a pesticide was unsafe for *any* use on or near food meant its total ban for food-related purposes. However, the FDA's authority extended only to residues left on fresh produce or commodities like eggs or milk, not to the use of nonfood pesticides, even if such use indirectly contaminated the environment or adulterated food. The FDA could, for example, regulate any chemical sprayed *directly* on apples, since it was considered a food additive, but had no authority over the use of herbicides to kill weeds below the tree. The USDA retained final registration authority over such nonfood uses, and FIFRA's provisions also allowed chemical makers to challenge any USDA registration decision, even if based on FDA advice.

The Miller Amendment passed with so little debate precisely because it theoretically balanced the needs of farmers, consumers, and chemical makers. Consumers were assured that the FDA now could prevent hazardous food additives, which was of no little consequence to food industry executives anxious to calm public fears. Because the USDA now worked with the FDA in setting tolerances, farmers theoretically no longer faced bifurcated federal authority that might spell unexpected FDA seizures of tainted crops. And the chemical industry once again stablized its market, for it now could promise farmers that the products purchased met both USDA and FDA standards.[58]

But, when all was said and done, the Miller Amendment perhaps benefited the agricultural and chemical sectors more than it appeared to do. The Delaney hearings may have publicized the potential dangers from *any* level of chemical additives in food, but, as Perkins observes, the law now "*legitimized* the use of insecticides by establishing legal doses that were considered insignificant. The FDA's extensive hearings were mooted by the legislation, because the Miller Amendment established a new method for obtaining a tolerance."[59] The FDA had argued all along that any additive should be considered harmful unless proven otherwise, but the law as passed reflected "the collective judgment of Congress that certain levels of the poisons could be tolerated as a daily part of our existence."[60] Congress thus assured the public that, while residues might remain in or on food, the levels were not sufficiently toxic to warrant concern. Such assumptions would be challenged in later disputes about food additives.

Conclusion

"With the end of the Delaney hearings and the passage of the Miller Amendment," Thomas Dunlap notes, "the controversy over the new pesticides seemed over, with the industry and the agricultural community the winners."[61] The FDA could now set tolerances for pesticide residues on agricultural commodities, but the overall registration of pesticides remained in USDA hands—the committees on agriculture saw no need to change FIFRA, despite the Delaney hearings. Additionally, the FDA still had to depend on research performed by the Public Health Service, which because of its research orientations continued to find little evidence of chronic danger to human beings in the new chemicals. The FDA gained a little, but the defenders of orthodoxy retained their dominance.

But, at a minimum, the Miller Amendment legitimized the *formal* inclusion of the FDA into pesticides regulation—a role never really provided by the 1938 law—and with this inclusion came a perspective about pesticides sharply distinct from the reigning paradigm. The agricultural community previously had dealt with the FDA on a less formal, more ad hoc basis, one in which the FDA usually found itself subordinate. Now the FDA—and, indirectly, the consumer—was central to at least one segment of the pesticides configuration. That it would be a disputatious partner soon would be apparent.

By mid-decade, then, the pesticides question once again approached a degree of tenuous consensus, at least regarding the procedures of federal policy. The pesticides subgovernment had faced a potentially crippling challenge, and emerged relatively unscathed—at least in the short run. The use of pesticides remained an agricultural matter first and foremost, even if it generated consequences for nonagricultural interests, and those others were defined in terms of the dominant theme. But the Delaney Committee's focus on public health obscured a facet of the pesticides equation soon to overshadow food issues and that would usher in a new, potentially revolutionary angle of policy vision. Nowhere did the committee discuss dangers to the "environment," a set of problems whose cumulative magnitude would surprise even those concerned about such matters.

Chapter Five

Creating a Public: The Eradication Campaigns

Problems and Their Publics

Suppose a farmer cannot sell corn to a local grocer unless it is free of worms and spots. The farmer's livelihood is at stake, so he sprays a chemical pesticide on the crop to make it conform to local consumer demands. The farmer has solved his immediate problem if the grocer regularly purchases the produce and if nobody (save the farmer) cares much about *how* the crop is grown. The transaction between farmer and grocer remains private, and the "problem" is resolved without making demands on those not immediately affected.[1]

Suppose, however, that the farmer's pesticide drifts onto neighboring properties and apparently kills other farmers' chickens. Actions taken to resolve a private problem now indirectly affect others. If those apparently affected perceive their woes to stem from that pesticide, *and* if they seek some systematic resolution to the perceived threat (such as a local spray ordinance), the problem becomes public. The "public," John Dewey argued, "consists of all those who are affected by the indirect consequences of transactions to such an extent that it is deemed necessary to have those consequences systematically cared for."[2] It is only when these "others" react to consequences generated by the original transaction that we can begin to speak of a "public problem." We speak incessantly about "the public," but the term may find little operational worth divorced from a referent. "Publics," like politics, may be defined by issues.

We have already considered one type of public problem, the possible adulteration of the nation's food by chemicals. Such concerns eventually prompted legislation, but, as was suggested, problem "resolution" apparently occurred with little attention by the "mass" public—that vast and amorphous community of citizens potentially affected by chemical residues on food. The "public" directly involved in the Delaney Committee and the Miller Amendment consisted of a relatively constrained set of professionals and an attentive elite, as so often proves the case with complex scientific matters. The broader public, the body of consumers, either failed to perceive the existence of a problem or discounted its significance. In either case, the average consumer failed to act upon what should have been a fundamental concern for personal health. It was, in large part, a public problem that lacked broad public concern. The public seemed to behave like the sort of unorganized and/or apathetic mass that is central to so many pluralist and elitist views.

The key to "creating" a public lies in shaping popular perceptions not only about the indirect consequences of human action, but also about their "directness," in how *acute* those consequences appear. Consumers during the 1950s apparently worried little about the potential health risks associated with using pesticides, perhaps because the more immediate benefits of clean and relatively cheap food far outweighed any remote potential harm from food additives. Combine these attitudes with the overriding faith in American business and science characteristic of the decade, and it is not surprising that the pesticides subgovernment was able to weather the spray residues episode with little short-term erosion in policy dominance.

The food additives dispute also failed to generate appreciable mass attention to the issue because it lacked the critical element of drama. Controversy existed, but the debate probably was so esoteric and technically arcane that relatively few comprehended either the nature of the dispute or the stakes involved. Delaney Committee hearings dealt primarily with chronic, or long-term, health effects—an impact whose immediacy or significance was (and is today) lost on those less concerned with that distant time known as "later." Besides, there was no suggestion that anyone actually died from "normal" pesticides use. A few unfortunates succumbed to the toxic effects of chemical "misuse," but nobody, not even the chemical industry's harshest critics, claimed that this new generation of pesticides posed grave immediate threats to health. If anything, when compared to the older formulations, they seemed rather safe.

Missing from the food additives controversy, to put the matter baldly, were bodies. Death is much more dramatic to the mass public than some uncertain claim about future effects—a fact understood viscerally by every newspaper editor. Evidence that pesticides *might* cause harm sometime down the road was not compelling enough for many, and print media coverage of the debate about food additives was limited to short wire-service items tucked far inside the newspaper by editors who apparently felt that, when compared to other news, this matter was of limited public interest. The average person's interest in personal health is supposedly a dominant concern, but the lack of compelling and dramatic evidence of a chemical danger may explain this paradoxical lack of mass attention. Public reaction to the food additives dispute proved in many ways a precursor of the later battle over smoking: in the face of contradictory evidence about a health threat, individual consumers will suspend their response until the preponderance of evidence builds on one side or the other. This is true particularly if the decision involves fundamental changes in one's personal behavior.

Roger Cobb and Charles Elder argue that issues vault most readily into mass public consciousness when they are socially significant, apparently nontechnical, defined broadly, and above all, emotional.[3] The food additives dispute certainly was socially significant, but it was far too technical, defined in rather narrow terms (for example, administrative coordination), and relatively low-key. This was a dispute confined largely to the committee room, one debated among experts in terms both technical and often arcane. Above all, there was little real drama.

The impact of chemicals on fish and wildlife, however, aptly fit Cobb's and Elder's criteria, and powerfully pushed the issue of pesticides into the public spotlight during the late 1950s and into the 1960s. The issue to come would seem far less technical and very broadly defined—dead birds, after all, are dead birds. And it would prove emotional, particularly because of how the issue would jump into the public arena. The problem of pesticidal impacts on wildlife crept imperceptibly along for several years, but would vault dramatically onto the public agenda because of two incidents that provoked a sharp backlash against those most ardently promoting the chemical eradication of pests. The irony would be that these actions—the massive spray campaigns against the gypsy moth and the fire ant—were both the pesticides subgovernment's finest hour and its eventual Waterloo. The eradication campaigns would show both civilization's scientific prowess and the reality of unintended consequences.

The Gypsy Moth Campaign

In 1957, with the debate about spray residues in abeyance, the USDA prepared to embark on ambitious efforts to eliminate the gypsy moth, a widely acknowledged threat to northeastern forest regions. Gypsy moths strip trees bare of leaves, depriving them of essential nutrients and sometimes causing them to die. Few questioned the moth's impact on the region's forests and related industries, and the USDA's Division of Plant and Pest Disease by early 1957 expressed confidence that it could now eradicate, rather than merely control, the pest that had mocked their efforts for decades. Division Chief M. L. Clarkson explained to the Whitten Subcommittee that a well-coordinated, regionwide spray campaign could extinguish the moth, and he requested $2.5 million for the project. "We have the tools to bring this to a final conclusion," Clarkson argued, proposing that the USDA, in conjunction with northeastern states, aerially spray infested areas with a mixture of DDT suspended in oil.[4] Such a broad-gauge approach would be more effective and cheaper than the traditional spraying of individual trees, a point sure to appeal both to subcommittee members' fiscal concerns and to their desire to display the department's pest control prowess.

Total eradication might take eight to ten years, Clarkson admitted, but he assured subcommittee members of final and total victory.

Whitten and his colleagues apparently shared that enthusiasm, for in March the USDA announced that some 3 million acres in four northeastern states would undergo aerial spraying between mid-April and late June, the period when moth larvae are most susceptible to chemical control. Federal appropriations for the campaign would be about $5 million (more than the USDA originally had requested), with additional monies provided by participating states—though the overwhelming portion came from federal funds, indicating the states' inability or unwillingness to deal with the pest unilaterally.[5] The USDA, in giving area residents short notice of the impending action, assured them that the spray program was safe, though a few fish in "shallow ponds" might succumb to the oil mixture.[6] Above all, the department emphasized in its press releases, the campaign would eliminate the moth threat. This was a statement contrary to the USDA's traditional caution about the permanence of any chemical control.

Elected officials in affected states expressed uniformly strong support, and the program began in earnest in April, as planes sprayed wide swatches of land. On May 8, however, a group of Long Island residents (described by some news reports as "organic gardening enthusiasts") led by noted ornithologist Robert Cushman Murphy filed suit in New York Federal District Court to enjoin the USDA from further spraying. Murphy and fellow plaintiffs complained that the DDT mixture in use was far too heavy for the intended purpose, that the program violated the USDA's own guidelines (planes flying too low, spraying during heavy winds, etc.), and that the spraying itself was a peril to gardens and local wildlife. USDA officials publicly responded that the spray mixture posed no such threats, but complaints from residents continued to pour in. New York agricultural agents heard from hundreds of farmers about crop damage, while other residents charged that spray planes flew too low, or that spraying left heavy deposits of the mixture on their properties. New York Governor Averill W. Harriman on May 22 stated that the spraying apparently had caused "significant" fish kills and other damage, and urged the USDA to exercise greater care to avoid spraying directly over water. Harriman did not, however, openly condemn the program itself.[7]

The Department of Agriculture grudgingly acknowledged some validity in the complaints, but did not halt the program because there was "no need to fear damaging reductions in fish populations."[8] Yet, simultaneously with such claims, New York residents reported massive fish mortality throughout the state, including some 600 trout killed in one upstate stream. Contradictions between USDA assurances and the apparently nonrandom pattern of wildlife mortality prompted the *New York Times* to question the government's moral and legal right to "spread poison" over private property without permission of the owner and in the absence of an overriding public health hazard. The editors, noting the U.S. Fish and Wildlife Service's apparent misgivings about the spray campaign, and citing the possibility of systematic chemical harm to local wildlife, urged future caution as well as the need for research on the relationship between pesticides and wildlife.[9]

Federal District Court Judge Mortimer Byers in late May denied the Long Island residents' request for an injunction against the USDA, arguing that plaintiffs did not show sufficient cause to warrant a halt. He ruled that Murphy and his colleagues lacked compelling evidence that DDT posed a clear and immediate threat to wildlife or human health—an objection to become painfully familar to anti-DDT forces in the future. Besides, Byers noted, the spray program was almost complete, making the complaint moot. However, he left the door open to requests for hearings on future such activities.

The gypsy moth campaign concluded in late June, but not before generating additional protests. Assistant Secretary of Interior Ross Loeffler in early June argued to reporters that the program had proceeded "without adequate information on the effects which the pesticides have on fish and wildlife resources."[10] The Fish and Wildlife Service lodged a strong protest with the USDA over its spraying of wetlands, which Agriculture subsequently agreed to avoid, but, as Robert Cushman Murphy noted sardonically to the *New York Times,* "private landowners appear to share no such privileges or rights."[11] The Department of the Interior at the same time proposed its own long-term research on pesticides and wildlife, a proposal that found little immediate support within Congress.

DDT Disturbs: The Emergence of Audubon

The gypsy moth campaign, redolent with charges of bureaucratic presumption and chemical overkill, did more than temporarily generate regional awareness of the pesticides issue. This campaign, and the fire ant program soon to follow, was a critical catalyst to the priorities of established conservation organizations, which up to this time had paid relatively little attention to the possibly deleterious impact of chemicals on wildlife. An important case in point here is the National Audubon Society, for whom the gypsy moth campaign transformed organizational goals and priorities.

Audubon's primary efforts during most of the decade aimed to preserve wetlands and forest areas as bird and wildlife sanctuaries. The society's president, John H. Baker, had cautioned about the proper use of chemical pest control agents as early as 1949, but Audubon's prevailing view throughout the postwar era was that careful use would minimize any possible harm to fish and wildlife. Too little was known about these new chemicals to make a definitive statement, but, as Baker urged members in 1952, "if serious harm to wildlife results from insecticide use, it is a matter of public concern and one that needs to be brought into the open with whatever factual evidence is available."[12] But the dearth of such "factual evidence," and the dominance of the pesticides paradigm within mainstream entomology, ensured that those worried about chemical effects on the environment had little ammunition with which to counter the reigning orthodoxy. Even when Audubon and other conservation groups became concerned about pesticides, their criticisms would be blunted by their lack of compelling evidence. This was to be a problem for years to come.

Little action on the pesticides-wildlife front occurred until mid-decade, when Audubon members in the field began to report alarming drops in certain types of bird populations. Most disconcerting was the sharp decline in the number of peregrine falcons, bald eagles, and ospreys, but concern broadened with additional reports of mortality and low fertility rates for birds throughout the nation. Few, if any, as yet linked these trends to pesticides; most observers blamed unseasonable cold snaps and the impact of urban encroachment on bird habitats. In mid-1956, however, Fish and Wildlife Service biologist Paul Springer detailed for Audubon members mounting evidence of a link between the heavy use of pesticides and the trends in bird mortality being observed. Pesticides were critical to life in modern society, Springer noted, but they nonetheless generated certain unfortunate consequences. "In problems of this type," he wrote pessimistically, "wildlife values are ignored as being of little significance. Or safeguards for [using chemicals] may be dismissed as hindrances to economic progress."[13] Demonstrations of "subtle effects" of chemicals on wildlife "points up the danger of using stable insecticides with high residual qualities," though Springer cautioned Audubon members not to condemn pesticides universally because some uses (or misuses) harmed wildlife.[14] The scientific community was fully aware of the possibly chronic dangers of pesticides, he assured readers, and "entomologists are just as open-minded as any group of people, and most of them are anxious to limit wildlife losses to a minimum."[15] Later assessments of entomologists' uncritical faith in chemical technologies would not be so charitable.

The gypsy moth campaign transformed the Audubon Society's ambivalence toward chemical technologies, its prior caution having been bred by uncertainty about the weight of scientific evidence and by a traditional belief that government officials were receptive to thoughtful advice. The spray campaign, and apparent USDA intransigence in the face of mounting criticism, turned caution into indignation. Evidence of this change emerged forcefully during Audubon's 1957 annual meeting. Whereas previous annual meetings focused on bird migration and other traditional society activities, this session was devoted almost exclusively to the gypsy moth campaign and to the relationship between chemicals and wildlife. The testimony of one member, Wilhemine Waller, of Westchester County, New York, typified the litany of complaints against the USDA: "I was assured both over the telephone and in press releases from the Department of Agriculture that no farmland would be sprayed; . . . the Department of Agriculture misinformed us twice—first in saying that no farmland would be sprayed, and second, in telling us that property would be sprayed but once. As a result of this dousing with DDT in oil base, we encountered . . . damage and destruction on our farm."[16]

The damage, Waller continued, included ruined garden produce, milk banned from sale because of high DDT residues, and high mortality rates among local birds and fish. Successive speakers condemned the USDA for alleged capriciousness, for program mismanagement, and for undertaking such a massive operation with little apparent understanding about its long-term impacts. "I have always been proud of being an American," declared Waller, the Westchester farmer, "but since the blanket spraying of personal property last spring, and the manner in which the U.S. Department of Agriculture carried out the spray program, I have been just a little less proud."[17] Veteran Audubon members could not remember hearing more acerbic sentiments.

The gypsy moth campaign often is credited with sparking the Audubon Society's transformation from a "bird club" to an "environmental protection" organization, one with pesticides placed atop the new list of the organization's priorities. Audubon already was going slowly in this direction as evidence about pesticides was painstakingly accumulated, but the spray campaign provided the catalyst, the "drama," that propelled a complex scientific issue into the spotlight and accelerated conservationists' fears that humanity itself was the primary enemy of the evironment. The campaign also provoked conservation organizations to question whether traditional public education activities were sufficient to promote their aims. And, if the gypsy moth campaign planted the seeds of doubt about pesticides and their patrons, the fire ant campaign would dispel any lingering illusions.

Fire Ants: The Political Imperatives of Pesticides

It would be ironic if we spent millions to fight communism and were eaten up by fire ants. The fire ant is by far the most dangerous importation yet.[18]

It is my opinion that there has been more misinformation, unfounded assertions, inaccurate reporting, and scare campaign propaganda, if not bold deception, fed to the American public in attempting to win support for the ill-conceived and irresponsible fire ant "eradication" program than for any other control program attempted in this generation.[19]

The fire ant migrated from South America through the port of Mobile, Alabama, in the 1920s, spreading gradually thereafter so that, by the mid-1950s, it covered parts of nine southern states. Like many insects, the ant delivered a painful sting, but one that one National Research Council report found to be no greater than of the stings of bees or wasps.[20] The ant's primary nuisance value appeared to lie in its propensity for building high mounds of dirt that interfered somewhat with haying or mowing operations; however, despite the rumors that would pervade southern folklore for years to come, there was scant evidence that the pest harmed human beings or wildlife. In fact, according to one 1959 study, the ant survived primarily on a diet of other insects—most notable among them being the boll weevil.[21] The "objective" nature of the fire ant "threat" appeared to differ sharply from that of the gypsy moth, yet the government's response would be identical, with profound implications for both the pesticides subgovernment and its critics.

The fire ant has been the target of the federal government's longest and costliest eradication campaign. The USDA's war against the ant, begun in late 1957, continues to this day despite vociferous and persistent opposition by biologists, environmentalists, and, increasingly, by residents within affected areas. The program survives nonetheless for the same reasons it began—politics. "More than anything else," Harrison Wellford argues, "the fire ant program is a monument to the power which key congressmen on strategic committees can exercise over environmental policy."[22] The origins of the campaign lay in the Whitten Subcommittee, which, in effect, became its permanent defender and coordinator.

USDA entomologists prior to 1957 worried little about the ant, confining their early actions to modest research studies that appeared to confirm the insect's minor status. Indeed, the USDA's own 1957 ranking of control priorities failed to mention the ant as a significant pest.[23] Early that year, however, Jamie Whitten received a request from the Mississippi Farm Bureau that the Agricultural Research Service look into the problem. The ARS also heard from a number of southern commissions of agriculture—who, it appears, had been prime movers of the fire ant program from the beginning. Says Wellford: "These commissioners are powerful men in Southern politics. They sit astride the flow of agricultural subsidies and benefits which amount to the largest source of outside capital for many impoverished Southern counties. They also control substantial patronage. The Southern commissioners have long been the most vocal lobbyists on Capitol Hill for fire ant control."[24]

Department of Agriculture studies later showed that few southern farmers controlled the ants on their own properties, that the pest did not particularly bother farm workers, and that land values were not affected by ant infestations.[25] The evidence ran contrary to claims made by southern agricultural officials, who, despite popular ambivalence about the pest, enthusiastically supported a massive federal eradication effort. Despite the uncertain nature of the alleged threat, it was to these officials, and to their congressional respresentatives, that Whitten and his colleagues responded.

The *New York Times* reported on March 19, 1957, that several southern House members had requested the Whitten Subcommittee to initiate a broad campaign against the fire ant. Witnesses cited "unsolicited" letters from constituents describing painful ant attacks against children and livestock. One Alabama farmer even argued that "the government should be building as big a defense against fire ants as they are against the Russians. The ants already have invaded."[26] Existing statutory authority, however, allowed the USDA to undertake control programs *only* under specific congressional authorization—a condition that, at least to Whitten and his colleagues, greatly hindered the swift interdiction of nascent pest outbreaks. Representatives in both chambers subsequently proposed to amend the Agricultural Organic Act of 1946 and grant to the USDA broad authority to control or eradicate insect pests, particularly the fire ant. The proposed legislation, both simple in language and sweeping in its scope, would allow the USDA to control pests without specific congressional authorization.[27]

The Senate Committee on Agriculture reported that, according to the USDA, treatment of the fire ant called for application of granulated dieldrin, chlordane, or heptachlor—persistent hydrochlorinated pesticides known to be at least forty times as toxic as DDT. Some 20 million acres across the South were involved, with treatment to cost approximately $5 per acre.[28] Yet, in the sole exchange on the Senate floor about the issue, Senator Allen Ellender assured his colleagues, "There is an existing contingency fund to take care of emerging outbreaks. No additional funds will be required."[29] But, in earlier testimony to the Whitten Subcommittee, Assistant Secretary of Agriculture Ervin L. Peterson had requested just such additional funding precisely because the department's contingency fund, estimated to be some $2 million per year at the time, was sorely inadequate for the proposed action. "For an eradication program to be successful," Peterson stated, "all infestations must be treated *without regard to location, land use, or ownership*."[30] Such a massive program was going to cost a lot, and all directly involved knew it, so one might reasonably suggest that program proponents shielded the true costs from inquiring eyes; it would take little effort to spot the obvious paradox between southerners' normal fiscal parsimony and their support for a massive and potentially expensive federal program. Those supporting the program played down its costs, and, as was noted earlier, the secrecy surrounding appropriations matters inevitably left outsiders in the dark about program commitments.

The Fire Ant Eradication Act (S. 1442) sailed through Congress with minimal debate or opposition in less than three months, a swiftness of passage made all that more notable by the simultaneous controversy brewing over the gypsy moth program. President Eisenhower signed the bill into law without comment on May 23, the same day that the *New York Times* editorialized on "poison in the air" and supported expanded federal research on the pesticides-wildlife relationship. The irony is considerable.

Controversy over the fire ant program began just as that over the gypsy moth began to wane. The USDA in late 1957 announced that with its new statutory authority it was about to embark on a joint federal-state fire ant eradication campaign that was anticipated to take at least three years.[31] Agriculture officials estimated the federal share to be less than 50 percent of total expenditures, but also noted that some 20 million acres required treatment. At the estimated $5 per acre treatment cost stated in the hearings on S. 1442 (estimates later found to run on the conservative side), and even assuming that the federal share remained at less than half of total costs (which it would not), it was obvious that the federal government would soon be pumping approximately $4–5 million annually into the effort. Yet, if documentary evidence is indicative, no opposition to the program immediately emerged from within Congress.

Late 1957 and early 1958 witnessed one of the more remarkable government publicity campaigns—critics soon would call it a "propaganda blitz"—in peacetime American history. Rachel Carson, whose chronicle of the fire ant program in *Silent Spring* later would infuriate both southern agricultural officials and Jamie Whitten, writes, "The fire ant program suddenly became the target of a barrage of government releases, motion pictures, and government-inspired stories portraying [the ant] as a despoiler of southern agriculture and a killer of birds, livestock, and man."[32] *New York Times* reporter John Devlin described the upcoming eradication campaign in terms almost straight out of USDA press releases. The U.S. federal government was helping the South in its fight against the "vicious" fire ant, Devlin wrote, and in "magnitude it will dwarf the spraying of DDT" that characterized the earlier gypsy moth program. Devlin's portrayal of the ant was almost identical to that given by the USDA in its report to Congress: "Farmers find the ant costly. Mowing bars hitting mounds jam or break. Because the ant tunnels several feet underground, poor soil is also distributed on top of the topsoil. Furthermore, the ants attack and kill calves, pigs, and chickens. They also bite farm workers."[33]

Critics later disputed the department's assertions that the fire ant (either singly or in numbers) could capably bring down a healthy calf, or even a chicken. At that time, however, reporters accepted the USDA's contentions at face value, and Devlin, writing for a large northeastern daily and with no personal experience with the subject, probably was no different. He went on to note that the department was conducting "an education program in the south to show the need for spraying. . . . Strong public support has resulted."[34] That the USDA conducted this education campaign to avoid a recurrence of the gypsy moth controversy seemed neither apparent nor worth commenting upon.

Central to the Department of Agriculture's education program, and one of its most enduring and controversial components, was a film produced for the USDA by Tulane University about the case of an eight-year-old Louisiana boy. The youth, who had acute allergies to all types of insect bites, died three days after being stung by several fire ants. This death became a rallying cry for eradication proponents throughout the South because, as one southern newspaper commented, "No one knows how many people have been killed by ants."[35] The USDA's Clarkson admitted to the Whitten Subcommittee in early 1958 that the youth's case was "extremely unusual, but it does illustrate the potential danger of the imported fire ant to some persons who happen to be extremely sensitive to the poison injected with its sting."[36] Whether the USDA intentionally used this case to drum up public support for eradication is a matter for debate, though there is little doubt that it became a strong justification for the program. And, as this excerpt from a youth magazine indicates, the fire ant fast was on its way to attaining mythic status. "This fire ant is only about a quarter of an inch long. But it packs a wallop like a tank! It gnaws a slit in the victim's skin and injects a dose of fluid that causes sharp pain and raises angry welts. Some victims of the fire ant may be hospitalized for weeks. A baby in New Orleans was killed by them."[37]

The publicity campaign, by all accounts, worked well. In early 1958 the *Times-Journal* of Selma, Alabama, thundered that the "deadly invaders" must be eliminated, while Representative Kenneth Roberts of Alabama called for all-out war on the "vicious, imported fire ants, intent on attacking and destroying if possible agriculture, cattle, dairying, and trucking in the several Deep South States."[38] Congress in 1957 appropriated $2.4 million to begin the fire ant program, Roberts noted, but this amount was "not much, considering the vast job to be done." Additional appropriations, he added presciently, would be necessary.[39]

If the educational campaign rallied southern opinion leaders to the fire ant cause, it also generated additional—and, from the department's perspective, undesirable—attention both from those already critical of federal pesticides programs and those newly alarmed by the magnitude of the proposed action. Two days after John Devlin's *New York Times* article, National Audubon Society President John H. Baker called the program a "chemical peril" to humans and to wildlife, and urged Secretary of Agriculture Ezra Taft Benson to curtail the action before it began. Dieldrin, Baker noted with alarm, is "one of the most deadly of the modern insecticides," and the USDA proposed to apply the chemical at approximately two and a half pounds per acre, a concentration sure to wreak havoc upon southern wildlife.[40] National Wildlife Federation President Ernest Swift, citing studies carried out by three Harvard biologists, warned that the program would serve only to devastate the ant's natural predators, and that too little was known about dieldrin's toxic or chronic effects to use it with confidence.[41] The *New York Times* editorial board, recalling the still-lingering effects of the gypsy moth program, chimed in: "There is good reason to get rid of the fire ant. . . . But the cure to which the Agriculture Department is suddenly committing itself on such an enormous scale may be definitely worse than the fire ant's bite. It is rank folly for the Government to embark on an insect-control program of this scope without knowing precisely what damage the pesticide itself will do to both human and animal life, especially over a long period. . . . Yet the Department plunges blithely ahead."[42]

Such criticism failed to deter the program's supporters. USDA Deputy Administrator Clarkson disputed charges of hasty planning, arguing to reporters that the campaign was designed to "do the least possible damage to wildlife and other insects."[43] Audubon President Baker, reporting on the society's survey of government officials throughout the nation, noted that "almost without exception, the defensive letters come from federal and state agricultural departments, state plant boards, and agricultural stations, all presently engaged in control.[44] Eradication supporters, driven as they were by the imperatives of both the pesticides interests and southern politics, plunged ahead, leaving their critics to find some other way to minimize the damage they feared would result.

Enter Interior: Institutionalizing a Third Perspective

The fire ant campaign not only provoked criticism by fish and wildlife groups but also spurred congressional action on pesticides research legislation that had languished since its introduction at the height of the gypsy moth episode. Fish and Wildlife Service officials, worried about the apparent impact of DDT on wildlife, had proposed extensive and long-term Interior Department research on the relationship between pesticides and the environment. Little such research existed at this time, both for scientific and fiscal reasons. Scientific concern about the effects on wildlife was only nascent, and what evidence did exist was largely a by-product of entomological research. The other obstacle was purely budgetary: biological research on chemical effects was greatly overshadowed by the millions spent by chemical companies and by the USDA to develop and test new products. Pest control already was a massive business, with industry producing some 500 million pounds of pesticides products in 1957. *Total* expenditures for pest control operations at this time neared the $2 billion mark, while the USDI was allocated a mere $56,000 for research on the effects of pesticides on wildlife.[45] This is not to imply that neither the USDA nor the entomological profession cared about such matters, but departmental priorities and entomologists' professional prestige were now inextricably tied to the success of chemical technologies. The FDA and the Public Health Service confined their respective research agendas to human health, and both relied heavily on the USDA for basic data—much of it generated by the chemical companies themselves. It was clear, at least to USDI officials, that the situation demanded a major research program by an organization without strong links to the pesticides subgovernment.

The USDI research bill (H.R. 783) was sponsored originally by Representative Lee Metcalf (D, Mont.), who in 1956 unsuccessfully proposed similar legislation following a massive and apparently DDT-induced trout kill in the Yellowstone National Park. The 1957 bill was referred to the House Committee on Merchant Marine and Fisheries, chaired by Representative Herbert Bonner (D, N.C.), who in turn sent the bill to the subcommittee on Fisheries and Wildlife Conservation, headed by Frank Boykin of Alabama. Both supported the USDA, and H.R. 783, like Metcalf's earlier bill, languished in subcommittee. Circumstances in 1957 seemed more dire, however, and Fish and Wildlife Service officials in July turned to Senator Warren Magnuson (D, Wash.), chairman of the Senate Committee on Interstate and Foreign Commerce. Magnuson faithfully served the interests of his state's apple growers (though never so closely as did Walt Horan on the House side), but since he also worried publicly about cases of bird mortality coming to light in Washington's apple regions, he agreed to sponsor the bill.

Magnuson held no hearings until March 1958. Reasons for the delay are unclear, but the record indicates that it took the fire ant controversy to move the process forward. It also appears that extensive behind-the-scenes negotiations delayed congressional action, because the bill to emerge finally from the hearings bore the marks of extensive compromise among those intent on pursuing such research and those seeking to maintain the USDA's dominance over federal policy. That the hearings on S. 2447 began almost simultaneously with the acceleration of the fire ant program only heightened the urgency to those concerned about the program's possible hazards.

Magnuson's Subcommittee on Marine and Fisheries gave critics of the USDA an institutional point of access previously untapped in Congress. Undersecretary of the Interior Hatfield Chelsant set the tone for the debate by lamenting the lag in research on chemical impacts on wildlife, which he said was "5 to 10 years behind the discovery, formulation, and field application of these chemicals."[46] Despite so little understanding about such effects, said USDI chemist James DeWitt, widespread use continued. "Probably the chief danger comes in the spraying operations which have been sponsored by other federal agencies," specifically the USDA.[47] The Sport Fishing Institute echoed DeWitt's charges, expressing doubt about the fire ant program's preparatory research and planning. "What the Department of Agriculture does not seem to know is what really may hurt America," the institute's spokesperson added darkly.[48] An official of the National Parks Association, Fred Packard, described the discovery of 800 dead birds soon after a USDA spray program near a Texas wildlife refuge,[49] while Audubon's Baker condemned such programs in general. Baker also recommended that the bill authorize $25 million for coordinated federal research but, he added, "We would recommend against lodging primary research responsibility in any government agency engaged in using chemical controls."[50] That Baker referred primarily to the USDA was apparent.

The sole public note of opposition came from NACA's Lea Hitchner, who, in a letter to Magnuson, argued for a more modest research program. The USDA had lengthy experience with pesticides, Hitchner argued, and any major expenditures for research of the type proposed were "unjustified." And, in a tone that no doubt infuriated the bill's supporters, Hitchner added, "In our opinion certain groups interested in wildlife are unduly worried about the use of these materials mainly due to the lack of knowledge of what has been and is being done. Sound research is always beneficial to all parties, and we believe would do much to allay the unfounded criticisms which appear from time to time in the press."[51]

The debate is all the more interesting because one would expect to find those interests most closely associated with the use of pesticides publicly defending their research prerogatives. But hearings alone are inadequate indicators of political activity, and it is likely that the opponents of the bill did not bother to appear because they did not have to—their work went on purely behind the scenes. The bill that emerged from the Magnuson Subcommittee already bore the marks of extensive compromise, for it was far more modest in scope and intent than was the original USDI proposal. Reality dictated that the bill's supporters would accommodate the wishes of the agricultural community, particularly with most key congressional veto points dominated by southern conservatives. The bill to finally emerge from committee posed no threat to USDA policy dominance, so the fact that Hitchner protested at all was more intriguing than any lack of industry or agriculture attendance. They needed no such show of "force," for the structure of bias embedded in Congress at this time definitely favored their perspective. The bill's supporters, however, needed to show both numbers and resolve to underscore their contention that the legislation was sorely needed. The legislative process has historically aided those seeking to prevent change, so making a show of force at committee hearings may in fact indicate relative weakness, not strength.

The maneuverings that produced a mild USDI research program failed to remove one major bone of contention. The original proposal left authorization levels unspecified in hopes that Congress eventually would appropriate "sufficient" funds for long-term research. A request for high funding levels over a period of years, as Audubon's Baker had urged, was sure to run into opposition from fiscal conservatives, so the USDI was content to leave matters open. The bill debated before the Magnuson Subcommittee, however, specified a $280,000 annual funding ceiling that Interior officials called "arbitrary" and grossly inadequate for the envisioned task. This "cap" originated with the House Committee on Merchant Marine and Fisheries, and was supported by the argument, "It is unlikely that the program could be started at a higher level than that set by the amendments" and that the ceiling "will insure that the Congress will be furnished with prompt and specific information as to the progress of the program if the Department wishes to expand beyond the maximum figure specified."[52] The USDI was not to get a blank check: House members, particularly Boykin and Bonner, envisioned a very modest research program, with the USDI primarily engaged in gathering and assessing data generated by others. The USDA, for its part, publicly did not oppose the bill, largely because department officials knew that they would be the USDI's primary data source.

The Magnuson Committee agreed with USDI protests and deleted the cap, and the bill passed the Senate without debate in May 1958. Interior Department satisfaction proved short-lived, however, as Bonner and Boykin immediately reinstated the cap and passed their version of the bill through the full House via the consent calendar in July. The weakness of the USDI position was made more apparent when the Senate accepted the House version without comment, with Interior and its allies apparently acquiescing for fear of jeopardizing even these modest gains. President Eisenhower took no part in the debate and signed the bill without comment in August.

Congressional defenders of pesticides later would rely heavily on such spending provisions to hamstring non-USDA research efforts. Spending caps usually reflect congressional desire for oversight or fiscal responsibility, but the practical impact in this case would buttress USDA dominance in pesticides research by keeping the USDI perpetually underfunded. And if those supporting expanded USDI research hoped to use the appropriations process to increase program funds, the committees on appropriations stood as determined bulwarks against such maneuvers. The USDI, according to Richard Fenno, lagged badly behind Agriculture in congressional prestige during the 1950s, and in fact suffered a disproportionate share of reductions in annual funding requests by the appropriations committees during the period under study.[53] That Interior trod lightly where the appropriators were concerned was not too surprising.

Congress did boost USDI pesticides research monies by $125,000 shortly after passage of the Pesticides Research Act, raising the figure to about $180,000 for 1958. A year later, as evidence mounted that the fire ant program was devastating southern wildlife, Congress raised the appropriations ceiling to $2.6 million per year, though the USDI still lobbied to eliminate it altogether. Nonetheless, USDI research funds increased only incrementally during the next few years. The department got little help from the Eisenhower administration, which instead was seeking to slash research funds as part of a general budget reduction. A frustrated Charles Gutermuth, of the Wildlife Management Institute, complained in early 1960 that the Bureau of the Budget had turned down Interior's request for additional funds even after Congress raised the funding cap:

> They got nowhere with that. They asked for additional money in the budget this time in compliance with this law that was passed by the first session of this Congress. They got nowhere . . . . I want to point out that in the Agriculture Appropriations [Committee] they are appropriating $2,400,000 and have for the last two years for fire-ant control work in the South, and still we get $280,000 for conducting all of the research which is being conducted on the effects of those powerful chemicals . . . . We contend, Mr. Chairman, that it is just about as important to keep from poisoning people as it is to balance the budget.[54]

Recurrent funding problems aside, the Pesticides Research Act at a minimum *informally* institutionalized a third perspective as part of the pesticides equation. USDI's modest research program seemed a minor matter to proponents of the status quo, which may account for the bill's passage in the first place, but later developments would magnify Interior's role and the centrality of its views in pesticides policy decisions. The significance of the research act, in retrospect, lay not in its scope but in the legitimacy it accorded to a new voice, however tentative, that ran contrary to views propounded by the USDA and its patrons.

Thus, by the end of 1958, three competing perspectives and three diverging sets of policy claimants, begin to characterize the pesticides debate. And, not unlike the three-headed Gorgon of mythology, the three perspectives perpetually lived in uneasy coexistence. The position of the USDA and its allies, favoring use, still dominated—the farm bloc still controlled the legislative route and the USDA the regulatory one—but the views expressed by the FDA and the USDI no longer could go ignored. The mass spray campaigns of the late 1950s strained any concord that may have existed among the three, and resulted eventually in congressional hearings on the administration of federal pesticides programs. Before turning to these hearings, which in many ways mark the end of pesticides "golden age," let us examine two additional incidents that proved influential, if only indirectly, in the struggle over the pesticides paradigm.

Court Suits and Cranberries

The furor over the USDA's fire ant campaign overshadowed a relatively minor drama being played out simultaneously in a New York federal district courtroom. The term "minor" only denotes the scant media publicity accorded the case, not the importance of the event itself to those vitally concerned about the future of pesticides use in the United States. To these actors the drama was compelling. It was to be repeated increasingly in future years.

Robert Cushman Murphy's group of Long Island residents, having failed to prevent the 1957 gypsy moth program, filed suit later that year to prevent future spray campaigns. To win their suit, the plaintiffs first had to show that the USDA had no authority to undertake mass spray actions without regard to land use or ownership—statutory authority implicit in the original campaign and later made explicit by the Fire Ant Eradication Act. Second, they had to prove that DDT was without doubt a health hazard, and that the plaintiffs were adversely affected by the spray. Without such proof, the USDA could not be prevented from acting as it did. Department officials, on the other hand, would be required to show that DDT was relatively benign and that the federal government had a right to undertake mass spray campaigns when they were deemed necessary and in the public interest.[55]

Murphy opened the nonjury trial in February 1958, testifying that the gypsy moth spray program and its kin upset nature's balance, and that it ultimately produced unintended consequences worse than the original pest infestation. Furthermore, he argued, the federal government had no right to apply such hazardous materials indiscriminately; the private citizen had a right *not* to be sprayed. Dr. Malcolm Hargraves of the Mayo Clinic testified that experiments on laboratory animals strongly suggested an erosion in human tolerance of DDT as absorption rates increased, which might cause jaundice and aplastic anemia. Biological and toxicological evidence presented by the plaintiffs showed pesticides to be apparent hazards to fish and wildlife. That they were hazardous to human beings would require a mere extrapolation from these data.[56]

The Department of Agriculture's case rested on testimony by Dr. Wayland Hayes of the Public Health Service. The PHS's findings, based on field studies, nearly always contradicted laboratory evidence about toxic and chronic chemical effects, and this time was no different. Hayes declared that PHS examinations of workers at DDT manufacturing plants, and of prisoners who received over 200 times the "normal" daily dosage of the chemical, showed no ill effects.[57] The amount absorbed by average citizens from spray programs, when compared to either test group, was "so small as not to be measurable."[58] Hayes also doubted openly that one could extrapolate human effects from laboratory tests on animals in the first place, so DDT, in his mind, was safe. The chemical undoubtedly caused undesirable effects to wildlife when applied improperly, but was not in itself a hazard to human beings.

Judge Bruchhausen apparently agreed, for on June 23 he argued that government's police powers allowed it to undertake spray campaigns on behalf of the public good. USDA evidence also showed the gypsy moth program to be a success—an end that justified the means. Furthermore, and more important for the future, Bruchhausen found no compelling evidence to declare DDT a health hazard, and ruled the plaintiffs' suit to be one of mere nuisance value. DDT was harmful neither to man nor wildlife when applied properly, the judge declared, and so long as the government carried out its programs with care, no harm should ensue.[59]

Thomas Dunlap, studying the history of DDT, notes, "In 1958 scientists were suspicious of DDT, but wildlife biologists were not willing to assert that the chemical was a danger to the environment."[60] Even the National Audubon Society, despite its fury over the eradication campaigns, refused to support the Long Island suit because of its uncertainty about the chemical's effects. Audubon's failure to press the court suit also was political; the organization still relied heavily on its traditional methods of public education to spread the good word about protecting nature. Neither Audubon nor any other mainstream conservation organization was as yet ready to use the judicial route, in no small part because of high hurdles erected by courts on the matters of establishing damage and cause. Given the state of the evidence and lack of support for the plaintiffs, the judge's decision came as no real surprise.

Murphy and his colleagues pressed their case to the Second Circuit Court of Appeals in 1959, and to the U.S. Supreme Court a year later. Both courts, noting that the gypsy moth campaign had ended years before, took the traditional view that the cause of the complaint was moot and therefore nonjudiciable. Plaintiffs in future spray program cases would encounter similar rulings, making it almost impossible to stop spray actions without temporary injunctions. And judges were loath to grant even those requests, both because of traditional deference to the expertise of executive agencies and because the evidence against pesticides failed to establish a compelling case. Where plaintiffs could show some evidence of damage, the courts just as invariably declared the cases to be moot, since most cases came to trial after the spraying had ceased. Either way, the judicial branch provided plenty of obstacles to those wishing to use that route in their fight against the use of pesticides.[61]

But, as so often happens in this tripartite structure of governance, the courts' reluctance to become involved in the pesticides issue was not copied elsewhere. The FDA in the late 1950s in fact would provoke one of the most heated controversies in the history of pesticides regulation, and would thrust the issue of spray residues squarely into the public view. The Great Cranberry Scare was an odd, yet in many ways telling conclusion to the golden age of pesticides. It was an episode that revealed clearly the dilemmas inherent in the chemical era. It also showed how policy in this system can be influenced indirectly through narrow niches in the decisional framework.

The Pesticides Control Amendment of 1954 allowed the FDA to set maximum tolerances for spray residue found in or on raw agricultural products—levels judged by the agency to be safe for human consumption. In March 1956, the USDA registered the herbicide aminotriazole for use on cranberry bogs after harvest, when no residues on fruit should occur. But, as always happened, some growers applied the chemical before harvest because it kept down weeds and boosted crop yields. Such practices prompted the FDA in 1957 to seize some 3 million pounds of cranberries for allegedly high levels of aminotriazole, even though it had yet to establish a tolerance for use on the fruit. FDA officials admitted their dilemma, but charged that residue levels found on the berries were sufficiently high to warrant caution. The USDA, upset over this turn of events, emphatically repeated its warnings about proper use, and the incidence of contamination fell sharply in 1958. The two bureaucratic units meanwhile negotiated on acceptable residue levels for the herbicide.

But economic imperatives proved more seductive than could any government cautions. To delay spraying the herbicide until after harvest meant lower yields, so many growers were willing to take their chances with the FDA rather than forego the certainty of greater profits. The cat-and-mouse game between the FDA and the growers, one developed over decades of confrontation, was on again, but the game would be more serious in 1959 because the agency would be armed with a new weapon.

Representative James Delaney became increasingly disenchanted with existing food and drug laws as the decade came to a close. His own hearings had paved the way for stronger FDA authority, but Delaney feared that alleged carcinogens still found their way too easily into the nation's food. The FDA could act on some pesticides residues, though largely after the products already were on the market, but completely lacked any premarket authority over such additives as dyes, sweeteners, or preservatives. The agency also bore the burden of proof when it attempted to ban the use of any substance—a process known to take years. Congress in 1958 finally acted on Delaney Committee recommendations for such a pretest provision, but the Food Additives Amendment applied this authority only to "intentional" additives. Pesticides residues were not deemed to be "intentional" when chemicals were not sprayed directly on the crop, and all "unintentional" additives were excluded from premarket testing by farm bloc representatives reluctant to impede farmers' ability to control weeds. The FDA still could set tolerances for chemicals applied directly on raw agricultural goods, but could not control indirect contamination. That authority resided with the USDA.

The Food Additives Amendment to emerge from the House Committee on Interstate and Foreign Commerce was the result of several years of hard bargaining among agricultural, food industry, and public health interests in Congress, and was perhaps the "best" bill that could have resulted under the circumstances. Food industry representatives as usual worried that any regulation would have a "chilling effect" on product development (and sales), but were willing to submit to premarket testing if only to quell mounting public anxiety about carcinogens. Farm bloc members were opposed to any sort of FDA premarket clearance, satisfied that FIFRA's registration provisions, when coupled to the FDA's tolerance-setting authority, adequately protected public health. A few "pro-consumer" representatives worried, nonetheless, that the Food Additives Amendment merely maintained the status quo, giving the FDA only reactive authority on pesticides. But, unlike earlier times, these members now had some leverage in shaping the pending legislation by virtue of the peculiar power resources to be found in the vagaries of the legislative process.

Delaney by this time was a senior member of the Rules Committee, strategically placed to hold H.R. 13254 hostage until the Eisenhower Administration and the Congress agreed to a simple change. His addition, thereafter known as the Delaney Amendment, stated only that "no additive shall be deemed safe if it is found to induce cancer when ingested by man or animal." This was one of those phrases that prove so tantalizingly ambiguous, deceptively simple in scope and intent, and so sweeping in its potential consequences. Delaney's amendment made no distinction between intentional and unintentional additives. It appeared to apply only to the intentional additives covered by the bill, and Delaney never really stated any intention to broaden its coverage (though he later admitted with pride that it did), but farm bloc legislators realized that the amendment could pertain to pesticides. More important, the amendment would now allow the FDA to ban any suspected carcinogen. Agricultural interests howled in protest, but for once they were unable to prevail. Too many others had worked for too many years on the Food Additives Amendment to risk failure, and these interests were willing to accept Delaney's amendment. So important was the bill to so many, including the Eisenhower administration (which actively pushed for its passage), that it sailed through Congress virtually intact. The strength of the coalition supporting the bill was most notable in the House, where a last-ditch effort to derail the bill failed when the chamber passed it by a two-thirds majority. President Eisenhower signed the measure into law in November 1958.

The Delaney Amendment stood until the mid-1960s as the only pesticides-related statute to pass through Congress over the objections of the agricultural subsystem.[62] Success came because the pending legislation was vitally important to a coalition of interests more encompassing than the farm bloc, and because the amendment's chief proponent controlled a crucial veto point in the legislative process. Those favoring stricter federal regulation of chemical additives for the first time were able to take advantage of congressional processes that so often had worked against them. Such lessons were no doubt learned at the hands of pro-pesticides legislators.

Relatively little time passed before the Delaney Amendment proved its potency to all concerned. In early November 1959 HEW Secretary Arthur S. Flemming announced suddenly that some cranberry products currently on the market were contaminated by residues of aminotriazole, which the FDA claimed caused cancer in laboratory rats. The cranberries in question were grown in the previous three years, and much of this fruit already was on the market in processed form. The FDA did not know the extent of the contamination, since the residues showed up in about ten lots of fruit produced in Washington and Oregon, but Flemming advised pre-Thanksgiving shoppers not to buy cranberries unless they were certain about the product's safety—which, of course, was impossible. The secretary based his action not on the tolerance-setting authority of the 1954 Miller Amendment, which pertained solely to raw agricultural goods, but on the Delaney Amendment, which declared *any* product to be unsafe if "tainted" with a suspected carcinogen. Thus products allowed on the market under a previous law now violated the FDA's interpretation of the new.

Flemming's action vividly illuminated the dilemmas faced by the FDA in dealing with spray residues. Aminotriazole was registered by the USDA as a herbicide, and sold well, but the FDA had warned in 1957 that the herbicide left alarming levels of residues on fruit shipped to market. The USDA revised its instructions to growers based on this advisory, while the FDA seized and froze tainted berries pending lab tests. These tests for carcinogenicity, using the chemical maker's own data, took two years, and showed that aminotriazole caused tumors in rats. The FDA interpreted these results as meriting a zero tolerance. Growers meanwhile continued applying the herbicide under USDA guidelines, so that when the FDA finally reached its conclusions in October 1959, several crops suddenly fell under suspicion. FDA officials argued that they had no ethical choice but to take action. Flemming acknowledged that his timing was disastrous for cranberry growers, but stated, "I don't have any right to sit on information of this kind; consumers had a right to know, and to know quickly, about possibly hazardous substances in their food."[63] The FDA acted precipitiously because it perceived little other choice.

The cranberry ban exploded into public awareness like no other food additive episode to date. The difference in 1959 lay with both the timing of the announcement—on the eve of the holiday season, when 70 percent of all cranberry sales occur—and because of the issue-expanding role played by the emerging medium of television, which, by this time, reached most American homes.[64] Public reaction to the announcement confirmed every worst fear about mass hysteria ever held by food industry executives: cranberry sales plummeted almost immediately, and several states and localities banned sales or urged consumer caution.

Criticism of the FDA rose almost as rapidly as the fall in sales, with the USDA leading the charge. Department of Agriculture officials flailed Flemming for failing to consult either with them or with food industry representatives, and noted that HEW gave but one hour's advance notice before going public. USDA officials also charged that the FDA read the food and drug laws too literally. Assistant Secretary of Agriculture Ervin L. Peterson argued to the Whitten Subcommittee shortly thereafter that USDA and HEW interpretations diverged on the literalness of statutory authority, stating that the HEW "holds that there is little room for scientific judgment on a given set of facts, whereas we in Agriculture believe that there is room for such judgment."[65] Flemming's actions, to USDA officials, were based at best on flawed assumptions, at worst on politics, but surely not on science—at least not on science as they saw it. Others agreed. The *New York Times* acknowledged that Flemming's job was difficult, but charged that he "went too far; . . . a warning would have been justified if there were clear proof that aminotriazole produces cancer in human beings—of which there is no evidence at all as yet."[66] The American Farm Bureau Federation predictably called the action "irresponsible," and several agricultural groups called for Flemming's ouster.

Administration officials and farm bloc representatives moved quickly to reestablish consumer confidence and save at least part of the 1959 sales season. The USDA, in conjunction with the food industry and members of Congress from producer areas, accelerated promotion of the fruit through ad campaigns. Vice-President Nixon gamely ate a generous portion of cranberry sauce for the cameras, as did his future presidential opponent, John Kennedy. Elected officials throughout the land followed suit. The HEW, under pressure from the White House, agreed to a short-term labeling program that would identify "safe" fruit for consumers, while Flemming assured the public that his family indeed would have cranberries for the holidays.[67] All of this was too late, however, as cranberry sales for 1959 fell by two-thirds, and losses to the industry totaled approximately $15–20 million.

Cranberry growers found not to be at fault for their losses later received some $10 million in indemnity payments from the USDA. Jamie Whitten was displeased that *his* department was footing the bill, but he and other farm bloc representatives felt obliged to support growers whom they saw as innocent victims of FDA capriciousness and hysteria-mongering. Whitten, during hearings on the indemnity program, worried aloud about the "chilling effect" on agriculture created by Secretary Flemming and his subordinates. He stated bluntly, "[Publicizing the matter] has caused most chemical companies to be hesitant to do their own future research, . . . they are afraid that they might wake up one morning and see in the press that Mr. Flemming has not only called it off the market but has destroyed the future of the commodity."[68] Flemming clearly broke the long-cherished USDA rule of reticence on contamination matters, and Whitten queried acidly, "I'm asking if there has been any understanding with Mr. Flemming that thereafter he will make every effort to handle this problem without unnecessarily inflaming the public?"[69]

Whitten's exasperation with the publicity surrounding the cranberry episode reflects the Achilles' heel of subgovernment politics. So great was potential public alarm about contamination that agricultural interests went to great lengths to avoid "inflaming" the public. The visibility of an issue is a key to the expansion of conflict, and public warnings of potential health hazards, no matter how carefully stated, were to create maelstroms of controversy that the pesticides subgovernment could not contain—to its detriment. The subgovernment successfully reimbursed many cranberry growers because of its unanimity on the matter, and because few in positions of authority sought to stall the payments. The Department of Health, Education, and Welfare was immobilized temporarily by the backlash to its action, and would be far more cautious during the next few years, but the episode nonetheless affected the pesticides subgovernment more fundamentally. This episode, coming as it did after successive controversies on USDA eradication programs, would force subgovernment actors to build bridges to policy stakeholders professing competing perspectives on the use and safety of chemical technology. The agricultural subsystem, by virtue of the USDA's authority to register pesticides, continued to dominate policy directions, but the mere fact that it had to take other views into consideration was a major change.

The End of the Golden Age

We are geared for a long, hard struggle against this tough and insidious pest, and it will call for the cooperation of every citizen in the South. But if we keep at it, we are pretty sure convinced that we can eventually get rid of the fire ant.[70]

It was with this zeal, at least as professed by the Assistant Secretary of Agriculture, that the USDA embarked on its eradication campaign, which, as Thomas Dunlap later wrote, "stimulated public and scientific interest in the effects of broadcast sprays on wildlife."[71] The 1950s began with enthusiastic and almost uncritical acceptance of the pesticides paradigm, but its end would reveal greater uncertainty, uncapping several wellsprings of criticism of both the USDA and its actions. The cranberry scare sparked momentary public awareness about pesticides residues in food, but such concerns, as happens so often, quickly dissipated. The fire ant program, however, would lay the foundation for fundamental transformation in national pesticides policy, though the changes would emerge some years down the road. More immediately, the campaign provoked deeper and more widespread criticism of USDA actions, and of pesticides in general, than had ever before been heard.

The department during 1958 treated some 900,000 acres in the South with granulated dieldrin and heptachlor, and reported optimistically about the program's success.[72] Assistant Deputy Secretary Clarkson acknowledged that a few "accidents" in the initial stages of the project resulted in some fish and wildlife mortality, but assured the Whitten Subcommittee that the USDA was working actively with fish and game interests in affected areas to minimize future damage. Some destruction of wildlife occurs inevitably during any such program, he noted, but not frequently. "We have been criticized for mentioning this aspect of the destruction of the imported fire ant," Clarkson complained; "We have tried to put the record in its proper perspective." However, he concluded, those interested primarily in defending fish and wildlife apparently were not interested in meeting the department halfway.[73]

The Whitten Subcommittee hearings of 1959 were all the more notable because those who hardly agreed with Clarkson's assertions came for the first time to argue against the program before its most ardent promoters. Roland Clement of the National Audubon Society argued that continuing field inspections of affected areas showed "alarming" rates of damage to wildlife. "In our considered opinion, and in light of our own observations," Clement stated, "there is no justification for continuation (and there was none for initiation) of the imported fire ant eradication program." He urged the committee to suspend appropriations.[74] Whitten promised to review the program thoroughly, but replied that the subcommittee had an obligation to implement "within reason" a program authorized by Congress. "I would not personally feel it is within our authority to veto a legislative act passed by Congress and signed by the President," he explained. "If the law were repealed, then, of course, we would not be faced with the obligation of implementing it."[75] It was obvious that no such appeal would come if Whitten had anything to say about it, and Congress quietly appropriated another $2.4 million for the program's third year.

The controversy widened perceptibly by late 1959, after another 800,000 acres had undergone treatment. Evidence of damage to fish and wildlife mounted, and the USDA's apparent intransigence in the face of such evidence infuriated state and national wildlife interests. The Audubon Society reported bird mortality in the 90 percent range for some species in Texas, while the Fish and Wildlife Service warned of the total decimation of quail populations in several states. Death to mammals, including rabbits, squirrels, and raccoons, apparently was widespread, with high levels of heptachlor residues found in the tissues of the dead animals. A south Georgia veterinarian reported the sudden deaths of over 100 head of cattle soon after a spraying of dieldin, and reports of similar devastation to farm animals filtered in from throughout the region. The Southeastern Association of Game and Fish Conservation Commissioners in October complained openly that the USDA had ignored its advice, though, according to a spokesman, "[The department] has repeatedly informed the public that its fire ant eradication program is cooperative with the States and with private landowners. Even so, areas have been aerially treated contrary to the wishes of property owners."[76] The USDA in mid-1959 did reduce the strength of the treatment per acre by one-third, the association noted, but this dosage still was highly lethal, particularly when applied indiscriminately by air.

Southerners, regardless of the wholehearted support maintained by their commissioners of agriculture, seemed less enthusiastic about the project as events unfolded. The fire ant program originally was designed as a cooperative venture, but by 1960 it was clear that the federal government alone would pay the freight. Georgian farmers by mid-1959 refused to contribute further monies to the program, while in Texas the USDA literally gave heptachlor away to any property owner willing to use it. The department assured the local public, "Heptachlor can be applied at no danger to livestock, pets, or birds," though it was considered widely to be far more dangerous to animals than DDT.[77] Finally, the state of Alabama, in which the fire ant first appeared, in late 1959 refused to appropriate funds for future participation in the program. State conservation officials calculated that up to 75 percent of Alabama's wildlife might be eliminated if the use of heptachlor continued, and state legislators apparently decided that further contribution to the program would be a bad investment. The USDA would continue to spray ants in Alabama, but without state help.

Amid this controversy, fought out largely behind the scenes, the USDA proposed to expand FIFRA's jurisdiction to include many classes of new chemical agents, including disinfectants, nematocides, and herbicides having entered the market since 1947. The proposed change was in scope, not direction, so the 1959 FIFRA amendment was rather minor relative to the overall pesticides question. One might have expected some degree of conflict over the amendment, given the ongoing spray campaign disputes, but such was not the case; the bill sailed through Congress with neither controversy nor opposition. USDA critics apparently focused their attention at this time more on the effects of pesticides on wildlife and the practices of the USDA, not on the registration process itself. Congressional critics of the campaigns also knew that they would have little chance to force policy change—it was not yet their moment. The FDA and USDI thus operated on the fringes of federal pesticides policy, while the USDA continued to dominate its core.

Yet the cranberry and fire ant episodes did disrupt the pesticides subgovernment, since they propelled the issue of chemical pest control into the uncomfortable spotlight of controversy. Threats to the hegemony of subgovernment policy still were rather undeveloped, yet the central actors of this policy community increasingly saw their paradigm encountering criticism from nonagricultural interests—a situation not particularly reassuring to those aware of agriculture's troubled future. Farm bloc representatives during the Whitten Subcommittee's 1960 hearings expressed undiminished and unabashedly strident support for the chemical option, and this tone intensified as the subgovernment was forced onto the defensive. Ervin L. Peterson of the USDA stated emphatically that although the department was careful to consider public health in its decisions regarding pesticides, its overriding responsibility was to the farmer, who "simply cannot produce enough safe, wholesome foods without chemicals."[78] Yet, Peterson noted with dismay, critics ignored the obvious benefits and focused instead on the unfortunate side effects that inevitably occur in the chemical age. The Department of Health, Education, and Welfare took a heavy beating from both USDA officials and members of the subcommittee for the cranberry episode, with Whitten particularly critical of HEW's literal interpretation of the Delaney clause and "capricious" exercise of administrative power. The chairman also wondered whether it would not be better to formally include HEW in any USDA pesticides research, "so that the research results will be partially theirs, and they will have to accept a part of the responsibility for such findings."[79] Whitten's intent was to co-opt the FDA into the research process, confident that inclusion would quell criticism and perhaps induce food and drug regulators to adopt the USDA perspective. Peterson, viewing the situation more in terms of bureaucratic turf, wanted little part of an arrangement that might erode the USDA's research dominance. The problem wasn't research, Peterson assured Whitten, it was simply that FDA officials really didn't support their decisions through sound scientific judgment. Everyone knew that pesticides produced some ill effects, but they were outweighed so heavily by benefits that the FDA simply did not take into consideration. USDA officials wanted nothing to do with interagency coordination of research efforts, particularly with those not sufficiently supportive of agricultural needs.

The question of interagency cooperation in pesticides research and registration arose again in 1960, but in a congressional decision arena not tied directly to the agricultural community. Rep. Leonard Wolf of Iowa, a member of the liberal Democratic class of 1958, in April introduced the Chemical Pesticides Coordination Act (H.R. 11502), which required advance consultation with the Fish and Wildlife Service by any federal agency preparing to embark on mass chemical controls. Wolf, apparently alarmed by wildlife losses stemming from various spray campaigns over the country, and with the cranberry incident still fresh in people's minds, argued that his bill also was in the best interests of farmers and the chemical industry. Otherwise, he added, "an indignant public will demand and get rigid controls."[80]

Wolf's bill was referred to the House Committee on Merchant Marine and Fisheries, still chaired by Herbert Bonner, and in turn went to Frank Boykin's

Subcommittee on Fisheries and Wildlife Conservation; as a result, prospects for any action looked dim. Below Boykin, however, were two Democrats of rising seniority, George Miller of California and John Dingell of Michigan. Miller and Dingell apparently lobbied hard on the bill's behalf, for the subcommittee held hearings in early May. Dingell, who in the 1960s would emerge as a leader in the House for stricter pesticides regulation, proved particularly active. The bill never had a chance to pass, since the farm bloc opposed giving the USDI any veto power over USDA spray programs, but it is apparent that Dingell used the hearings to get publicity and perhaps to establish some momentum that might pay off later down the road.

Conservation groups, state and federal fish and wildlife officials supported Wolf's bill, all stressing the need to control federal pest control programs and avoid repetitions of the fire ant fiasco. The Audubon Society's president Carl Buchheister, for one, charged the USDA with overriding the concerns expressed by conservation interests:

> USDA officials, when questioned, say that, yes, the fire ant operations have been carried out "in cooperation with" or "after consultation with" wildlife officials. Upon investigation we have found that in general, in the operations of the fire ant program, the recommendations of the wildlife agencies have been ridiculed publicly and usually ignored. Such consultation as has existed has usually been after the fact.[81]

The USDA came under heavy fire for its fire ant program. Perennial critic Dr. Clarence Cottam called the USDA's "education campaign" one of "misinformation, unfounded assertions, inaccurate reporting, and scare campaign propaganda, if not bold deception,"[82] while Charles Kelly of the Alabama Game Department called it "a glaring example of riding roughshod over the responsibilities of other public and private agencies."[83] Alabama's previous withdrawal from the program no doubt buttressed Kelly's charges. Critics argued overall that current legislation was inadequate to protect wildlife and public health against pesticides; the only answer was a national policy on pest control requiring the USDA to consult with the Department of the Interior. Both Dingell and Miller expressed their opinions of the USDA in no uncertain terms. The USDA "does not seem to care about the foods that we eat insofar as the toxic things that we get in our food," Dingell charged, adding that Agriculture officials had always "looked down their noses at the Department of Health, Education, and Welfare." Miller noted, "Every once in a while we find an agency of Government . . . that takes unto itself certain infallible qualities and does not want to connect with others." The USDA, Miller concluded, "has been rather highhanded when it comes to dealing with what it considers the minor interests of Government—fish and wildlife particularly."[84]

The departments involved, however, wanted nothing to do with formal coordinative arrangements. USDA officials, not surprisingly, saw the Wolf bill as completely misplaced and unnecessary, largely because it emphasized the protection of fish and wildlife over national welfare and security. Besides, argued USDA Deputy Administrator William Popham, the congressional committees on appropriations annually reviewed USDA spray programs as part of the budget process; mandatory interagency coordination would simply be redundant.[85] Extensive cooperation in research and implementation did exist, he claimed, arguing that the department had consulted widely with others in planning the fire ant program. Some damage to fish and wildlife was inevitable, but, he added, "Evidence is lacking to indicate that widespread permanent damage to the wildlife resources of this country have resulted from the use of pesticides in connection with Federal-State eradication and control programs."[86] Agriculture's real fear was that the USDI would delay vital control programs, though critics noted that the fire ant war took a full year to plan. So long as cooperation remained purely advisory, the USDA could argue that it informed other agencies about its actions, regardless of the extent of such cooperation or how readily the department considered objections or advice. Mandatory cooperation evidently would force the department to balance competing angles of vision.

More intriguing was unexpected opposition expressed by both the USDI and HEW, departments that stood theoretically to benefit from the proposal. Sports Fisheries and Wildlife Director Daniel Janzen explained that the USDI appreciated the growing congressional concern with and support for wildlife matters, but he could not as yet recommend passage of an interagency coordination act: "The progress of research . . . has not yet reached the stage where we can make positive recommendations on how to modify many of the control programs. At this time we can offer but little effective information, and, therefore, it is not feasible . . . until more complete data are gathered on the immediate and long-term effects of each control agent."[87]

The Interior Department at that time was capable of only informal cooperation with the USDA, Janzen argued, even though such arrangements did have severe complications, and pointed to the fire ant program as a prime example. Interior simply lacked the research capabilities or data base to make valid contributions, and therefore did not want to be forced into an unequal working relationship. HEW officials similarly opposed the bill, noting that many pest control efforts were emergencies that might go unanswered if the USDA needed USDI clearance. Besides, HEW's Dr. Samuel Simmons pointed out, the Interdepartmental Committee on Pest Control had since 1945 provided for the free exchange of information and cooperative planning "where there is a joint interest."[88] This little-known, underutilized, and ad hoc committee was purely voluntary, with no authority of any sort. Yet it appeared that both USDI and HEW preferred to stick with the inadequate present rather than plunge into a risky future.

That the USDI itself declined to be part of a mandatory review process served in the end to buttress USDA arguments that informal processes served the bureaucratic actors well (though unified opposition to the bill in no way indicated agreement about pesticides activity and regulation). The Wolf bill did pass both the House Committee on Merchant Marine and Fisheries and the Senate Committee on Interstate Commerce, indicating strong feelings at that level, but went no further. Objections by both the USDA and congressional farm bloc members assured the bill's eventual demise, and it never even went up for a floor vote. USDA critics in Congress could do little more than to shake loose a few more dollars for further USDI research, which farm bloc representatives were quite willing to grant if it would dissipate any heat on pest control programs.

The pesticides subgovernment was wounded only slightly in the short term by the fire ant and cranberry episodes; in 1960, backed by the Eisenhower administration, it was able to exempt pesticides used to preserve fresh produce from food and drug labeling provisions. The 1960 amendment to the FDCA required labels on shipping containers to indicate the presence of pesticide residues on produce, but no such declaration would be necessary for commodities displayed for retail sale. Supporters of the exemption argued that pesticides on harvested produce technically were not "preservatives," a view disputed vigorously by the FDA. The cranberry scare, however, had mobilized farm bloc members in Congress to defend "farm interests," and consumers would remain unaware about residues on their produce.

Coda: The Precipice of Change

As the 1950s began, the pesticides paradigm was regarded as the undisputed panacea for the nation's agricultural woes, and the USDA as its unchallengeable keeper. By 1960, however, both assumptions faced growing skepticism, a trend that increasingly alarmed those whose interests lay in defending chemicals and growers' unrestricted right to use them.

The eradication campaigns of the late 1950s, and their apparent dire consequences for both public health and the environment, for the first time catalyzed the view among scientists and the attentive public that those who planned and executed federal pesticides programs too rarely considered nonagricultural values. Gradual accumulation of data and greater scientific understanding about pesticidal effects had some impact, but the magnitude and apparent capriciousness of the gypsy moth and fire ant campaigns provided still uncertain critics with more compelling evidence that both pesticides and the USDA needed greater scrutiny. One wonders what might have been the course of events had the USDA not undertaken these projects; perhaps opposition might never have coalesced as it did. Yet, in another sense, these programs were inevitable: a traditionally aggressive and client-oriented department, supported by and indeed prodded on by equally aggressive congressional patrons who shared a supreme faith in pesticides, could only have created such programs. The "bigger and better" syndrome, integral to the postwar American psyche, inevitably led to a desire to eradicate unpleasantness in a single and compelling stroke. The dispositions of subgovernment actors, particularly those in Congress, could lead to no other outcome at this time. It is ironic, then, that the same programs inaugurated with so much hope would also provide critics with enough rope to construct a noose.

By the decade's close, the embryonic "environmental" community began to shift its attention to the issue of chemicals versus nature, but their involvement was tempered both by the uncertain state of scientific information and their continued reliance on traditional methods of public education and low-profile "lobbying." The courts, as the Long Island DDT suit showed, still were out of bounds, or, at least, trying to win through litigation was not worth the effort because of the stringent standards for claiming damage still evident at that time. Only one conservation organization supported a professional lobbyist in Washington during most of the 1950s; the majority shied away from more active lobbying roles for reasons ranging from traditional disdain for "political advocacy" to fears, already noted, that such behavior might lead to revocation of their precious tax-exempt status.[89] This quiescence, in contrast to concerted activity and publicity on the part of the agricultural community and the chemical industry precluded anything more than modest alterations in federal policy, those deemed relatively inoffensive by defenders of the status quo.

The snail's pace of change in participatory behavior during this decade mirrored perfectly the stability—or state of rigor mortis, if you will—displayed by the nation's ruling institutions. Congress changed little during most of this period. Some faces changed, but power continued to be held by committee chairs dominated by conservative farm bloc supporters. And in the case of key committees relevant to pesticides, particularly in the House, the same faces as ten years before controlled the positions of policy leverage. The central role played by Whitten and his subcommittee as the fulcrum of the pesticides subgovernment would go unchallenged for many years to come.

Congressional structures and rules buttressed, and perhaps magnified, the influence of those most faithful to the pesticides paradigm, but Congress as an institution was on the edge of change as the fifties faded. The shifts as yet were imperceptible, for, as so often is the case with that institution, change bubbles up slowly from the bottom, from new cohorts in the membership. This loose coupling between demographic shifts and institutional transformation runs through cycles in American history. The changes to come would trace their roots to the influx of liberal Democrats after the 1958 congressional elections and to the subsequent creation of the Democratic Study Group, an intraparty caucus of unparalleled organization that would give liberals a base from which to counter the conservative coalition in the House. These forces, begun in the fertile soil of social transformation, would bear fruit some years later.

The presidency in 1960 stood as another defender of the pesticides status quo, or, at least, as a nonparticipant in the entire matter. Subgovernments thrive in the absence of high-level executive intervention, and, for most of the 1950s, mid-level USDA careerists were left pretty much alone to carry out their prerogatives.

The Eisenhower administration, though it supported some changes in food and drug laws, consistently sided with the agricultural and business sectors on pesticides-related issues, particularly with respect to residues. The cranberry incident was an example of independent action by a cabinet member, not of presidential policy initiative. The fact that cranberry growers eventually were indemnified for their losses without much controversy testifies to Eisenhower's reluctance to meddle with the agricultural community.

Deep within the executive establishment, however, the strains of diverging attitudes on pesticides were beginning to show. The FDA could set spray tolerances and the USDI perform some research, but the USDA still dominated regulation. The paradigm, guarded jealously by its government patrons, still reigned, but few believed that the dissonant voices now being heard could be quieted. The doubts would intensify in the years to come, but the transition from one reigning paradigm to some other perspective would not be smooth. Those long entrenched in positions of policymaking leverage would utilize every instrument at their disposal to maintain the status quo and to disparage competing claims. They would succeed for many years, despite massive shifts in social attitudes, which underscores the existence of the innumerable veto powers made available in the fractionated U.S. political system to those smart enough, or well placed enough, to use them. The fortifications manned by the pesticides subgovernment were cracked, albeit slightly, in the late 1950s. But they were cracks nonetheless, and were soon to widen perceptibly.

Chapter Six

Sea Change in the Sixties

New Leaders, New Environs

While new leaders and leaders in general make some difference, particularly at certain times, the impact of elections tells us as much about environmental change as about personnel change. Leaders matter, but always within certain parameters.[1]

John Kennedy, who pledged to get America "moving again," came to office during what Samuel Huntington calls "a shift from complacency to commitment" that would transform American society and its politics during the next decade.[2] The new president was in many ways a product of tidal change, a decade of activism to contrast dramatically with the 1950s and its self-assured affirmation of the American Century. It was a nation already in flux as a new generation, not content with that previous "atmosphere of greater serenity and mutual confidence," began to challenge traditional assumptions about virtually everything. Not even Kennedy, whose administration sparked the decade of "reform," could have foreseen its future depth, scope, or strains on the body politic.

Huntington pictures a nation in "creedal passion," a condition that surges to the fore periodically throughout American history. General support for mainstream social and political institutions gave way to skepticism as Americans of many stripes questioned the credibility of national ideals apparently at odds with reality. A common perception that the American dream had its warts would unleash reformist impulses reminiscent of previous "passion" periods (such as the 1890s), and the 1960s would remind scholars once confident about the "end of ideology" that in the United States ideas, and ideologies, were anything but consensual. A vast generation of postwar youth, and a good number of their elders, Huntington writes, participated in "a dramatic renewal of the democratic spirit in America. . . . The spirit of protest, the spirit of equality, the impulse to expose and correct inequalities. . . . The themes of the 1960s were those of Jacksonian Democracy and the muckraking progressives; they embodied ideas and beliefs which were deep in the American tradition, but usually do not command the passionate commitment that they did in the 1960s. . . . It was a decade of democratic surge, of the reassertion of democratic egalitarianism."[3]

This "democratic surge" took many forms. Government institutions, and institutions in general, suddenly faced a generation wary of organizational structures and power. Voter turnout plummeted, while more direct modes of political action blossomed, as a highly educated and mobilized postwar generation disdained the ballot and turned to more "efficacious" routes to reform. "The circumstances of the 1960s and early 1970s," Richard Brody notes, "gave rise to an interest in protest marches and demonstrations as media for the expression of political opinion."[4] From Freedom Riders to those protesting the Vietman War, the "Sixties Generation" expressed its discontent through new channels.

The participatory impulse emerged in the new political organizations that proliferated almost exponentially. Overwhelmingly, these groups dared to defend the so-called public interest—backing issues and concerns not traditionally represented in the spectrum of recognized interests—and to speak for social groups heretofore largely ignored. Jeffrey Berry notes that fifty-two out of eighty-three public-interest organizations surveyed in 1972–1973 were formed *after* 1959, and most emerged late in the 1960s.[5] This explosion in public-interest groups came about, Andrew McFarland argues, because of an increase in middle-class political participation, the corresponding increase in the politics of issues (as opposed to party politics), a growing skepticism about the utility of existing politics, technical advances in communications, economic prosperity, and, finally, initial success bringing more success.[6] The rise of public-interest organizations, ranging from consumer groups to legal defense funds of every sort, would expand and transform the universe of organized interests that had been studied by pluralists in the 1950s. It would be, by 1970, a broader and more variegated universe, that would have a dramatic impact on both public policy and political science.

New impulses found new technologies. Television emerged in the 1960s as the primary instrument for political discourse and exposure, a medium particularly suited to the spirit of reformism at hand. Nightly news programs, once the poor stepchild of network programming, by the decade's end regularly introduced Americans to legions of incidents, issues, and interests. And the smart political actor fully realized the power of the now dominant medium. "Not until the 1960s," Theodore White observes, "did television become the atmosphere of politics in which politicians breathed or suffocated to death."[7] The politics of Sam Rayburn gave way to the "new Nixon," and television influenced mightily who made it and who didn't, and on what terms.

Television fed, and was fed by, a systemic impulse toward open government that transformed an already public forum into one of seemingly transparent visibility. Reformers now could communicate *their* issues and *their* views directly to the public, and television, when combined with such changes as the Freedom of Information Act, dramatically affected the nature of political power in America. A medium capable of making more issues more visible to more people in less time would itself be a counterweight to the institutions of government and business. The news and information media in general act as "custodians" of national values, Huntington argues; "They expose and denounce deviations from these values; they bestow legitimacy on individuals and institutions who reflect these values. In creedal passion periods, the media have no recourse but to challenge and expose the inequities of power."[8]

Postwar affluence had produced a generation reared in relative comfort, one now in search of "postmaterial" values long deferred by their elders. Attitudinal disparities lay at the heart of the much-discussed "generation gap," but even older Americans sought change. Once-dominant economic concerns gave way to "superior" goods, those not necessary to human survival but increasingly regarded as essential to the overall quality of life. Americans increasingly cared more about civil rights, foreign policy, and government reform; not until the mid-1970s would the economy regain its position at the apex of the public issue agenda. Postwar affluence freed Americans to elevate new values—often, ironically, at the expense of the very institutions responsible for that affluence.

One of those institutions was agriculture, increasingly ignored by a more urban society on the one side, and embattled by an expanding range of detractors on the other. Farm population dwindled to but 8.7 percent of the nation's total by 1960, and would plummet to half that in the next ten years (see table 2, above). Almost one million farms folded, and those that were left grew, accelerating anew the trends toward fewer and larger begun in the late 1940s. Farm income rose in absolute terms, but remained low compared to nonfarm earnings, while the capital required to farm successfully more than doubled. The trends were unmistakable and ominous to those who cared, particularly to the congressional farm bloc. Jamie Whitten, in a 1961 speech to the National Agricultural Chemicals Association, accurately linked diminished farm population to farmers' weakened political influence: "For the future, the problem of reduced influence seems likely to become worse rather than better. Interest in and understanding of agricultural problems is decreasing as our rural population declines and our young people turn to other occupations and ways of life; . . . there will be few legislators in the next generation who will have any knowledge of agriculture, and the voice of agriculture in the Halls of Congress will indeed be weak."[9]

Whitten's fears were borne out in succeeding years, as both Congress and the executive underwent profound changes: Democrats in 1958 picked up forty-eight new seats in the House and fifteen in the Senate, with most newcomers elected from former GOP strongholds in the urban North. Republicans made modest gains in 1960, but the Johnson landslide of 1964 further undermined the conservation bloc, as liberal Democrats grabbed thirty-eight new House and two new Senate seats. The changing demographic character of postwar America had begun to be reflected in Congress, and the influx of new blood would soon shake that institution.

Such demographic trends, oddly enough, were aided by the structure of bias built into southern electoral systems. The conservative coalition, with its roots in the New Deal and its base in the single-party bastions of the South, aged dramatically as the decade progressed. The same social and political system that consistently returned southern Democrats to office now worked perversely to ensure that little new blood trickled in from that region. Southern and border-state Democrats in 1960 made up three-fifths of the fifty most senior members of the House, but below them waited lay a vast cohort of younger, more liberal representatives.

The effects of age on a stagnant representative population were profound. Southern Democrats in 1960 controlled thirteen out of twenty House standing committees, twelve out of seventeen such committees in the Senate, and they dominated both appropriations committees.[10] By 1970, however, southerners controlled only seven out of eighteen House committee chairs and six out of fourteen chairs in the Senate—and the trend would continue into the next decade. The farm bloc held its own far better on the Republican side of the aisle, but its influence there was blunted by Democratic control in both chambers.

Structural changes also made a difference, particularly in accelerating the urbanization (or suburbanization) of the national legislature. Rural populations had traditionally been overrepresented in the House in part because of the power of rural-dominated state legislatures to draw district boundaries. Malapportionment was so widespread by 1960 that a two-to-one disparity between the most and least populous districts existed in *every* state, giving some sparsely populated rural areas for greater representation in the House than was enjoyed by many cities. These artifices of rural power were dismantled by the Supreme Court between 1958 and 1964, culminating in the 1964 decision in *Wesberry* v. *Sanders* that required congressional districts to be of equal population size. Within two years of *Wesberry,* 258 House districts in 31 states were redrawn, the cities and suburbs being the chief beneficiaries.[11]

These intersecting trends were first manifested in the weakening of two key levers of the conservative coalition's power. Judge Smith's House Rules Committee, the secretive scourge of liberal Democrats, was truncated in early 1961 by a Rayburn-led and Kennedy-advocated expansion in its membership that altered the committee's ideological character. The Senate filibuster, instrument of so many conservative stands, declined gradually in importance as liberals increased in number and occupied key postions.[12] The trends were to amplify in subsequent years; Congress by the end of the decade bore little resemblance to its 1960 counterpart, and more sweeping changes were yet to come.

Finally, Valerie Bunce argues, "New leaders mean new policies and old leaders mean a continuation of old priorities," but always within certain contextual parameters.[13] The ascent of John Kennedy accompanied, and was borne by, a nation in flux, a society facing the ambiguous transition from "old" to "new." The Kennedy era began the recalibration of the structural disequilibrium that had long supported traditional sectors of power in American society, and this shift accelerated through the decade. John Kennedy was important less for his substantive accomplishments than for a change in tone, a departure from past orientations. At a minimum, his administration "established a climate of *greater receptivity* to the demands for environmental improvement that had been growing during the Eisenhower years."[14] Whether Kennedy fostered this shift, or was merely part of it, is both difficult to say and immaterial; his administration was a lens that refocused the gaze of society and government away from one set of dispositions and toward others. It is within this perspective of executive initiative, seeing the presidency as agenda-setter, that we examine pesticides politics at the dawn of the new decade.

Executive Action: Kennedy and the FPCRB

> I am hopeful that consistent and coordinated Federal leadership can expand our fish and wildlife opportunities without the present conflicts of agencies and interests . . . [including] one agency encouraging chemical pesticide that may harm the songbirds and game birds whose preservation is encouraged by another agency.[15]
>
> —John F. Kennedy

The president, John Kingdon notes, like "no other actor in the political system has quite the capability . . . to set agendas in given policy areas for all who deal with those policies."[16] James Reston is more concise: "The president of the United States is the one who can get the attention of the American people."[17] John Kennedy wholeheartedly supported economic progress, but early on indicated that he would not sacrifice America's natural resources in the process. Conservation groups widely applauded his appointment of Stewart Udall as secretary of the interior and of Abraham Ribicoff to head HEW, as well as his appointment of leading conservationists to a number of advisory panels to investigate important problem areas. Other issues still dominated the agenda of government, but the new president at least offered to conservationists strong verbal and symbolic commitments to their goals, as well as the promise of far greater budget flexibility than had existed under Eisenhower.[18]

Congress in 1960 had failed to pass the Pesticides Coordination Act despite the cranberry and fire ant controversies, and John Dingell reintroduced the measure early in 1961, hoping that the new administration would support stronger pesticides regulation. Kennedy, however, still smarted from the House Rules Committee fight, and, anxious to avoid more conflict with southern Democrats, ordered the relevant agencies to hammer out their differences. Dingell was not particularly enthusiastic about this administrative route to policy change, but agreed to withdraw his bill and give the administration time to work out the problem internally.

The bureaucratic response was the Federal Pest Control Review Board (FPCRB), constituted to coordinate and monitor all federal pesticides activities.[19] The agreement for the first time gave the USDI a *formal* voice in federal pesticides policy, realigning the department's informal role found so wanting by Dingell in 1960. FDA participation also expanded beyond its statutory authority to set spray residue tolerances—the cranberry scare having convinced the new administration that the agency was ill-positioned to prevent such problems in the first place. FPCRB members, according to the late 1961 memorandum of understanding, would review proposed pest control programs and "pass judgment" but, to the dismay of the USDA's critics, lacked statutory authority to veto such actions; the FPCRB essentially was to be a formal advisory board, and little more. Nonetheless, a new administration joined together what Congress had been unable (and unwilling) to wed.

USDA support for the new board hinged on a desire to avoid mandatory policy coordination. "The advantage of formal over statutory coordination," notes John Blodgett, "was that formal coordination was essentially voluntary and compliance to other departments could not be forced."[20] The Department of the Interior and HEW also wished no statutory encumbrances, but both saw the FPCRB as an avenue for regularized access to policy decisions not previously available to nonagricultural viewpoints. Absent the authority to enforce collective judgments, the panel was better than no access at all.

The obvious problem with the FPCRB, or with any coordinative mechanism of this sort, was that its members were put in the awkward position of advocating their respective bureaucratic views while simultaneously deigning to judge the plans of others.[21] Interagency "coordination" sounded good in theory, but conflict resolution depended totally on voluntary cooperation. No single party could force agreement or enforce collective decisions, and there was no mechanism for welding three competing viewpoints into common, or compelling, agreement. That the FPCRB during the next two years proved unable to prevent continued USDA campaigns against the fire ant, despite repeated protests by other board members, painfully emphasized the arrangement's flaws. Lack of a credible action-forcing mechanism made it a toothless tiger.

Critics of the USDA, not surprisingly, viewed the FPCRB as a patina of formal cooperation covering the status quo, and later renewed their efforts to legislate more stringent regulation. Administration officials urged patience, arguing that too little time had elapsed to judge the board's efficacy. However, by mid-1962, events were to make patience a scarce commodity.

Silent Spring and the Wellsprings of Environmentalism

The June 16 edition the *New Yorker* carried the first of three articles by Rachel Carson, the noted biologist and nature writer whose earlier works were read widely and had been uniformly praised. Carson's thesis was both simple and powerful: pesticides were used excessively, with little proper regard for their impact on either human health or nature. The results, she argued ominously, were environmental degradation and a more widespread threat of cancer and other chemical-induced maladies.[22] Carson nowhere argued that the use of chemicals should cease entirely—a point usually disregarded by her most vitriolic critics—but simply that humanity neither understood nor appreciated the *environmental* effects of its headlong plunge into the chemical age. Human action, she concluded, had become the dominant environmental influence, if not an outright hazard. This was a view fundamentally counter to American society's traditional faith in science and technology as panaceas for national problems.[23]

Three factors, in retrospect, contributed to Carson's major impact on the national consciousness about the side-effects of the chemical age. First, the articles came hard on the heels of the widely publicized thalidomide tragedy in Europe and its subsequent ban in the United States. Carson's timing could not have been more fortuitous to those critical of federal pesticides policy, for her writings intersected with heightened sensitivity to the grievous errors science is capable of. Second, the articles struck a sympathetic chord within the highest levels of the Kennedy administration, and among members of Congress not previously involved in the pesticides debate; the influx of new congressional actors begun in the late 1950s proved fruitful soil for receptivity to Carson's thesis. Third, and probably most important in the long run, Carson's original writings were quoted, discussed, and publicized by other media on a scope not previously experienced for an "environmental" matter. That initial *New Yorker* series reached a limited audience, but particularly explosive pieces of journalism tend to have marked multiplier effects— a pattern of attention first on the part of political leaders and the "attentive public," and later the mass audience, as other media organs report on the original story. The first two factors influenced short-term government response; the third helped *Silent Spring* to spark the modern environmental movement.

National media attention to the articles began with the *New York Times,* which on July 3 editorialized,

> Miss Carson will be accused of alarmism, or lack of objectivity, of showing only the bad side of pesticides while ignoring their benefits. But this, we suspect, is her purpose as well as her method. . . . If her series helps arouse enough public concern to immunize Government agencies against the blandishments of hucksters and enforce adequate controls, the author will be as deserving of the Nobel Prize as was the inventor of DDT."[24]

The *Washington Post* a week later commented, "Carson's negative case is virtually as powerful as the poisons she deplores."[25] Houghton-Mifflin announced the September publication of a book-length version of Carson's attack, to be known as *Silent Spring*.

Department of Agriculture officials meanwhile wavered between private outrage and public moderation in their response to Carson. The *Wall Street Journal* reported, "Secretary [Orville] Freeman squelches trigger-happy underlings who itch for a quick rebuttal of Rachel Carson's magazine articles on safety of chemical insecticides. The Agriculture Department builds a careful defense of its encouragement of insecticide use. An indirect reply: The Department pushes work on non-chemical war against bugs."[26] USDA representatives on the FPCRB reportedly "alternated between angry attacks on *Silent Spring* and nasty remarks about Miss Carson,"[27] while the department's public statement praised Carson for a "lucid description of the real and potential dangers of misusing chemical pesticides."[28] Carson, the USDA reply added, "does not advocate halting all use of chemical pesticides. She does advocate the use of best judgment in selecting the right control method. . . . The USDA will continue to exercise strict control over pesticides for sale in interstate commerce and will constantly seek better methods to control pests."[29]

The USDA's public ambivalence in no way resembled "the greatest uproar in the pesticides industry since the cranberry scare of 1959."[30] Industry spokesmen admitted the veracity of Carson's facts, but nonetheless sharply criticized the author's "imbalanced" views. Monsanto Corporation President Peter Rothberg called Carson "a fanatic defender of the cult of the balance of nature," while the National Agricultural Chemical Association announced a major public relations program to offset Carson's expected (or feared) impact and to reaffirm the unalloyed benefits of chemical use.[31]

Public criticism of *Silent Spring* eventually coalesced into two broad categories. Chemical industry and agribusiness executives viewed the problem largely in educational terms: the public needed to be reminded about the benefits of pesticides to society so that such attacks could be weighed properly. Their response was to flood the public consciousness with "experts" who loudly refuted Carson's claims—or, at least, her overemphasis on the hazards of pesticides. Manufacturers expanded advertising campaigns to emphasize that a gloomy future loomed should *all* pesticides be banned, a "worst case" scenario traditionally posed by defenders. An industry that viewed all problems through the lens of public relations naturally went that route.

But Carson's sharpest detractors were those in the scientific community who saw *Silent Spring* as a direct and dangerous assault on their professional reputations. Carson had charged many scientists with trading professional objectivity for obeisance to the needs of industry and of their own research funding—criticism that struck at the heart of much of the pesticides research performed both in industry and in major universities. The author also portrayed a National

Academy of Sciences–National Research Council panel on pesticides (formed in 1960) as dominated by industry-sponsored experts. She revealed that most panel members in fact worked either directly for chemical companies or performed industry-funded research. Carson's attack, in this sense, was not only on pesticides but on reputed "scientific neutrality;" her thesis challenged scientific professionals to defend both their views and their integrity as scientists.

Industry scientists largely derided Carson's thesis, as did many members of the NAS-NRC pesticides panel. George C. Decker, chair of the panel's subcommittee on Evaluation of Pesticide-Wildlife Relationships, and who had long identified with the industry, reviewed the book for *Chemical World News* and called it "science fiction."[32] Fellow panel member William J. Darby said simply that the book "should be ignored; [however,] the responsible scientist should read this book to understand the ignorance of those writing on the subject."[33] The attacks hit not at Carson's data but her values, which challenged the core tenets of scientific orthodoxy. To these critics *Silent Spring* was but a reactionary attack on technology, if not on "progress" itself, and Carson was dangerous because she challenged science's uncritical view that "progress" was inherently good. Accepting Carson's thesis, Darby warned darkly, meant "the end of all human progress, reversion to a passive social state devoid of technology, scientific medicine, agriculture, sanitation, or education. . . . Indeed, social, educational, and scientific development is prefaced [*sic*] on the conviction that man's lot will be and is being improved by a greater understanding of and thereby increased ability to control and mold those forces responsible for man's suffering, misery, and deprivation."[34]

The debate sparked by *Silent Spring,* like so many later disputes over technological issues, centered less on "scientific truth" than on values, for at its root Carson's book challenged prevailing assumptions about American social behavior and economic practice. If man *could not* adequately control nature without disastrous side-effects, then of what value was the entire structure of power supporting the nation's scientific and economic institutions? Carson's thesis was a dagger pointed at the heart of mainstream assumptions about technological progress, and the defenders of that paradigm stridently rejected both the thesis and its implications as only so much reactionary twaddle. Scientists, especially those representing the pesticides interests, later would endure harsh criticism for their reactions: "No one denies that these scientists had a perfect right to review the book," wrote John Blodgett. "But by associating themselves with interests rather than weighing the book on its merits they projected an image of clashing values and a lack of objectivity which policy-makers think that scientists must avoid if their advice is to be useful to the non-specialist in making decisions."[35]

Whatever the nature of the debate yet to come, Carson's articles prompted John Kennedy, in August 1962, to order the President's Science Advisory Committee (PSAC) to study the issue. Kennedy's choice of his personal scientific advisors over the National Academy of Sciences is instructive. Academy members were viewed increasingly as reflexive supporters of the pesticides paradigm, while PSAC members had fewer direct ties to subgovernment interests. Kennedy also may have avoided the traditional NAS-NRC route because the academy's own panel on pesticides-wildlife relationships was embroiled in a sharp internal fight over its long-delayed final report, which would not be released until mid-1963. Choosing the PSAC appeared the least controversial route.

Attacks on Carson, meanwhile, mounted noticeably as *Silent Spring* neared publication. The Manufacturing Chemists Association distributed a cartoon skewering Carson to over 50 daily and 800 weekly newspapers, while Monsanto parodied *Silent Spring* in its company newsletter, a move widely criticized after the satire was leaked to the general press.[36] Velsicol Chemical Company officials threatened to sue Houghton Mifflin for libel over Carson's allegedly "inaccurate and disparaging" remarks about chlordane and heptachlor, two persistent pesticides manufactured solely by the company.[37] A letter to Houghton Mifflin from Velsicol's general counsel, Louis McLean, aptly summarized the views expounded by Carson's sharpest critics, views to be repeated regularly thereafter:

> Unfortunately, in addition to the sincere opinions of natural food faddists, Audubon groups and others, members of the chemical industry in this country and in Western Europe must deal with *sinister influences,* whose attacks on the chemical industry have a dual purpose: (1) to create a false impression that all business is grasping and immoral, and (2) to reduce the use of agricultural chemicals . . . so that our supply of food will be reduced to east-curtain parity. Many innocent groups are financed and led into attacks on the chemical industry by these sinister parties.[38]

Houghton Mifflin, to be on the safe side, sought independent review of the disputed material and methodologies. The toxicologist supported Carson's conclusions, and Velsicol dropped its threatened lawsuit, though not its objections.

The specter of the scientific, corporate, and government establishments publicly ganging up on the demure Miss Carson propelled the controversy into the realm of a human interest story, the sort of news eminently attractive to large segments of the mass public. Heavy-handed and well-publicized attacks no doubt caused many otherwise uncertain or uninterested citizens to sympathize with the author, whose status as underdog appealed to a rich vein in the American populist psyche. The entire prepublication brouhaha prompted *New York Times* writer Walter Sullivan to comment that *"Uncle Tom's Cabin* would never have stirred a nation had it been measured and 'fair.'"[39] Audubon's Charles Callison observed simply, "What probably scares the industry most is the effect on public opinion which the writings . . . may create."[40]

Public response soon proved Callison's prescience. *Silent Spring* became an instant hit, selling almost 500,000 hardback copies in just six months. Conservation groups, consumer organizations, and the Book-of-the-Month Club promoted the book to a suddenly interested reading public, and the Columbia Broadcasting System (CBS) announced late in 1962 its intent to air a special on the book the following spring. Local protests to government spray programs jumped noticeably in several regions, and over forty bills calling for stricter regulation of pesticides hit various state legislatures by the year's end. Several states in fact instituted pesticides review boards and new local standards, though these measures tended to have little real effective power.[41] Audubon Society membership in 1962 grew by 24 percent, much of it credited to Carson, and other conservation organizations reported similar growth.[42] At the federal level, Connecticut Senator Thomas Dodd (D, Conn.) proposed to create a Federal Chemical Council that would formulate *national* pesticides policies and standards. In the House, John Dingell announced that his patience with the FPCRB was at an end, and that he would reintroduce his Pesticides Coordination Act in early 1963. Rachel Carson, Secretary of the Interior Stewart Udall recalled, clearly had "awakened the nation by her forceful account of the dangers around us."[43]

Public awareness reached new heights when CBS on April 3, 1963, aired "The Silent Spring of Rachel Carson," a prime-time special that for one hour graphically portrayed her thesis. Television would dramatically escalate the pesticides debate, as conservationists hoped and the industry feared. So controversial was the issue that the network received over 1,000 letters prior to the airing, most demanding that CBS not proceed. Three network sponsors—Standard Brands, Inc., Lehn and Frank Products (Lysol), and Ralston Purina—abruptly withdrew their ads, citing "incompatability" between the show's content and their products. Standard Brands, one writer noted somewhat sarcastically, nonetheless sponsored that evening's showing of "Mr. Ed," a comedy about a talking horse.[44]

Program transcripts underscore the network's efforts to offer a diversity of views on the pesticides issue. CBS producer Jay McMullen simply allowed those representing competing views to state their cases, interspered with narrator Eric Sevareid's short informational bridges. Despite this attempt at balance, viewers apparently were impressed more by Carson's simple and compelling themes about unintended side effects and chemical hazards than by scientific experts' more arcane discussions about relative risk, tolerance levels, or carcinogenicity. Sevareid's script aptly summarized the crux of the debate:

> In "Silent Spring" Miss Carson stresses the possibility that pesticide chemicals may be working harm to man in ways yet undetected—perhaps contributing to cancer, leukemia, genetic damage. In the absence of proof, her critics concede that these are possibilities but not probabilities and they accuse Miss Carson of alarmism. Yet few scientists deny that some risk may be involved.[45]

This last sentence constitutes the core of Carson's theme, one both simple and irresistibly powerful. Television compresses complex issues into more easily comprehensible snippets of information—or tries to do so. *New York Times* writer Jack Gould commented that the Carson special "was largely a symposium of experts giving their views. . . . But at times the TV medium illustrated points in its own distinctive way. There was a moment of striking contrast in scenes of small children covered with filthy insects and later of majestic mountains and streams in all their unspoiled loveliness."[46] In the case of "The Silent Spring of Rachel Carson," as with so much that television does, the most easily comprehensible theme repeatedly stood apart from other, more complex arguments—we know not the harm we may face. The effort by CBS to achieve a balance in presenting the issue probably mattered little in the end, because the simplest theme still rose above the cacaphony of dissonant voices. That theme, as defenders of the pesticides paradigm feared, was Carson's.

Television is a powerful political tool because it helps to define issues *for* its viewers. Newspapers may lay out issues in more thoughtful or complex ways, but arguably are less successful as devices for expanding public awareness because readers must take a more active part in defining the issue. Television magnified Carson's definition of the pesticides problem for a public not used to thinking in "environmental" terms, and thereby redefined pesticides *as an issue*. No longer were pesticides seen purely in agricultural terms; they now were matters affecting human health and the environment—considerations far more central to a society increasingly devoted to "superior" social goods. Carson defined pesticides as an environmental issue, and it stayed that way, no doubt to the deep dismay of those devoted to promoting chemical technologies. The pesticides subgovernment long had controlled how the public saw pesticides, but now that capability lay squarely in the hands of its most persistent critics.

Those who define an issue most successfully also predetermine the parameters of valid debate, and losing that power proved a bitter blow to those staunchly defending pesticides; they now had to fight on terms not of their choosing. Carson redefined the issue, CBS amplified the message, and a new issue clambered onto the public agenda.

Studies and Counterstudies: Science in the Crossfire

An agenda, John Kingdon notes, is the "list of subjects or problems to which governmental officials, and people outside of government closely associated with those officials, are paying some serious attention at any given time."[47] Mass attention to a public issue tends to ebb soon following some response to the perceived problem, regardless of that action's true success.[48] Political leaders, however, tend to watch the issue far longer, until they are satisfied about its resolution (or, perhaps, until they give up). The public's issue awareness surges and decines with succeeding "crises," but policymakers are likely to display more consistent attention to an issue since they tend to be more knowledgeable about and committed to the question at hand. Such would be the case with pesticides after *Silent Spring*.

The "post–*Silent Spring* era" spans the years 1963–1969—a period that witnessed a fascinating proliferation of "expert" studies, each claiming to be the definitive analysis on pesticides and debunking claims propagated by rival studies. The nearly half-dozen major studies to emerge in this period may be classified into two dichotomous perspectives. The first held that federal pesticides policy inadequately deals with mounting human health and environmental dangers and therefore requires major revisions; the other argued that public fears about pesticides are misplaced because the alleged hazards are minuscule compared to the benefits, and no overhaul is necessary. This babel of panels, each fortified with scientists of all stripes and methodologies, would serve to muddle further any public understanding of the real stakes involved, and would reveal science split asunder. Into this breach stepped the political actors, bringing with them their own particular ideational baggage. By the decade's end, scientists themselves would become more visible *political* actors, but within a context that blunted the force of their knowledge and compelled them to take a back seat to the nonexpert. Being so split, science became less powerful as an independent force, and the reputation both of science and of scientists suffered. The politicization of science ironically made it "less scientific" in the eyes of other stakeholders.

The CBS special sparked a new flurry of public and private activity among state and federal policymakers. Newly elected Senator Abraham Ribicoff of Connecticut (formerly secretary of HEW) announced immediately that a subcommittee of the Committee on Government Operations would investigate administration of federal pesticides programs. Senate rules authorized that committee to investigate the implementation of *any* federal program—a sweeping oversight role that would give critics of the USDA a highly visible and authoritative avenue for access. Ribicoff's hearings began in May, simultaneously with the administration's release of the long-awaited report by the President's Science Advisory Committee.[49] Kennedy's emphatic endorsement of the report, and his subsequent orders to executive agencies to implement its recommendations, seemed to mark a new era for those critical of the pesticides orthodoxy. But bureaucratic intransingence, particularly within the USDA, later would convince critics that more than presidential directives were needed. Audubon President Carl Buchheister in late 1963 lamented, "I am afraid some of us may have been lulled into complacency by that report. For surely, if a White House committee makes such recommendations, one would expect the executive bureaus to take heed. Such has not been the case."[50]

It is easy to see why conservationists expected great changes. The PSAC panel acknowledged the benefits to society from chemical use, but overall concluded that "decisions on safety [were] not as well-based as those on efficacy," and that "until publication of *Silent Spring* by Rachel Carson, people generally were unaware of the toxicity of pesticides."[51] The panel recommended a series of administrative and legislative changes that cumulatively would tighten federal regulation, beginning with stronger action-forcing mechanisms linked to interagency coordination and cooperation. PSAC members found the FPCRB inadequate for the task at hand, noting that its objections consistently failed to prevent massive USDA eradication campaigns. More research, both on the effects of pesticides and on nonchemical control technologies, was essential, as was a federal monitoring program to track the production, sale, and use of the vast quantities of chemicals applied annually.

On the legislative front, the PSAC report proposed to eliminate FIFRA's "protest registration" provision, calling it a loophole without merit—a view long held within the USDA itself. Recommendations that product labels carry federal registration numbers also gained support from the department and many chemical firms, which hoped that such a provision would quell consumer fears about product safety. Finally, panel members recommended that FIFRA be amended to include fish and wildlife as "useful" vertebrates and invertebrates—a status previously enjoyed solely by farm animals—so that chemicals known to harm "useful" fish and wildlife could be kept off of the market.

Such recommendations mattered largely to policymakers and a narrow range of attentive interests; organizational fine-tuning and limited legislative change are not the stuff that excite public opinion. But the panel, in language that clearly vindicated Carson, concluded, "The accretion of residues in the environment can be controlled only by orderly reductions of persistent pesticides," and stated that the federal government should phase out all uses of chemicals like DDT and heptachlor.[52] This recommendation, and its potential impact on pest control programs, had prompted USDA and PHS attempts to delay the report's release, and would be at the center of the sharpest debates to come.

Reaction to the report differed starkly from that greeting *Silent Spring,* in no small part because its authors were eminent scientists who consulted with a wide array of experts in both the private and public sectors. One also suspects that the language of the report, written in a style typical of scientific panels, exuded that aura of "objectivity" allegedly disdained by Carson. Many who earlier had vilified Carson now applauded the PSAC report for its "temperance" and "balance," presumably because the panel, unlike Carson, acknowledged the social benefits of chemical use. *Silent Spring* publicized the potential dangers in the widespread use of pesticides and was called biased; the equally critical PSAC report was "thoughtful." Some industry scientists challenged the panel's data, supplied by federal agencies, but few argued that the report misrepresented the pesticides dilemma. "Because of the poorly balanced nature of most of the previous government statements," Frank Graham comments, "the PSAC report seemed revolutionary to both the conservationist and the pest control interests."[53] The medium indeed may be more important than the message.

Proponents of stricter federal regulation thereafter used the PSAC report to support contentions that the USDA was inherently ill-suited for pesticides regulation. Defenders of the status quo would counter that the panel nowhere explicitly advocated transfer of USDA registration duties, but merely recommended programmatic changes and internal reorganizations. These arguments are correct, but the panel did emphasize the need to endow the FDA with clearer, more central policymaking roles. The panel recommended that the FDA coordinate a federal monitoring program, vastly expand its research, and enjoy more inclusive authority to set residue tolerances. What was more important, and most controversial, the panel proposed that the FDA register all pesticides used for non-agricultural purposes—a proposal that struck clearly at the heart of USDA registration prerogatives because the line between agricultural and non-agricultural chemicals was thin indeed. Thomas Dunlap comments that the panel's intent was to make a fundamental shift in federal authority and the overall pattern of chemical use: "To shift responsibility for pesticides registration to the FDA, which then only controlled the level of pesticides in products reaching the market, would seriously upset agricultural and industrial power to influence registration. It would also remove congressional power over pesticides policy from the agricultural appropriations subcommittees."[54]

The PSAC report in many ways was more dangerous to the pesticides subgovernment than was *Silent Spring*. Rachel Carson could be dismissed by her critics as an eccentric, or, worse, as a naive neurotic, but PSAC members were not so summarily ignored. Where the former appealed largely to an uninformed mass public, the latter carried tremendous influence with government insiders. Proponents of the pesticides paradigm were clearly in a quandary: publicly, at least, they needed to be seen as both cooperative and willing to accommodate the views of their critics, lest their intransigence generate new controversy. Few in or out of Congress believed that the most fundamental legislative changes called for by the PSAC would pass, since farm bloc members still wielded tremendous influence both within Congress and in an administration anxious to avoid debilitating battles with rural representatives. Obvious attempts to block reform would be dangerous, however, since a display of brute power might well sway the administration to the side of the rapidly expanding cohort of conservation-minded individuals both in government and in the electorate.

This quandary was exacerbated by the scathing reception encountered by the NAS-NRC's long-delayed final report on pesticides and wildlife. The academy's first two reports were released in 1962, but only after two years of embarrassing internal squabbling. The final report was delayed yet another year.[55] The problem, said the critics, was that the panel was dominated by scientists with public ties to industry or agriculture, many of whom were also Rachel Carson's harshest opponents. A minority of panel members privately dissented with majority views, but panel rules forbade minority reports or inclusion of alternate opinions. This public dispute heavily tarnished the academy's image, and only amplified critical disdain of its conclusions. Frank Engler, writing in *Atlantic Naturalist,* described the NAS-NRC report as "unscientific" and "written in a style of a trained public relations official of industry out to placate some segments of the public that were causing trouble."[56] Audubon biologist Roland Clement was a bit less scathing, but just as critical:

> Given their authoritative backing, these two reports are extremely disappointing. . . . The whole is a generalized and undocumented statement that, far from coming to grips with the problem, seems to disregard much important evidence and does no more than offer a gentle admonition to pesticides users to be more careful in the future. The result is either a mark of ecological incompetence in the committee or, more likely, evidence that the viewpoints of the advocates of pesticides use within the committee prevailed almost entirely.[57]

Central to the dispute were the academy's arguments that persistent pesticides were fundamental to American health and bounty, that biological controls were impractical or did not meet farmers' needs, and that damage to wildlife by pesticides was minimal and caused only by misuse. Additional research was needed, as was better interagency cooperation (a point about which few disagreed), but, overall, the panel advocated no real changes. Audubon's Frank Graham called the report a "whitewash," charging, "The fact that the entire study was prompted by the USDA's disastrous policies and procedures is studiously ignored."[58] The final report only worsened the academy's reputation for objective and independent analysis, for it was surrounded by charges of doctoring at the hands of its executive director, a former employee in the USDA's Insect Control Division. The result was a rather desultory discussion of future research needs that—worse yet—contained absolutely no documentation.

The juxtaposition of the PSAC and NAS-NRC reports, combined with the public attacks by academy members on Rachel Carson, arguably did more to besmirch the image of science and scientists than did any other prior incident. The same experts who called Carson "unscientific" themselves were ridiculed for *their* biases and public attachments to vested interests. Conservationists thereafter wanted little to do with the academy, while defenders of the pesticides paradigm would use its reports to justify their claims. This condition of "subjective science" would seriously undermine the credibility of many experts, and would bring science directly into the political realm.

One should note in all fairness that the pesticides paradigm itself was undergoing a fundamental transition. Knowledge about the effects of persistent pesticides on human health or on wildlife lagged far behind understanding of the efficacy of pesticides, as both Carson and the PSAC report noted. The technology to assess long-term effects simply had not yet caught up with common understanding of immediate chemical benefits. Once it did, however, technological change would prompt shifts in prevailing scientific assumptions, both about chemicals and about the roles played by industry and government in protecting the public interest. Labeling provisions that may once have "protected" public health did not conform to the new knowledge, and the gap between technology and policy yawned anew. "Moreover," notes John Blodgett, "as the technology changed and affected new interests, these new interests joined the policy making process."[59] Or at least they tried, with varying degrees of success.

Even so, the rigidity of dominant scientific views and the ties publicly cultivated by many experts to major economic and political interests meant that the pesticides paradigm in crisis had to be ardently defended. Paradigm shifts, Thomas Kuhn argues, involve a dismantling of the dominant scientific view and its replacement by an equally compelling alternative.[60] The pesticides paradigm was under clear attack, but no alternative of comparable scientific or economic power as yet compelled a whole generation of scientists to renounce the accepted orthodoxy. Defenders of the pesticides paradigm would denigrate its detractors—more often than not exposing their own institutional, political, or economic stakes in the issue than their challengers' scientific failings. Such stakes, however, provided powerful reasons to resist change.

The remainder of the decade reveals a protracted series of conflicting scientific assessments about pesticides, a pattern of "increasingly strong and more sophisticated attacks on the position of the USDA as preponderant maker of pesticides policy."[61] Defenders of the paradigm countered those attacks with their own studies, attacking their opponents in increasingly strident tones. Evidence supporting the critics eventually emerged as more credible, but policy change did not follow smoothly. Science was but an advisor to the seats of power; policymaking was still the purview of the politician.

Subgovernment Response, Subgovernment Adjustment

The transformation in the pesticides debate was so swift that those promoting the use of pesticides were unprepared to deal with new sets of policy claimants. The twin blows delivered by *Silent Spring* and the PSAC report made it obvious to all save the most recalcitrant defenders of the status quo that *some* change would emerge as the debate moved from the front pages to the committee rooms and the federal office buildings. How the various subgovernment actors dealt with the new challenges reveals a great deal about the potential depth and effectiveness of that change.

The new debate intensified first in Congress as more legislators perceived statutory reform to be the answer to apparent regulatory shortcomings. The 88th Congress (1963–1964) in fact held almost a dozen separate committee or subcommittee hearings on pesticides matters—a record number. Six bills aimed at such reform suddenly emerged in the House during 1963, and three in the Senate, and this legislation was unlikely to evaporate into thin air. Reform proposals were of two general types: one sought to eliminate loopholes in FIFRA and strengthen federal registration; the second demanded advance USDA consultation with both the Department of the Interior and HEW on all registrations and pest control programs. Several in Congress, led by Senator Abraham Ribicoff and Representative John Dingell, sought complete transfer of USDA registration duties to the FDA, though few thought such direct attacks would succeed at that time. The preferred route was to require prior USDA consultation, to do legislatively what the FPCRB apparently failed to achieve administratively. Proponents of reform also sought to reorient federal pesticides regulation from a mere labeling process to one emphasizing premarket testing of all chemicals.

The path taken by reform legislation during the 88th Congress illuminates the extent to which subgovernment actors were willing to allow change while not surrendering their core prerogatives. Senators Ribicoff and James B. Pearson (R, Ky.) in May 1963 proposed to eliminate FIFRA's "protest registration" provision and require federal registration numbers on all product labels. Both provisions were recommended in the PSAC report, and in fact were proposed by the USDA.[62] The Ribicoff-Pearson Bill (S. 1605) also allowed manufacturers to challenge USDA regulatory decisions before a National Academy of Science review panel. This proposal was designed to deter protracted and costly court battles that might ensue in the absence of protest registration. Appeals to a panel with unblemished scientific repute, at least among subgovernment participants, also seemed preferable to a trial before an "amateur" judge. The right to appeal agency decisions in local federal courts of appeals remained, though the bill's language limited claimants to those "adversely affected," terminology that, at this time, pertained solely to farmers and pesticides makers. Any claimant who could not show direct and adverse harm still was denied access to the courts, though the meaning of "adversely affected" later would undergo expansion by the courts themselves.

Critics wanted more—much more—but were unwilling to press their case too far. *Some* change, clearly, was better than none at all. The Kennedy administration proceeded cautiously, in part because of interagency wranglings over the PSAC recommendations and, more critical, because administration officials worried about provoking the ire of farm bloc representatives. The Johnson administration would proceed just as cautiously in 1964. Change could come only from within the committees on agriculture, a fact of life keenly apparent to proponents of reform. They would acquiesce to minor legislative changes, but seek "true" change through other avenues.

The 1964 FIFRA amendment passed with little controversy primarily because subgovernment participants acceeded to the need for some change and the critics took what they reasonably could get. Federal agricultural officials supported changes that tightened the more gaping holes in the regulatory framework and finally shifted responsibility to registrants for proving product efficacy and safety. Chemical companies for the most part recognized a public relations coup in the making and used congressional hearings both to support the amendment publicly and to reassert their commitment to "sound" regulation. Protest registration had seen little use, but its image as an escape route for recalcitrant producers gave industry in general a bad reputation, and most company officials would just as soon have substituted a more publicly respectable appeals process. Industry also supported the registration number requirement because it would assure growers and users that the product carried the federal seal of approval. Overall, as in 1947, the industry hoped such changes might forestall, or at least reduce, the threat of state-by-state registration that began to reemerge in the wake of *Silent Spring*.

The agricultural community's support hinged primarily on the protection any changes offered to farmers against ineffective and hazardous products. The registration number provision meant to farmers that *both* the USDA and the FDA approved the chemical in question: if so cleared, the product *must* be safe, effective, *and* free of residue problems. Revocation of protest registration likewise offered assurance to farmers that no renegade products littered the market. Few in the agricultural community objected to the changes, though many representatives worried about granting too much authority to the USDA—showing a traditional chariness about bureaucratic discretion. The science advisory board and appeals court provisions adequately protected chemical companies from bureaucratic arbitrariness, while simultaneously offering to constituents assurance that their purchases were safe and effective.

The only controversy of note involved the question of proprietary information—data and information submitted by a producer to support registration. Ribicoff's bill provided that only "trade secrets" and chemical formulas be kept secret, allowing for public access to all health and safety data. FDA officials naturally supported this provision, seeking to free themselves from USDA control over information, and were backed by the USDI and the major conservation groups; all wanted to place health and safety data in the light of public scrutiny. The USDA and its industry allies, not surprisingly, opposed it. Chemical companies regarded any such data to be "trade secrets"—a very broad definition of proprietary information—and the specter of public access to their research gave industry lobbyists fits. House farm bloc members, fearing that public access might "chill" new product development, sided with the industry. Nobody seemed willing to compromise on this matter, as would be the case for decades to come.

The Senate passed Ribicoff's bill largely intact in October 1963, but the House Committee on Agriculture immediately tightened information dissemination in line with the industry's and the USDA's concerns. The House version eventually became law, defenders of pesticides ceding expendable chunks of the regulatory framework, critics winning a few, largely symbolic, changes. These changes, in retrospect, did little more than plug a few onerous holes in the registration process, but the debate itself at least introduced health and environmental values, if only tangentially. Society and Congress alike were changed enough by 1964 to pressure for some policy alteration, but those challenging the subgovernment view were not yet strong enough to force fundamental shifts.

The litmus test of opposition strength—or weakness—came not with FIFRA but with attempts to force the USDA to cooperate with the USDI. Critics argued that USDA control over registration guaranteed policy rigidity and lax enforcement, since the department's primary role as promoter of agriculture worked inherently against other goals. The PSAC panel, hoping to inject health and environmental perspectives into federal policymaking, explicitly recommended that the USDA consult with both USDI and HEW prior to undertaking pest control programs or product registration. Agriculture obviously opposed such changes, but PSAC chair Jerome Weisner argued to the Ribicoff Subcommittee that Interior's approval of registrations was essential where the product in question affected fish and wildlife, in the same way that the FDA already had formal authority to set tolerances on residues. The FPCRB was composed of departments with disparate pest control duties, Weisner noted, which "restrict[ed] the Board's effectiveness in reviewing the programs of member agencies."[63] Weisner carefully avoided criticizing the USDA, but Secretary of the Interior Udall was not so reticent. Referring to his department's running feud with Agriculture over the fire ant program, Udall argued for mandatory interagency coordination and a mechanism "where we could get down to cases and be heard."[64] The FPCRB was better than nothing at all, Udall added, but something more action-forcing than an advisory board was critical to environmental health. Interior's call for a more central policy role contrasted sharply with its earlier reluctance, a change of heart attributed both to shifts in departmental leadership and to a heightened sense of urgency about the problem. Lower-level careerists probably doubted USDI's ability to play such a role, but Udall and other high officials displayed no such reservations.

The fight to require USDA consultation was led by John Dingell, whose bill (H.R. 4487) would require the USDA to transmit to the USDI all information pertaining to the effects of pesticides on wildlife, order all such information to be indicated on product labels, and authorize broader dissemination of pesticides information to the public. The bill would also expand USDI research on pesticides and wildlife, removing the $2.6 million appropriations cap in place since 1959. Dingell's proposal went to the subcommittee on Fisheries and Wildlife Conservation, on which he sat, while Oregon Senator Maurine Neuburger's companion bill (S. 1251) went to Warren Magnuson's Subcommittee on Merchant Marine and Fisheries. Both held hearings in June 1963, but progress stalled thereafter.

Other PSAC recommendations similarly ground to a halt, largely because required clearance by the appropriations committees dragged on through 1964. On the administrative front, the USDA in September 1963 proposed new guidelines for warnings on product labels, while the FDA requested that the National Academy of Sciences review the chemicals aldrin and dieldrin, and began to move against the pesticide chlordane later in the year. There was little apparent progress on interagency coordination, however, as agency representatives wrangled bitterly over departmental jurisdictions and the authority of a reformulated FPCRB to override agency programs.

Yet another highly publicized incident apparently broke these logjams. The State of Louisiana in late 1963 called in the Public Health Service to investigate a phenomenal fish kill on the lower Mississippi River, the latest in a series of annual disasters now reaching epidemic proportions. Public Health Service researchers, aided by the Fish and Wildlife Service, subsequently diagnosed the cause to be acute insecticide poisoning, apparently from residues of the pesticide endrin migrating downstream from a Velsicol Chemical manufacturing plant in Memphis, Tennessee. The PHS, traditionally cautious in its assessments on the causes of mortality, verified its conclusions with five teams of researchers, including a private research firm, before releasing its full report in March 1964.

Publication of the report indicting endrin and Velsicol sparked a major renewal in the pesticides controversy, at least in Washington. Ribicoff resumed his subcommittee hearings; Secretary of HEW Anthony Celebreeze called for a multistate conference to determine the causes of the accident; the USDA began its own investigation. Representative Jamie Whitten blamed the fish kill on unidentified "others"—among them neither farmers nor the USDA—and ordered *his* subcommittee staff to examine the entire pesticides issue. Whitten later accused those seeking to curb the use of pesticides of undermining agriculture and of scaring the public unnecessarily: "One of the things I am most disturbed about is the headline seekers who rush to the President with big scare stories. Frequently they are worded in the negative. With that attitude, unless we find the answers before the troublemakers make it impossible to use what we have, the American people could go hungry."[65]

Department of Agriculture officials in June 1964 announced that their findings contradicted those of the PHS, with one stating that "none of the evidence presented either at the hearings or at the conference (conducted by the PHS) was scientifically adequate to justify withdrawal of endrin, aldrin, or dieldrin from farmers."[66] Velsicol executives charged the Public Health Service with gross errors, charges echoed by Senator Everett Dirksen (R, Ill.), who condemned the PHS for "wild accusations" that "unjustly crucified Velsicol and the chemical industry in general."[67] Public Health Service supporters replied that Velsicol had both a history of insensitivity toward the environment and was in the forefront of the sharpest attacks on Rachel Carson. Senator Ribicoff made the most cogent summary of the entire conflict:

> The PHS conference in New Orleans has been compared to a "kangaroo court" by a chemical industry trade magazine. The USDA hearings in various parts of the country seemed to lack adequate efforts at thorough fact-finding. When the hearing concluded there were no specific findings by the hearings examiner, and the general conclusion was reported to the public in the middle of a USDA press release regarding the establishment of a pesticides monitoring program by that agency.[68]

The fish kill controversy, Ribicoff's subcommittee later argued, underscored the problems of agency jurisdictional disputes and industry intransigence, exacerbated by technical ineptitude and parochialism by all involved. Such weaknesses seriously undermined the federal government's ability to collect valid information and to redress apparent problems.[69] This incident revealed to all, even Jamie Whitten, the need for stronger interagency cooperation and coordination of programs. The dilemma, as usual, was the form that such arrangements would take, and whether these changes would be made administratively or through Congress.

The Johnson administration moved more quickly, largely because Congress was locked in a stalemate over the Dingell bill. The Departments of Agriculture, Interior, and HEW in June 1964 replaced the FPCRB with a Federal Committee on Pest Control (FCPC), later described by the Ribicoff Subcommittee as "the most significant step taken by the Government since the pesticides problem gained wide attention."[70] The agreement provided that the FCPC would review all federal pest control programs, "with particular reference to possible harmful effects," and would advise member departments on desirable program changes.[71] Each department was to inform the others fully about new developments and research. The USDA would supply weekly reports on registration proposals and related matters, while the FDA would report on all matters pertaining to spray residue tolerances. Committee members would review all current and proposed actions, and any matters left unsettled would go to the department secretaries for final resolution. But, like its predecessor, the FCPC was purely advisory; committee decisions would not impede an agency's statutory obligations.

The administration also requested an additional $29 million to develop alternative technologies for pest control (a project that had all but been abandoned in the late 1940s) and safer, less persistent pesticides. Such funds would boost USDA research appropriations to almost $69 million for fiscal 1965. Jamie Whitten complained that his subcommittee already had granted an $18 million boost in research funding, but most of that money was earmarked for production and marketing, *not* basic research.[72] Whitten grumbled loudly that the USDA had played both ends against the middle to get more funds (which it did), but his subcommittee allocated $27 million for new research, as well as over $1 million for the USDA's share of FCPC costs.

Farm bloc representatives gave ground grudgingly on new research funding and interagency cooperation, but legislative attempts to mandate USDA consultation with the USDI met with failure—though not without a fight. Dingell's Pesticide Coordination Act emerged from Merchant Marine and Fisheries only slightly modified from the original version. The irksome appropriations cap remained, but the committee did raise its ceiling from $2.6 to $3.2 million for fiscal 1965, and to $5 million annually thereafter. Otherwise, The bill still required advance USDA consultation with Interior, mandated that all information regarding the effects of chemicals on wildlife be placed prominently on product labels, and expanded public education efforts. The administration took no public stand on the bill, but the USDA was widely known to oppose all undertakings but increased research. Both HEW and the USDI supported the entire package.

The Senate companion bill, however, emerged from the Commerce Committee shorn of all but the research funding hike. Warren Magnuson argued during the short floor debate that the recent FCPC agreement adequately ensured interagency coordination and that the labeling provision was being worked out administratively. "This is not an enforcement measure," Magnuson said of the emasculated proposal, "nor is it designed to regulate an industry."[73] The bill instead simply called for better research. Thus assured, the Senate passed S. 1251 without dissent.

Passage of H.R. 4474 was far different, and also was one of the more notable, if fleeting, defeats yet for the staunchest defenders of USDA policy. Merchant Marine and Fisheries Chairman Bonner moved to suspend the rules to prevent floor amendments and require a two-thirds majority vote for passage—a true litmus test of House intent—but nonetheless advocated passage. Provisions in the bill were already being carried out administratively, Bonner noted, but, he added, "I see nothing wrong with Congress adopting this policy through legislation."[74] Besides, the bill in no way affected USDA statutory authority over registration. Recent fish kills in North Carolina apparently tempered Bonner's traditionally wholehearted promotion of chemicals, and he was now convinced of the need for better agency coordination.

Jamie Whitten was not so moved, arguing that the FCPC adequately served the intended purposes and that specific legislative decrees might rob administrators of the flexibility necessary to operate effectively. Unstated, however, was Whitten's fear that product labels including data on the effects of chemicals on fish and other wildlife might give the USDI "control" over the registration process. Whitten defended his environmental record, but stressed, "I do not want you to make the American people go hungry because we started giving specific authorities to various departments where, with time, each could veto, in effect, necessary action by the other."[75] Farm bloc loyalists feared that the label provision would muddle agency jurisdictions if Agriculture and Interior fought over label language. Representative James Jones (D, Okla.) also argued that the dissemination of information to all "interested persons" would only bring out more "scare stuff" to the uninformed and the excitable; such fears had been commonly expressed since the days of fire ants and cranberries. To even suggest that Agriculture disseminate registration information to a wider audience unnerved those most protective of the department's tradition of secrecy.

Those supporting the bill succeeded in the end, as the chamber passed H.R. 4487 by the required two-thirds: 235–119 (87 not voting). The vote breakdown reveals clear regional splits within the Democratic party: northern Democrats voted 104–05 in support; southern Democrats only 49–35. Republicans also split, 83–70. Southerners thus divided for the first time on a pesticides issue. The massive fish kills throughout the South and the ongoing fire ant controversy had apparently convinced many that interagency coordination was both worthwhile and relatively painless. Many farm bloc representatives no doubt believed that the agricultural perspective would prevail anyway, but also felt that better coordination might prevent messy interdepartmental conflicts and costly missed signals. Clear cracks thus appear for the first time in the conservative coalition's solidarity on pesticides—fractures generated apparently by mounting doubts about the unalloyed benefits of the chemical appraoch, as well as by a desire to protect farmers from the effects of future administrative turf wars.

Those staunchly supporting USDA prerogatives eventually prevailed, however, as the two chambers failed to reach a compromise. Senate sponsors, unwilling as yet to interfere with the new FCPC arrangement, insisted on their bill, while House conferees split on the matter. Farm bloc members opposed *any* bill, while Dingell and his supporters insisted on Senate acceptance of H.R. 4487. The result was a stalemate, and both bills died with the 88th Congress. Expanded USDI research was approved in 1965, and again in 1968, but research funds never would reach the $5 million mark set by the 1965 law.[76] Proposals to reform various aspects of federal pesticides policy averaged three per session in each chamber between 1964 and 1968, but only the two extensions on USDI research ever passed. In the legislative arena, at least, policy changed stalled, largely through the ability of farm bloc representatives to use their positions as bulwarks against policy modification. Change would come, but its genesis lay elsewhere.

The Battle for DDT: Playing by the Rules of the Game

The post–*Silent Spring* reform wavelet ran its course by mid-decade, with the farm bloc fractured, but not shattered, by the controversies of 1962–1964. Direct attempts to overhaul FIFRA died regularly in the committees on agriculture, while more circuitous routes to policy change inevitably were sidetracked in the nether realm between committee and floor, or in the chambers of congressional appropriators. Jamie Whitten and like-minded colleagues accelerated fire ant eradication efforts despite the continuing controversy (and waning USDA enthusiasm for the program), while simultaneously gutting proposed increases in USDI research. A congressional sea change was under way beneath the committee and subcommittee chairs, but these most pivotal junctures in the legislative process remained firmly under the control of those most defensive about pesticides and their centrality to American society.

The executive branch proved equally exasperating for those seeking policy change. The Federal Committee on Pest Control mechanism gave cautious USDI and FDA bureaucrats formal access into pesticides policymaking, but always with the knowledge that the USDA could proceed despite objections. Proponents of the pesticides paradigm continued to dominate the middle levels of administrative discretion, their power backed mightily by long and friendly ties to congressional appropriators. If many within the USDA privately questioned the continued sanctity of pesticides orthodoxy, few openly voiced their doubts, and fewer still supported any structural or institutional shakeups that might rob their bureaus of policy jurisdiction. At the higher levels of administrative power, the issue was shunted aside in favor of more pressing business such as commodity subsidies and farm income. Pesticides legislation constituted a political graveyard for the Johnson administration, since "the environment" as a generic issue area had yet to displace other matters at the top of its agenda for action.

Conservation groups, with growing membership lists, broadening organizational horizons, and greater confidence in their scientific evidence, chafed at their apparent inability to translate truth into action. They were unhappy especially about what they perceived as inequities built into the federal tax code; in their minds, these biases simply buttressed unequal access to government. Audubon's Charles Callison in late 1967 admitted that the organization's public education and quiet lobbying activities were far too passive to blunt the onslaught of corporate and USDA public relations campaigns, but added that the society could do little more.

> Why doesn't the National Audubon Society organize write-your-congressman campaigns on pending issues? Tax laws are one reason. If "substantial" funds and effort were expended for such work, the Society would forfeit the status under which contributors can now list their gifts as income tax deductions. . . . It is unfair that profit-seeking enterprises can write off costs of lobbying campaigns as business expenses, but that non-profit institutions are penalized when they try to tell their side of the story.[77]

How much the tax code retarded greater activism among conservationists during the decade (or earlier) is uncertain, but there is no doubt that nonprofit organizations knew the pitfalls inherent in their status. And, if they needed reminders about their fiscal Achilles' heels, they needed only to look as far as the Sierra Club.

The Sierra Club by mid-decade had abandoned its regional orientation and, under the direction of David Brower, rose in national prominence as a militant proponent of environmental protection. Sierra's primary focus was on land preservation, its efforts aimed especially at fighting proposals for new water projects and other types of development throughout the Southwest. In June 1966 the club purchased a series of full-page advertisements in major newspapers urging the public to fight the proposed Grand Canyon dam, a project supported heavily by southwestern utilities, industries, and congressmen. The ads contained forms that readers could clip and send to their legislators, which was an obvious attempt by the Sierra Club to stir up opposition to the project. The IRS had ignored previous Sierra ad campaigns, but now agency officials suddenly decided that this current version constituted "substantial" lobbying, encouraged no doubt by letters from southwestern legislators. Even at the risk of being charged with bureaucratic arbitrariness and caving in to congressional pressure, the IRS revoked the Sierra Club's 501(c)(3) status, making contributions to the organization no longer tax deductible. Internal Revenue officials argued that Sierra's actions during 1964–1966 indicated that the organization's cumulative effort had "not been a casual or incidental or sporadic concern, but a regular, formal, purposeful part of [their] functions during the period," and that its behavior was sufficient reason to revoke the group's tax status.[78] If this weren't bad enough, insofar as conservationists were concerned, proponents of the dam were able to deduct *their* regular and purposeful lobbying expenses as "ordinary and necessary" costs of doing business.

The action clearly chilled conservationist attempts (actual or planned) to promote their goals more actively. The IRS in 1970 would also go after newly emerging public-interest law firms and their patrons in the private foundations, but this attack was blunted by growing congressional and public indignation over apparent bureaucratic capriciousness. The IRS backed off under congressional pressure, but, as the *New York Times* later editorialized, "Private interests still get the tax breaks and the public interest organizations still live in uneasy apprehension that the IRS may swoop down on them if they are suspected of lobbying."[79]

The Internal Revenue Service's assaults on tax exemptions for public-interest law firms stemmed directly from the success these new forces began to enjoy in the late 1960s. Success came through the courts, as activists learned to bypass the more entrenched centers of economic and political power. Much of this success came in campaigns like those for civil rights and social welfare, but some of the greatest impact wrought by these law firms concerned environmental protection. Conservation groups would turn to the judiciary, despite its traditional defense of private economic interests, because that branch was the only avenue actually left open to them. Fear of losing tax-exempt status after the Sierra Club incident was one reason; the others lay in the realities of congressional power and administrative prerogatives. "The only way out," Thomas Dunlap notes, "was to find some forum in which the defenders of [pesticides] could not use their advantage in excellent public relations, impressive scientific figures, and powerful political support."[80] William Butler, now with Audubon, recalls:

> When one petitioned the Department of Agriculture for restrictions on pesticides use, nothing happened. Therefore, a few lawyers and scientists decided that the best way to get at this problem was through the legal process. That is because the legal process was the great equalizer. It took only one person with a good argument to have a major effect on a court, whereas before an administrative agency or in Congress the odds against those in favor of stricter registration were much longer.[81]

The environmental road through the courts began, not surprisingly, in defeat. However, it was a defeat far different from earlier ones. In spring 1966 Victor Yannacone of Long Island filed suit in Suffolk County Court to restrain the county from spraying DDT to control mosquitoes. Yannacone, a lawyer, persuaded several local research scientists to back his suit, and obtained a temporary injunction against the spraying based largely on their affidavits. The plaintiffs could have settled out of court, since county officials were reluctant to press the issue, but instead proceeded for two primary reasons. First, Yannacone wanted scientific evidence inserted into the official court record, even if he lost—evidence that might persuade others to act. Second, what was no doubt discomfiting to local officials was that Yannacone wanted publicity; county officials would need to defend their use of pesticides, which might embarrass "the polluters" and inform the public about potential dangers. Yannacone's clear intent was to use the courts, regardless of outcome in this case, as a device for building a reservoir of evidence and for expanding public issue awareness.[82]

The National Audubon Society, through its recently formed Rachel Carson Fund, provided partial financial support for Yannacone's suit, and later agreed to record and publish the trial proceedings. Frank Graham, writing in *Audubon,* explained the society's new orientation: "Conservationists have learned that it is not enough to complain to the world at large. Their most effective weapon against pollution is a well-substantiated case aimed at a specific target."[83]

The target was DDT. Despite a strong scientific attack on the pesticide and its effects by Yannacone and his experts, and a poor defense by county officials, the judge ruled that prohibiting DDT use was a political, not judicial, question, and remanded matters directly back to the county. Loss in court was in fact victory, because the attendant publicity prompted county officials to suspend future DDT use. Furthermore, as Dunlap argues, Yannacone and his scientific allies discovered through this experience a potent weapon against pesticides—"an alliance of legal tactics with scientific information in a forum outside the government agencies or Congress."[84]

Out of the Long Island case emerged the Environmental Defense Fund (EDF), which for several years would disdain traditional membership appeals and congressional lobbying strictly in favor of the judicial route. This approach conserved EDF resources, allowed it to focus entirely on a single policy arena, and also enabled EDF to retain its tax-exempt, tax-deductible status. Tax-deductible organizations *could utilize the court system without jeopardizing their status,* a lesson quickly appreciated by established conservation groups. The Sierra Club soon formed a legal defense fund, a tax-deductible wing for the no longer tax-deductible organization, while other organizations, such as the Natural Resources Defense Council (NRDC) and Environmental Action (EA), would emerge in the early 1970s exclusively to pursue judicial pathways. Audubon took longer in this respect, but provided EDF with its seed money and continuing support from 1967 onward. The period of environmental advocacy through the courts had begun in earnest.

The EDF failed twice more in court during 1967, being unable to establish proof of DDT danger according to the tough standards of proof erected by the judicial system. But if EDF lost the battles, it was winning the war, because the publicity surrounding each suit sparked public attention and the defendant's subsequent prohibition of DDT—just as it did in Long Island. These results were, in themselves, gratifying, but the war was both frustrating and expensive. Yannacone and his colleagues desperately sought some way to establish their case in a judicial setting that did not require the harsh standards of proof demanded by the courts, but which also could convict DDT.

They found their venue in Wisconsin, where state law contained a novel quasi-judicial mechanism that allowed any state resident to request a state agency to rule on the applicability of a particular set of facts to any rule enforced by that agency. Upon receiving such a request, the agency would hold public hearings, subject to the standard rules of evidence and cross-examination, that would clarify administrative procedures. Plaintiffs in these cases need not establish judiciability, meaning that they could focus entirely on some broader question rather than a single case. Furthermore, rules for these hearings provided that witnesses could appear *without* declaring themselves on either side in the dispute. This requirement in regular courts had long dissuaded scientific experts from testifying for fear of jeopardizing their "objectivity." These rules, Thomas Dunlap explains, were perfect for the EDF: "In the hearing room the EDF could define the issue—was DDT safe for man and wildlife? Although it and the intervenors could raise other issues that went outside the legal case, . . . the main thrust of the hearing was the issue the EDF wanted to discuss. The hearing, in effect, allowed it to choose its ground, set the terms of discussion, and force the industry to come and defend itself."[85]

Critics of DDT for once held the trump cards, since they as the plaintiffs—not the court—defined the issue to be discussed. The judicial rules of evidence would lend credibility to a successful "prosecution" of the case, while the lenient rules about witnesses allowed previously cautious scientists to testify with far less fear of being categorized as pro-business or pro-environment. The forum was perfect for the EDF, and the late 1968 hearing on DDT use in Wisconsin caught both industry and the USDA completely unawares. The defendents arrived expecting just another court case, but instead found themselves and DDT on trial according to the EDF's own definition of the issue. Yannacone and "his" experts simply shredded the traditional defenses proffered by industry and USDA representatives, established the strength of the evidence damning DDT, and also established for the record the apparent isolation and intransigence of the USDA respecting the effects of pesticides and public access to the registration process. Especially damning to the USDA was its grudging admission that it made no independent checks on industry-supplied registration data. Interior Department scientists called by Yannocone, now protected (theoretically, at least) by the rules of the hearing, powerfully buttressed his attacks, while a hastily formed industry "task force" put up a markedly feeble defense.[86]

The hearing panel in early 1969 declared DDT to be a hazardous pollutant and, though it could not enforce its declaratory judgment, recommended strongly against further use of the pesticide in Wisconsin. The decision was an unparalleled defeat for defenders of DDT and, by extension, for the pesticides paradigm as a whole. The "rules of the game" this time worked against the pesticides subgovernment, and the weight of the scientific evidence apparently supported the EDF case. The administrative hearing, designed as a device for bureaucratic accountability, proved a powerful ally in the hands of knowledgeable lawyers against more traditional forms of economic and political power. This lesson would be repeated in the 1970s as similar adjudicatory mechanisms evolved at the federal level. For the time being, however, the Wisconsin ruling sent shock waves throughout the agricultural community, the chemical industry, and government at all levels—impacts that went far beyond Wisconsin.

The Pace of Change: Prelude to Policy Transformation

The EDF "victory" in Wisconsin sparked a flurry of activity on DDT at both the state and the federal level, a trend coinciding with a sudden surge in overall concern about the environment in the wake of the Santa Barbara oil spill, the battle over the Everglades jetport, and other highly publicized incidents. A society previously sensitized to its problems of war overseas and inequities at home perceived increasingly that environmental degradation posed apparent dangers to national health and the quality of life. "The environment," that conglomeration of interconnected conservation, health, and preservation issues, became by mid-1969 a powerful force on the public agenda. Comments Dunlap: "*Silent Spring* had prepared many Americans to accept the idea that human actions could be a significant factor in the global ecosystem, to believe that technology could be harmful, and to distrust the experts. By 1969, the accumulation of incidents had turned suspicion to active discontent."[87]

The events of 1969 were but a prelude to the changes to come, yet they had powerful impacts on subsequent policy scope and direction. Richard Nixon's avid pursuit of the American middle class and his efforts to undercut Senate Democrats like Edmund Muskie (D, Maine), prompted the president in May to form a cabinet-level Environmental Quality Council to advise him and to foster interdepartmental cooperation on environmental matters. Congress, for its part, passed the landmark National Environmental Policy Act (NEPA), a bill originally opposed by Nixon (largely because its sponsors were prominent Democrats) but later embraced publicly by him in order to regain his place as *the* dominant environmental advocate in government. NEPA simply required federal agencies to consider environmental matters before undertaking new actions, and also created a cabinet-level Council of Environmental Quality (CEQ) to develop national environmental policies. NEPA on its face seemed merely the government's symbolic affirmation that it cared, and the law passed with little controversy or public attention; few in Congress, or in society, understood the breadth or depth of change that the Muskie-Dingell proposal would generate. This "symbolic action" would, however, place into the hands of environmentalists and other citizen activists a potent action-forcing weapon—the Environmental Impact Statement—which would prove to be the bane of many a developer or government official. Reams of print now cover NEPA and its aftermath, so it is sufficient here to suggest that the law helped to catalyze the environmental decade to come.[88]

The battle over possession of the heart and soul of federal pesticides policy flared with unprecedented intensity amid this emerging national concern about the environment. The momentous "defeat" for defenders of DDT in Wisconsin spawned a host of concerted attacks on both persistent pesticides and their patrons. Bans on DDT began to dot the national landscape, resulting in a "crazy quilt" of state and local regulations disconcerting both to the chemical industry and environmentalists. Michigan, for example, banned the use of DDT in April 1969 after the FDA seized several thousand pounds of apparently tainted coho salmon, a move designed to protect lucrative commercial fishing and tourist industries. Ohio, however, did not ban DDT use, which effectively subverted the effect of any ban by its northern neighbor. National Audubon Society officials at midyear decried this growing hodgepodge of local standards, calling for both a uniform national policy and a national ban on DDT. Defenders of DDT opposed any ban, but supported uniform national standards for traditional commercial reasons.

Any national DDT ban obviously would not come through Congress, though legislative activity on pesticides issues blossomed to proportions exceeding those in 1963–1964. Thirty-one separate proposals to either ban certain pesticides or rework federal regulation were referred to the House Committee on Agriculture *alone* in the 91st Congress (1969–1970), but no action beyond hearings occurred. Environmentalists and their congressional allies fared little better in the more sympathetic climes of the Senate. Even so, the congressional tide began to flow distinctly toward some fundamental reordering of federal priorities. The General Accounting Office late in 1968 and early in 1969 severely criticized USDA registration and enforcement procedures, charging that the Pesticides Regulation Division (PRD) of the Agricultural Research Service failed to take any enforcement actions against some 562 major violations of FIFRA in 1966 alone. Even worse, the GAO reported, "There have been no enforcement actions taken by ARS to report violators of the FIFRA for prosecution in 13 years, even in instances where repeated violations of the law were cited by the agency and when shippers did not take satisfactory action to correct violations or ignored ARS notifications that prosecution was being contemplated."[89]

USDA officials acknowledged the validity of the GAO's findings, but argued that budget and statutory limitations were at the root of the problem. There is evidence that the USDA itself had worried about this problem for several years, and, as department officials argued, congressional budget constraints did not allow them to publish judgments on registrations. Funding for all FIFRA activities had increased from around $1 million in 1960 to about $3.8 million in 1969, but even this level of funding apparently proved insufficient to allow the ARS to keep track of some 60,000 products now on the market.[90] Nonetheless, the GAO reports and subsequent House hearings established clearly that the Agricultural Research Service consistently allowed violations to go unprosecuted and unreported, that it apparently ignored HEW and USDI objections to a number of pesticides registrations over the years, and that the FCPC mechanism fared little better than did its predecessor as a device for enforcing the agency's cooperation. The House Subcommittee on Intergovernmental Relations also found that FIFRA itself failed to provide for the pretesting of new pesticides, nor did it allow the ARS to suspend use of certain pesticidal products without laborious court proceedings. The USDA and FIFRA were under the congressional microscope in 1969, and neither appeared all that appealing to legislators increasingly concerned about keeping "bad" chemicals off the market. The Wisconsin case and the GAO reports neither improved the USDA's reputation nor settled nagging concerns about the pesticides paradigm itself. Despite a renewal of congressional attention that, at a minimum, created an atmosphere more conducive to reform, the legislature did nothing. Change again would have to come through indirection.

The Food and Drug Administration's seizure of Lake Michigan salmon in March 1969 provoked howls of protest from five Great Lakes state governors, who feared that the pall cast by the action might cripple inland fishing and tourism. In response, hoping to finally establish some agreement about the future of DDT and other persistents, Secretary of Health, Education and Welfare Robert Finch in April announced the creation of an independent Commission on Pesticides and Their Relationships to Environmental Health, to be chaired by Dr. Emil Mrak. Finch's action did not please environmentalists, who felt the issue of DDT to be a dead letter after the Wisconsin case. "That still another study was deemed necessary," declared William Rodgers, "demonstrates the continued rigidity of the forces opposing effective regulation of chemical pesticides."[91]

This may have been true, but creation of the Mrak Commission also illustrated the intensity of the dispute under way throughout the scientific community. A decade of "experts" and scientific studies had failed to resolve the question to anyone's satisfaction. Staunch supporters of pesticides continued to claim that they did no harm to human health, while environmentalists denounced these chemicals in increasingly strident tones. The National Academy of Science–National Research Council in May released its latest study on persistents, only to find itself ridiculed and the study excoriated by critics as "ecologically incompetent." The academy, Audubon officials argued, buried all evidence damning to DDT deep into the report, while it simultaneously stated that the pesticide posed no threat when properly used.[92] Secretary of Agriculture Clifford Hardin, on the other hand, called the NAS-NRC report "reasonable and balanced."[93] The entomological community itself, once united firmly behind the pesticides paradigm, was at the decade's end nearly split on the issue of persistents, and yet another study would probably do little to quell the acrimony.

The Mrak Report, released in early November 1969, probably was the most comprehensive study to date on pesticides and their impacts. The panel recommended against blanket suspension of pesticides, which few advocated anyway, but concluded that there was "adequate evidence concerning potential hazards to our environment and to man's health to require corrective action."[94] The report called for a phased elimination over two years of all but "essential" uses of DDT, and recommended a national shift away from all persistent pesticides. Panel members also advocated changing the zero tolerance clause embedded into law by the 1958 Delaney Amendment because rapid technological advances made the idea of "zero residues" simply untenable. With respect to the administration of federal policy, the panel argued that "a single agency should take the initiative to insure the effective monitoring of the total environment," but stopped short of recommending specific shifts in existing agency jurisdictions.[95] Far closer interagency cooperation was essential, as usual, which called for a *new* interagency agreement. But, the panel added, if such agreements again fail to protect public health, "it will be necessary to amend the FIFRA" and statutorily enforce interagency cooperation.[96] Above all, the Mrak Report proved influential and unique because it "was the first recommendation by a group of such high standing to set a deadline for ending uses of DDT and to define an operational criterion for determining what uses might be continued. This criterion of unanimous agreement among USDA, HEW, and USDI in effect meant the last two would have veto power over USDA decisions; this would have represented a major realignment of authorities."[97]

Nothing was likely to sway the most hard-line defenders of the status quo nor please the most extreme elements in the nascent environmental movement, but the Mrak report did influence those most unsure about the purely benign character of pesticides. And if the legislative road to policy change remained apparently blocked by agricultural committees, the administrative route was not. On November 12, Secretary Finch, backed by President Nixon, announced a total federal phase-out of all but "essential" uses of DDT over two years. The FDA ban of course applied only to uses that left residues on foodstuffs, and did not apply to nonagricultural uses, but Finch's announcement nonetheless marked a watershed in federal regulation of pesticides. The stated intent to cancel the use of DDT marked the first time that the government would forbid use of the pesticide *not* because it was mislabeled, adulterated, or misused, but because *in itself* it posed potential dangers. The issue no longer was use versus misuse. The chemical was not as yet proven to be harmful to man, Finch noted, but "prudent steps must be taken to minimize human exposure to chemicals that demonstrate undesirable responses in the laboratory at any level."[98]

Total cancellation of DDT use could come about only through the USDA, with its statutory authority to register pesticides. In previous years—the 1959 cranberry scare comes to mind—the department and its allies may have resisted, and prevailed, but the tide had turned. And, to emphasize the shift in federal orientation, President Nixon intervened on the side of the critics. Whether he did so purely for "political" reasons is both difficult to know and irrelevant. Whatever the motivations, the effect was to force the USDA into line with the FDA's action, despite howls of protest by farm bloc representatives and the chemical industry. Secretary of Agriculture Hardin soon announced additional limits, canceling approximately 33 percent of all DDT registrations, and stated his intent to cancel all but "essential" DDT use by late 1970. Hardin also announced that the USDA would soon enter into a new interagency agreement with USDI and HEW that would give them virtual veto power over all future decisions on DDT and other persistents.

Environmentalists naturally hailed the administration's action. Frank Graham, writing in *Audubon,* exclaimed euphorically, "The sudden capitulation in late November of the stubborn, narrowly positivistic, industrial bureaucratic forces which have sustained DDT through 20 long years of bitter scientific controversy opens up a new era."[99] The *New York Times* editorialized simply that "the warning sounded by Rachel Carson in *Silent Spring* is finally being heeded in Washington."[100] The critics, it appeared, finally had won a decisive victory over those long entrenched in positions of policy dominance.

Or had they? The announced ban on DDT came at a time when that chemical was no longer so essential in the battle against pests. Domestic use of DDT dropped by 50 percent during the 1960s as it gradually gave way to new, more specific, and less persistent (though more acutely toxic) formulations. Concerns about residues on produce figured modestly in this reduction, since growers sought formulations that left no residues purely out of economic self-preservation. The primary reason for DDT's fall from grace, however, lay simply in its decreased effectiveness in an ecosystem that had been subjected to millions of pounds of the pesticide during the previous twenty years. By the late 1960s some eighty-nine species of insects had developed a resistance to the chemical, and the prospects for future DDT use looked rather unprofitable for its manufacturers.[101] Besides, getting rid of DDT domestically would spur the sales of other, more expensive formulations. In this sense, then, DDT was a relatively easy target.

Still, the announced ban on DDT was a major apparent change in the direction of national policy, particularly because the chemical maintained powerful supporters in agriculture and industry. Some 70 percent of *all* domestic DDT use in the late 1960s (approximately 20 million pounds) went solely to cotton, so the ban was sure to rile southern legislators—the staunchest defenders of all. Additionally, some 79 percent of all U.S. DDT production went for export, and, while the ban did not include exports, any major action by the United States was sure to affect actions taken by other governments. Canada already had banned most DDT use, and many other nations were likely to follow suit. The significance of the administration's action against DDT therefore meant much more than an environmentalist "victory" over a single perceived chemical hazard. Victory this time gave both legitimacy and momentum to the emerging environmental movement, and the decision to cancel the domestic use of DDT sent a powerful symbolic message that a new era had begun. Any substantive victory claimed by environmentalists was in fact rather small, since the power to determine federal policy remained largely in the hands of the stoutest proponents of pesticides. These actors now were on the defensive, but the power inherent in their positions and numbers guaranteed that fundamental policy change would require the intersection of greater numbers of well-positioned advocates of policy change.

The pesticides policy community expanded in the 1960s, and some change occurred—new actors, new directions. But fundamental change would require more than just new actors; it would require clear shifts in the structural and institutional patterns of policymaking. New policy claimants might influence the edges of policymaking—regarding DDT, for example—but the direction of central policy still lay with the USDA and the congressional committees promoting agriculture. Yet a transformation in the actual structure of power that might produce significant shifts in policy direction was in the offing.

Chapter Seven

Environmentalism, Elite Competition, and Policy Change

The Rise of Environmentalism

A miracle of public opinion has been the unprecedented speed and urgency with which the ecological issues have burst into the American consciousness. Alarm about the environment sprang from nowhere to major proportions in a few short years.[1]

Public opinion does not emerge like a cyclone and push obstacles before it. Rather, it develops under leadership.[2]

The Gallup Poll in mid-1970 asked respondents to select the domestic problems that government should address. Most chose crime first, which was not unusual, but this time some 53 percent of those polled also cited air and water pollution—a dramatic shift from days when such questions barely made the slate of national priorities.[3] Clearly, or so it seemed, environmental concerns had vaulted to the forefront of the national consciousness. That these attitudes should have surfaced just shortly after Earth Day 1970 (April 22), probably one of the most effective instances of "demonstration democracy" in modern times, is no surprise. "Demonstrations," Amitai Etzioni argues, "are a particularly effective mode of political expression in an age of television, for underprivileged groups, and for prodding stalemated bureaucracies into taking necessary actions."[4] Publicity-generating techniques honed in civil rights and antiwar demonstrations now proved particularly useful for those seeking to reorient national priorities.

Mass public opinion tends to influence the agenda of government, the order of priorities laid before (or forced on) policymakers, more than it does actual policy choices. "Government officials," notes John Kingdon, "may pay attention to a set of subjects partly because those subjects are on the minds of a fairly large number of people."[5] Earth Day indeed set the government agenda, but, unlike the way that natural cataclysms suddenly structure official priorities, it was no accident. Earth day was the brainchild of Senator Gaylord Nelson (D, Wis.) and Representative Paul McCloskey (R, Calif.), two legislators long accustomed to waging a sometimes lonely fight for stronger national conservation efforts. These two entrepreneurs of opinion mobilization sensed (as did many others) the emerging strength of environmental concern, an as-yet amorphous support for new initiatives awaiting some sort of catalyst, some type of focus. Backed by a small Conservation Foundation grant, the two organized the nationwide "teach-in" that surpassed most observers' expectations. Nelson, at least in retrospect, knew it would succeed. "I saw Earth Day as a chance to create a political dialogue. It was an agenda-setting device. I wasn't surprised at the response, because I had a sense that the people were way ahead of the politicians on this."[6]

Such is the stuff of elite perceptions about mass opinion. The broader question here is whether Earth Day in fact "created" the environment as a major national issue or merely benefited from mounting public concern about perceived environmental ills—the "chicken-and-egg" argument. In retrospect, Earth Day and its impact probably worked symbiotically with a rising tide of public awareness. Nelson's assertion that the people were "way ahead of most politicians on this matter" may be true in rather abstract terms, but one might argue plausibly that public opinion lay largely nascent and unformed *until* Nelson and other policy entrepreneurs crystallized its dimensions around an identifiable and compelling event. Public opinion in favor of federal action was becoming a powerful force in 1970, but lagged noticeably behind already intense elite competition, particularly between Richard Nixon and Senate Democrats, over who in fact would tap the *potential* benefits of an "environmentally aware" public. Nelson succeeded, and most other political leaders quickly jumped onto the bandwagon.

We can never resolve this causal dilemma; however, there is no doubt that the emergence of an "environmental public" eventually produced government policies congruent with the direction of mass opinion. Some policy change in fact preceded the crystallization of mass opinion—for example, NEPA and the DDT decision in 1969. Did these actions "create" public opinion? To some extent, yes, but it is more likely that political leaders *anticipated* public attitudes and acted accordingly. In general, Page and Shapiro argue, "It is reasonable in most cases to infer that opinion change was a *cause* of policy change—at least a proximate or intervening factor leading to some government action, if not the ultimate cause."[7] Such was the case with the bundle of environmental issues.

Public opinion frequently gets exercised about some problem or malady, only to ebb quickly, and with no perceptible impact on public policy. Public concern about chemical pesticides soared with the publication of *Silent Spring* and during the Mississippi fish kill episodes of 1962–1964, but these concerns were apparently sated by incremental policy change and largely symbolic administrative adjustments. To compare the early 1960s to the early 1970s, however, is to compare two vastly disparate sets of conditions; changes occurring in the intervening years led to profound policy change. One major difference was increased social awareness of the environment *as a problem,* which reflected a slow growth of popular comprehension over the previous decade. The spate of scientific studies on pesticides convened during the 1960s may have muddied the waters of scientific consensus on this issue, but, if nothing else, these professional disputes spilled over into greater public awareness that something was indeed wrong. Additionally, visible public concern about some problem accelerates *prepolicy* activity by political leaders, ranging from more frequent symbolic posturing to "more and stronger proposals through the policymaking process."[8] Broader public awareness and accelerated prepolicy activity had intersected by late 1969, propelling the issues across a critical threshold toward policy change. Earth Day 1970, in this sense, marked the critical transition point between the slow gathering of public opinion and its blossoming into a potent political force.

More Actors, More Patrons, More Access

"What was different in 1970," Charles O. Jones notes about the Clean Air Act, "was the lobbying by environmentalists."[9] The difference lay in the *expansion in the number of participants* in policy development, *active competition* among elected officials to produce and be credited with strong legislation, and *constant media attention* to policy developments.[10] To his list I add a fourth factor, one not normally included in pluralist analyses. In addition, critical *structural changes* created opportunities for access to the policymaking game for those previously excluded. The first condition, more players in the game, may in fact depend powerfully on the last.

Earth Day, true to the "disturbance theory" of group mobilization, begat a host of new environmental organizations.[11] Environmental Action, formed by Nelson and McClosky to organize Earth Day, outlived its original mission and moved on to lobbying. The Natural Resources Defense Council (NRDC), formed in 1970 by several northeastern lawyers, emulated the increasingly fruitful judicial tactics pioneered by the Environmental Defense Fund, while the League of Conservation Voters found its niche as the political action arm of the movement. Mainstream conservation groups expanded existing efforts, accompanied in many cases by radical alterations in organizational mission and strategies. Many created legal defense funds or non–tax deductible lobby wings to complement traditional activities, adapting to new political opportunities in ways that allowed environmentalists to expand both their legal and legislative efforts. Conservation lobbyists were virtually nonexistent in the 1950s, but, by 1970, thirteen separate environmental organizations were so registered, and their ranks expanded thereafter.[12] By 1971, almost seventy environmental organizations had established some type of office in the capital, creating in the process a sense of greater permanency for the movement, a sense of credible "presence" previously absent.[13]

Presence in Washington accompanied, and was aided by, growing organizational strength, both in membership and resources. Sierra Club enrollment rose from 48,000 in 1967 to a national membership of more than 130,000 by late 1971.[14] National Audubon Society rolls in 1970 alone grew by almost 70,000 new members (to 148,00), to double again by 1973.[15] Much of this expansion in fact came after major policy breakthroughs, lending credence to the old saw that success breeds even greater success.

Numbers alone cannot account for environmentalists' subsequent policy success. Equally if not more crucial was the emergence of well-positioned patrons both inside and outside government. The minuet of policy initiative and credit-claiming between a Republican president and his Democratic rivals (real or feared) accelerated noticeably after Earth Day. All sides promised to be first with the most on the environment, which created a condition of unparalleled receptivity to environmental interests that greatly expanded the opportunities for participating in decision making. This competition among political leaders—based as much on electoral considerations as on any real concern for the environment, and sharpened by deepening distrust among these putative partners in national governance—quickened the pace of policy activity and broadened its eventual scope. The period 1969–1972 was, in retrospect, a protracted game of high-stakes political stud poker, with Nixon and his rivals raising the ante ever upward. All players in the game sought to call the others' bluff, hoping that somebody would fold, and all envisioned large political payoffs to the winners.

Schattschneider's observation that the scope of conflict powerfully conditions policy outcomes was borne out during Nixon's first term. Page and Shapiro note that "on issues about which the public has more well-defined opinions and shows more concern, where the scope of conflict is broad, policy tends to move in harmony with public opinion."[16] The scope of conflict at its core depends on the visibility of the issue and competition among leaders, conditions that characterized the early 1970s. Heightened public awareness fanned intense conflict among policymakers. The effect, argues Charles O. Jones, was to sidetrack "normal" incremental policy styles and inject into the process spasms of "speculative augmentation" on the part of all involved:

> Normally decision makers are expected to refine existing policy by determining what is technically and administratively feasible, as well as what is within the limits of acceptability to those being regulated. In speculative augmentation, however, the basis of decision making changes. Feasibility is less important than estimating what will be acceptable to a rather indistinct "public" perceived to be demanding strong action.[17]

Such dynamics produce notable policy breakthroughs, but they also multiply the possibilities for access to decision making and influence by outsiders. Policymakers now actively courted the advice and support of "legitimate" environmental interests, those apparently speaking for a concerned public. Interests traditionally represented in the game found the policy arena suddenly crowded with competing claimants. They also found the arena itself bathed in the limelight of intense public scrutiny, an uncomfortable condition that made traditional patterns of mutual accommodation appear dubious, if not patently antidemocratic. Above all, business and producer interests were ill-prepared for the new reality. David Vogel observes that the friction that developed between business and government during this period "has no parallel outside the United States, traditionally considered the most conservative or pro-business capitalist democracy."[18] So swift was the surge in the influence of public-interest groups, and so accustomed were business interests to having no credible competition for the trust of the public and control over policymaking, that the latter would take years to regain some of their lost leverage.

Environmental groups found a more congenial reception within government, but might not have been able to capitalize on their opportunities were it not for other patrons, particularly among the nation's private philanthropic foundations.

Table 6. Foundation Grants to Environmental Organizations, 1974–1984

(in thousands of dollars)

	Environmental Defense Fund	*National Audubon Society*	*Natural Resources Defense Council*	*Sierra Club*[a]	*Total*
1974	$ 37.2	$210.0	$ 85.0	$ 86.9	$ 418.2
1975	222.0	625.0	180.0	62.5	1,089.5
1976	391.0	163.2	545.0	222.5	1,321.7
1977	386.3	187.5	885.0	135.0	1,593.8
1978	210.0	155.0	2,052.7	125.0	2,542.7
1979	624.0	795.0	583.0	305.0	2,307.0
1980	497.0	81.0	577.5	85.0	1,240.5
1981	438.5	543.0	275.0	146.0	1,402.5
1982	509.0	357.3	637.5	220.0	1,723.8
1983	521.0	163.0	1,661.5	276.5	2,622.0
1984	471.0	715.0	565.0	178.5	1,929.5

Source: Foundation Grants Index (New York: The Foundation Center and Columbia University Press, 1974–1985).

Note. Figures cited were compiled from multiple editions, with some donations not appearing in print until two or three years after the fact. This lag time accounts for the incompleteness of the amounts cited for the years 1982–1984. The figures indicated, particularly for later years, are incomplete; they represent an *idea* of foundation largesse to environmental organizations. Figures indicate the year in which a grant was awarded; some grants were for several years.

a. Includes the Sierra Club Legal Defense Fund.

The still-fragile environmental movement in fact survived its tumultuous birth pains and was able to mature more because of foundation largesse than mass public support. The Ford Foundation played a particularly critical role in nursing the EDF, the NRDC, and other "public-interest law firms" to health and success, and in many ways also bankrolled the entire public-interest law movement that was to revolutionize American jurisprudence. Ford Foundation grants to these firms during the 1970s would total almost $21 million, with major chunks granted to environmental organizations.[19] Table 6 partially indicates the extent to which foundation money subsidized the environmental movement, ensuring that it would be more than a blip on the screen of history.[20]

Foundation patronage of public-interest law outraged those defending the status quo, and Congress in 1969 passed a Tax Reform Act barring such grants unless the foundations assumed "expenditure responsibility" over the recipients. The IRS in 1970 started to investigate both the firms and their patrons, but these actions were stopped dead in their tracks by adverse publicity and the strenuous objections of Senate liberals. The Nixon administration eventually issued guidelines to shield both the foundations and the law firms from IRS penalties so long as the firms do indeed serve the "public." Foundation support soon flourished, but the tax law remained on the books, an uneasy reminder to public-interest groups of the limits to their activities.[21]

The Pathways to Change: Reforming the Structure of Bias

John Kingdon argues that in addition to changes in mass public opinion and the "balance" of organized interests, the nature of the "political stream" depends on government's own internal dynamics and the positions held by key actors within that structure of governance. "Agendas," he adds, "are changed because some of the major participants change, and agendas are markedly affected by the drawing of jurisdictional boundaries and by battles over turf."[22] The "national mood" and the "power" of interest groups are potent forces for change, but *who* the key players are and the rules they play by influence which issues find a place on the public agenda—not to mention what alternatives become feasible.

The reweaving of the American political fabric accelerated through the 1960s, and in many ways reached a climax as the new decade began. The same reformist impulse so critical to earlier social movements focused increasingly on the structure of the system itself; reform was redirected toward the parties, the election process, and virtually every form of political organization in the structure of governance. It was a logical next step. Organizations are hierarchies of power bound together by rules and run by those who control key decisional junctures. Change the rules that bind the organization and you likewise alter its power structure—which, one assumes, redirects the organization itself. The most logical step of all for those intent on broad-scale change was to attack the very center of national political power—Congress.

Organizations must adapt to their external conditions to survive, or at least to remain credible. Congress during the early 1970s was buffeted by tremendous external cross-pressures, ranging from Nixon's perceived assaults on legislative prerogatives to social demands that Congress "open" itself up to new claimants.[23] A legislature characterized by very little turnover during the preceding three decades (and increasing seniority for those blessed with safe seats) witnessed between 1969 and 1975 an unprecedented influx of new members—new cohorts dominated by liberal Democrats intent on restructuring the institution's internal allocation of power.[24] The intersection of these pressures first generated the Legislative Reorganization Act of 1970, which instituted guidelines for media coverage of committee sessions and installed electronic-teller voting in the House. Reformers, disenchanted particularly by the disproportionate leverage wielded by more conservative senior members from the South, openly challenged the hallowed institution of seniority. The breadth of the changes to come are too numerous to discuss here, but their cumulative impact produced by the end of the decade a legislature far more open, far less hierarchical, and probably far less predictable than Clarence Cannon and his contemporaries could ever have imagined.[25]

Environmentalists joined with others to push such institutional "reforms," changes dedicated to giving "public interests" greater access to decision making and to countering the power wielded by "iron triangles" over public policy. Such changes came grudgingly, however, and environmentalists feared that their opportunities might fade before they could make substantive progress. Congressional reform and statutory change therefore became long-range strategies; for immediate impact, they pursued the increasingly successful judicial route. The Environmental Defense Fund and its emergent kin, seeking to repeat the apparent success in the battle over DDT, expanded their litigation against government and business activities alleged to have detrimental environmental effects. The Sierra Club alone pursued some forty state and federal lawsuits in 1969–1970.[26] Accelerated environmental litigation coincided with a seemingly fortuitous transformation in the courts as instruments of policy change; however, fortune had less to do with the reformers' success than did the fact that environmentalists consciously pursued the most promising avenue in the system. One can understand these perceptions only by examining the changes in the courts that, by the early 1970s, placed them squarely at the center of the policy debate.

First—and perhaps most critical—was a marked lowering of the barriers to judicial access. The 1965 *Scenic Hudson Preservation Conference* v. *Federal Power Commission* decision had broadened the judicial concept of "standing" beyond traditional economic criteria and toward a broader notion of what constitutes an "aggrieved party," and this concept was expanded even further by the Supreme Court in 1970.[27] Plaintiffs by the early 1970s no longer needed to exhibit direct and adverse injury to obtain standing; they needed only to argue plausibly that the interests in question were covered by some relevant statute or constitutional guarantee and were in some way affected by the actions of the defendants. *Sierra Club* v. *Hickel* (1970) established the radical (for common law) concept of class action, allowing plaintiffs to file suit on behalf of entire sectors of society, even if those "represented" knew nothing about the dispute. Plaintiffs no longer needed to name each party to the suit; they could now claim damages on behalf of society itself. Lowered barriers now gave public-interest lawyers unparalleled access to the courts. They were not guaranteed to win, but at least they could play the game. However, as Shep Melnick argues, environmentalists "would have had little reason to use their newly acquired access to the courts if they always lost their cases on the merits."[28] This caution is valid, since—at least regarding pesticides litigation—the courts consistently had sided with the pesticides producers and the USDA.

But this tendency also was changing. The courts by 1970 increasingly interpreted existing statutes "to guarantee a wide variety of groups the right to participate directly in agency deliberations as well as to bring their complaints to court."[29] Judicial insistence that federal administrators incorporate environmental concerns into their actions began most noticeably with the pathbreaking *Calvert Cliffs* decision of July 1971, in which federal district court judge J. Skelly Wright halted federal licensing of a nuclear power plant after determining that the Atomic Energy Commission failed to consider its environmental impact.[30] The purpose of the National Environmental Policy Act, Wright declared, "was to tell federal agencies that environmental protection is as much a part of their responsibility as is protection and promotion of the industry they regulate."[31] *Calvert Cliffs* put real teeth into NEPA, and the onus increasingly fell on administrators to prove in court that they had weighed all relevant criteria prior to starting new projects. Successive court decisions showed the judges increasingly willing to second-guess bureaucrats on both procedural and substantive matters; this was a challenge to administrative expertise that was a striking departure from judicial tradition. As the "generalists" now questioned the "specialists" on key scientific, technical, or economic questions, the courts took on a new and powerful role that no doubt sent chills down the spine of many a bureaucrat.

Thus, by the mid-1970s, the Third Branch played a major role in federal policymaking. But, as Melnick reminds us, the courts in many ways preceded the executive and legislative branches in riding the environmentalist wave: "Their efforts to increase the government's role in protecting the health and safety of its citizens and to decrease the influence of industry in regulatory policymaking preceded those of Congress and the White House. The Courts did not follow, they led. And even when they followed, they did so with surprising eagerness and aggressiveness."[32] And, in the process, they created inviting opportunities for those claiming a stake in environmental issues.

These facets of the "political stream"—national mood, more actors, elite competition, structural change—all intersected in the early 1970s to alter significantly the fundamental structure of bias that long had favored narrowly "private" interests over those of the more vaguely defined "public." Schattschneider argues that "procedures for the control of the expansive power of conflict determine the shape of the political system."[33] Those procedures for conflict containment were, in a rather short time, dramatically refurbished in a way that *allowed* conflict to more easily expand. The result, as Schattschneider might have foreseen, was to change the nature and outcome of the political game.

These factors also coalesced in a way that created inviting "windows of opportunity" for those promoting new and stronger environmental policies. These windows would remain ajar largely until 1973, when the "energy crisis" displaced the environment on the nation's agenda for action. Until then, however, nearly a dozen major statutes, covering virtually the entire expanse of the "environmental problem," came into law. Policies, like problems, often travel in bundles, and success in one area (such as the 1970 Clean Air Act) bred success in related policy areas. This context of vast and widespread advocacy for fundamental change regarding environmental protection almost guaranteed some change in FIFRA. The "inevitability" of policy change may in fact have been accelerated by the last, and perhaps most critical, structural change to be considered here: creation of the Environmental Protection Agency (EPA). The circumstances surrounding that agency's birth tell us a great deal both about the power of executive initiative and about the impact on policymaking wrought by fundamental changes in the structure of governance.

Enter EPA: The Environmental Advocate

Passage of the National Environmental Policy Act was a signal by Congress—though most legislators failed to appreciate its significance—that the federal government must get its own environmental house in order before turning to the larger problem. But, as Mazmanian and Nienaber argue, "Policy changes must filter down to the operating levels of government, to the departments and the agencies which not only administer programs but apply and interpret and hence in effect make policy."[34] Avoiding such a "leakage of authority" often requires modifying organizational goals, bureaucratic behavior, or both. One might institute some novel supradepartmental planning mechanism (for example, a new budgeting system) or, more commonly, one could reorganize the structures and process of entrenched bureaucratic power. After all, suggests Hugh Heclo, "Between the politicians' intention and the government's final action passes the shadow of the bureaucrat," and every political executive seeks to make that shadow as nonthreatening as possible.[35]

Scholarly analysis of Richard Nixon's "administrative presidency" tends to focus, for good reason, on the president's attempts to "infiltrate" the career bureaucracy with political loyalists or on his second-term proposals for massive structural realignments within the executive establishment.[36] These efforts failed largely because of the Watergate scandal, though, as William Safire later noted,

"The infiltration and reorganization which now seems so villainous will be carried out by more principled people under the banner of reform."[37] Less frequently analyzed years after the fact is one of Nixon's most enduring administrative legacies—the EPA. History will (or should) note that the reorganization spawning the new environmental agency supported two key notions about political power: (1) the ability to manipulate the structure of the executive establishment is critical to "successful" governance (or at least political control), and (2) fundamental alterations in that structure profoundly affect both government goals and policy outcomes, often unintentionally. Such intervention, as a result, also powerfully disrupts existing configurations of political influence, if only for a while.

The genesis of an independent environmental agency, as with so much of the Nixon administrative strategy, lay in the President's Advisory Council on Executive Organization, led by Roy Ash. The Ash Council at first considered consolidating all federal environmental and natural resource policies in an existing agency, but encountered the usual obstacles thrown up by bureaucratic turf battles and jurisdictional aggrandizement. Department of the Interior officials lobbied hard in 1969 for total jurisdiction over natural resource matters—a proposal with some merit—but were opposed strenuously by those in both the USDA and HEW. Enhanced interdepartmental coordination seemed equally doomed; the disputatious partners in pesticides policy were unlikely candidates for marriage of any sort. Yet the haphazard expansion in federal environmental activities during the late 1960s clearly demanded some type of coordinative mechanism, a more efficient and effective structure for policy consolidation and operational control. The latter concern was critical to Nixon, since operational control, when wedded to some type of strict sorting out of administrative function, meant tighter White House control over policy. Or so the theory went.

A new and independent environmental agency was the path of least resistance, both with respect to the bureaucracy and, particularly, to Congress. Nixon's June 1970 reorganization transferred antipollution programs to the new agency in a manner that spread the pain among all bureaucratic "losers," thus preempting inevitable attempts by members of Congress to preserve selected bureaucratic domains. The reorganization succeeded where earlier, more segmented, attempts had failed, primarily because of its breadth, its executive-led imposition, and its timing. Nixon clearly challenged his rivals to "put up or shut up," stealing much of the thunder on the environment away from Senator Edmund Muskie and his Democratic colleagues. Muskie earlier had proposed such an agency, so Nixon's action forced Democrats onto the defensive just as the environmental wave hit its crest. Nixon in essence gave Congress an ultimatum: accept the package as it is or vote the whole thing down. No middle ground was acceptable.

The reorganization achieved in one stroke what more than a decade of administrative wrangling and aborted legislative efforts had not—a fusion of federal pesticides policies. USDA registration functions, HEW tolerance-setting authority, and USDI research efforts—all were to be combined into EPA's new Office of Pesticides Programs (OPP). This was a shotgun marriage of longtime bitter rivals that, as we shall find, had profound consequences for early EPA strategy. Pesticides manufacturers, agricultural groups, and congressional farm bloc representatives bitterly opposed the plan, maintaining that final authority for approving new products properly lay with the USDA. Those already on the defensive in the battle over DDT saw Nixon's plan as an even more ominous shift in political influence. However, a department wracked by charges of regulatory laxity, conflict of interest, and mismanagement was ill-positioned to defend its prerogatives in the eyes of an increasingly skeptical and demanding public. Defenders of the status quo complained bitterly, but few could oppose the consolidation openly under the current political circumstances. The EPA officially came to life in December 1970.

The Environmental Protection Agency is important to our concerns here almost purely in terms of its impact on the traditional structure of relations in the pesticides subgovernment. The new agency, simply put, gave environmentalists somewhere to go. The USDA contained an uneasy duality of purpose, responsible for both promotion of key clientele interests and regulation of ill effects; environmentalists found that, when push came to shove, clientele interests always came first. EPA, on the other hand, had a single organizational goal—to protect the environment. William D. Ruckelshaus, the agency's first administrator, stated the EPA's mission with brutal clarity. "EPA," he told the House Committee on Agriculture, "is an independent agency. It has no obligation to promote agriculture or commerce, only the critical obligation to protect and enhance the environment."[38]

EPA was to be an environmental *advocate,* fulfilling a role perhaps not foreseen entirely by a president more concerned with political control over the bureaucracy, but it went in this direction nonetheless. In doing so, the EPA, regardless of its later successes or failures, gave environmentalists an institutional point of access comparable to that traditionally enjoyed elsewhere by business and producer interests. Succeeding administrations might have sought to restrict group access into policymaking, but few could ignore environmental perspectives without denying EPA's basic mission. In this respect it is useful to note that the EDF, one day after the EPA's official birth, filed suit in federal district court to require that Ruckelshaus cancel the organochlorines aldrin and dieldrin. That action, one former EDF staff counsel later observed, was "to help EPA get its agenda set."[39] Environmentalists may have found a bureaucratic patron, but they clearly were taking no chances.

The agency's impact on federal policy is well documented. One point, however, bears some comment here. EPA was the product of executive initiative, but it still depended on Congress for its operating funds. Creation of a new administrative unit often results in reallocation of program jurisdiction among congressional working groups, since prior jurisdictional lines may no longer make much sense, but such was not the case with pesticides. The rural sector's overall political power may have waned considerably by 1970, but it still enjoyed considerable leverage among the most senior and powerful members of Congress. The committees on agriculture, as a consequence, retained sole legislative authority over pesticides considerations, maintaining that duality of roles—promotion and regulation—just recently excised within the bureaucracy. Pesticides still were agricultural matters to the ruling committees of Congress, and the EPA would have to deal with "nurses" that were indeed hostile to the patient.

Equally important, however, were simultaneous shifts in appropriations jurisdiction. The chairman of the House Appropriations Committee, George Mahon (D, Tex.) stunned EPA officials and environmentalists by assigning *all* appropriations for environmental and consumer protection programs to his kindred spirit, Jamie Whitten. The most vocal congressional opponent to stricter federal pesticides regulation continued to control the purse strings. This was a fact of fiscal life that caused environmental advocates numerous headaches in the years to come. Creation of EPA jarred subgovernmental dominance within the bureaucratic side of the configuration of power, but in Congress the old central players still held away.

Imperatives for Policy Change: The Byzantine Ban on DDT

True change, a fundamental reorientation of existing national priorities and procedures, normally occurs irregularly, unlike the more incremental sorts of policy tinkering provided by routine openings in the political process. Such change occurs most readily when policies are widely perceived to be obsolete, when most (if not all) parties attached to an issue acknowledge some discontinuity between government intent and its statutory ability to meet those goals. Such perceptions began to emerge respecting FIFRA. Those concerned about chemical effects long had demanded stronger federal regulation, but, by 1971, even those promoting chemicals recognized that extant policy no longer met evolving technological, scientific, or social needs. The question pondered by all was, of course, where to go.

A number of forces pushing for policy transformation intersected soon after the creation of EPA. Prior to this moment, and despite mounting claims professed by nonagricultural interests, pesticides as an issue traditionally had been defined in rather simple terms. The 1947 law was a "truth in packaging" law and little more, devoted to smoothing the flow of pest-control technology from the factory to the farm. By 1970, however, the pesticides issue had evolved from a rather simple problem to one now fractionated into multiple and often contradictory clusters of policy claims. A regulatory framework predicted on a rather simple bipolarity—honest versus dishonest claims—now was polycentric, and was found wanting. The degree to which FIFRA had failed to accommodate the surge of competing claims was masked for years by the USDA's regulatory dominance, since departmental perspectives had tended to define the issue essentially into a "food versus famine" dichotomy that made swift approval of new products an overriding goal. FIFRA now was in the EPA, an organization less burdened by structural conflicts of interest and dedicated to values not traditionally locked into the reigning regulatory framework. The EPA also trod in areas the USDA had long avoided—using FIFRA as a mechanism to cancel certain uses of pesticides. These efforts exposed the statute's inherent legal weaknesses and procedural biases. Widespread acceptance of the need for statutory change in fact emerged simultaneously with the EPA's efforts to ban use of DDT. A brief foray into the case helps us understand the imperatives for policy reform.

Our discussion in the preceding chapter ended with the apparent cancellation of several uses of DDT by Secretary of Agriculture Clifford Hardin and his promise to end all but "essential" uses of DDT by late 1971. Many environmentalists, and particularly the general public, assumed that the books were all but closed on the famous pesticide. But, as they were to discover with exasperation, the battle had just begun. First, announcement of intent to cancel applied only to the uses stated in the ban, *not* to the chemical itself, and applied only to products in interstate commerce. Canceling a specific use under FIFRA forced the manufacturer only to indicate forbidden uses on the product label, but the product itself remained on the market. Whether prospective users actually adhered to these "paper bans" is, of course, anyone's guess, but experience shows old habits die hard. Harrison Wellford, long a critic of DDT and FIFRA, noted, "Farmers may continue to use DDT on tobacco, a prohibited use, under pretense of buying it for cotton, a permitted use."[40] Besides, the federal government had to rely on USDA county agents to enforce such bans, and such agents were known to give dubious support for stricter regulation.

Second, announcement of intent to cancel did not immediately stop the use of products if the manufacturer objected. The only way the EPA could immediately halt use was to suspend all manufacture and use of the product because of some "imminent hazard" to human health or the environment. Suspension did not, however, prohibit *sale* of the suspect product. This was a quirk of regulatory life that made the effort somewhat quixotic. Cancellation, on the other hand, set into motion a rather long and convoluted appeals process that was designed originally to protect the manufacturer from possibly capricious bureaucratic action. Additionally, *each separate use* was guaranteed such an appeal, which, in the case of DDT, meant over 300 possible separate proceedings. Regulatory fine-tuning thus was problematic; the choice between the blunt, abrupt suspension order and the elongated cancellation proceeding was not a happy one to federal administrators. Such a choice may in fact have been one reason why the USDA never bothered.

Finally, though manufacturers had to establish both the efficacy and the safety of a product during initial registration, in cancellation proceedings the burden of proof was paradoxically shifted to the federal government. Each use, in essence, was considered innocent until proven guilty, and manufacturers could produce and sell the product until all appeals were exhausted, which could take years. To be sure, the manufacturer had to prove the nonhazardous nature of the product with each appeal, but the byzantine nature of the process no doubt gave federal regulators tremendous pause before initiating action. The process benefited the manufacturer, which was exasperating for the EPA as it sought to carry out the letter of the law.

The battle over DDT did, in fact, take years, and not until December 1973, when the last court finally ruled against the industry, did the battle end (though manufacture and sale of DDT continued unabated in the interim). The reason for the length and the ferocity of the battle was clear: DDT no longer formed the cornerstone of the nation's war against pests, but it still was a powerful symbol of the pesticides paradigm. It was an affirmation of humanity's ability to control the environment and to ameliorate life's irritations. Defenders of the paradigm feared that a ban would signal open season on all persistent pesticides, and statements by some of the more zealous environmentalists did little to calm the Cassandras among those favoring the use of chemical products. The appeals process embedded into FIFRA allowed the determined defenders of DDT to delay actual cancellation, but they could only put off the inevitable. An alarmed (if somewhat amorphous) public, courts that sided increasingly with environmentalists, the passage of the NEPA, and the creation of the EPA—all coalesced to shift the tide against those traditionally in control of pesticides policy making. The DDT battle forced those concerned about pesticides—an ever-expanding pool of claimants—to accept the need for statutory change. Environmentalists and their allies long had favored stricter federal regulation, but, by the early 1970s, they were joined by their longtime enemies.

Those traditionally in control of pesticides policy now feared the growing influence of "outsiders," worrying that these actors neither understood nor cared about the unique needs of the farm economy. Farmers feared losing long-favored pest control technologies, while chemical manufacturers envisioned lost markets. Both feared a destabilized regulatory environment, since state governments by 1970 had begun to respond to the question of pesticides in varied and often unpredictable ways. The impetus behind statutory change in 1947—uniform regulatory standards—proved even more powerful this time. Farmers and homeowners left uncertain about product safety or legality made for unstable customers, and the possibility that a product might be suspended abruptly by the EPA left chemical makers even more determined to refurbish existing law. Regulation was a necessary evil, but it had to be expeditious, authoritative, and—most of all—less likely to make products on the market overnight pariahs. To do otherwise simply would maintain market uncertainty and reduce incentives for future chemical research and development.

Those in both EPA and USDA by the early 1970s advocated wholesale regulatory change out of a common perception that existing statutory authority made little regulatory sense. The USDA's unhappy experiences in the late 1960s convinced department officials that, in the words of Dr. Ned Bayley of the USDA, "The law is not adequate in either its enforcement provisions, or in providing the ability to control pesticides use in a way that minimizes hazards."[41] The USDA's perceptions of the pesticides problem underwent notable change: most department officials now worried that farmers might suffer disproportionately unless federal policy accommodated both traditional needs and the new environmental ethos. Above all, department officials sought mechanisms to fine-tune regulatory decisions and avoid a reprise of the messy DDT battle, a fight that in their minds served only to traumatize farmers and freeze progress toward new and better technologies. While still defensive about the needs of agriculture, USDA officials now recognized, if reluctantly, the views professed by their longtime critics.

EPA officials, for their part, understood the need to balance their primary organizational mission with the interests of pesticides makers and users. Environmental protection was a dominant national goal, but it was not the only one, and the EPA's future mission constantly to weigh multiple and competing national objectives. Hence, argued William Ruckelshaus, "If we could control much more carefully ultimate use, we would be in a much better position to permit beneficial uses to continue without taking drastic action."[42] Caught between the desire to protect the environment and the need to promote the judicious use of pesticides, EPA officials from the outset recognized that FIFRA offered too few regulatory choices. Cancellation was too prolonged and ineffectual to remove a "proven" hazard; suspension seemed too abrupt, likely to harm farmers by forcing them into precipitous and uninformed searches for alternatives. Both were "back-door" options; what was needed was a premarket clearance process that would make the choice between suspension and cancellation unnecessary.

Finally, congressional bloc representatives, though even more adamant than the USDA in their defense of the pesticides paradigm, recognized that current law only hurt their constituents. The few regulatory options open to the EPA were ungainly cudgels, and the DDT experience showed the need for more delicate and expeditious instruments, else farmers find themselves whipsawed in the battle between regulators and the regulated. FIFRA's archaic standards and ponderous procedures no longer suited constituents' needs for smooth and authoritative registration.

Above all, those concerned about the pesticides problem, save for the most recalcitrant defenders of the status quo (such as cotton growers), recognized that existing law no longer balanced now-dominant social interests with traditional needs. A common perception of the need for change in no way meant agreement on answers, and the battle to come would show the implacability of complex national problems. In many respects, the pesticides problem would reveal an exasperating complexity of goals and means, perhaps more so than any other environmental issue. Unlike the air or water pollution battles preceding it, the effort to reformulate national pesticides policy had no easily caricatured villains, no commonly agreed-upon "sinners" against whom to rally the environmental believers. To install more stringent federal regulation over chemical pesticides was to seek a balance among equally powerful national values—abundant and cheap food, insect-free communities, public health, and a safe environment. It was a "choice among equals" that always had existed, but now all the choices were out in the open and more equally represented in the political arena.

The Path to Statutory Change

House Action: The Subgovernment Prevails

The path toward statutory change officially began with Richard Nixon's Message to Congress on the Environment of February 8, 1971 (see figure 2). Nixon's proposed major rewrite of federal pesticides policy surprised no one; congressional activity and administrative changes during 1970 had propelled that likelihood along. That year saw over twenty separate FIFRA reform proposals within the House, about a dozen in the Senate, with only the administration's assurance of forthcoming legislation apparently halting further action. Some type of movement during 1971 seemed certain.

Particulars about key provisions in both the administration and House versions of the Federal Environmental Pesticides Control Act (FEPCA) are provided in table 7. The EPA bill completely rewrote FIFRA, with environmental criteria now prominent. The proposed law called for a tripartite classification system: (1) "general use," comprising those products deemed by the EPA to be no demonstrable threat to man or the environment; (2) "restricted use," pertaining to more toxic materials that were to be applied only by certified applicators; and (3) "use by permit only," those products deemed the most persistent and environmentally hazardous. This last category was to include products subject to use by prescription, and only for demonstrated critical needs. The era of unlimited and little monitored use of pesticides would end.

Equally important, FEPCA proposed to radically alter procedures for registration and cancellation. Prevailing practice allowed the petitioner to decide what information and data would support an application, but the proposed law enabled the EPA to make that determination, including, if necessary, the product formula itself. The aim was to make registration more credible and demanding, a process that regulators hoped would screen out potentially hazardous products. Revisions in the cancellation process sought to give the EPA the discretion and flexibility essential to deal with complex products and problems in a more straightforward manner. No longer would it be required to convene *both* a science advisory committee and a public hearing during cancellation appeals; the former would be discretionary, the latter streamlined. The EPA administrator also could convene "quick" review on products slated for suspension if so requested by the manufac-

Figure 2. House Action on the Federal Environmental Pesticides Control Act (FEPCA)

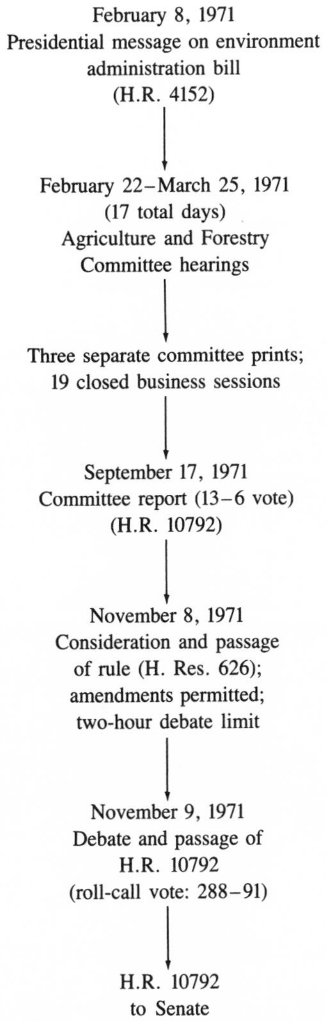

February 8, 1971
Presidential message on environment
administration bill
(H.R. 4152)

↓

February 22–March 25, 1971
(17 total days)
Agriculture and Forestry
Committee hearings

↓

Three separate committee prints;
19 closed business sessions

↓

September 17, 1971
Committee report (13–6 vote)
(H.R. 10792)

↓

November 8, 1971
Consideration and passage
of rule (H. Res. 626);
amendments permitted;
two-hour debate limit

↓

November 9, 1971
Debate and passage of
H.R. 10792
(roll-call vote: 288–91)

↓

H.R. 10792
to Senate

turer, and would enjoy the authority to suspend all product sales during the cancellation process itself. Registration information would be open to the public, save for those data classified by the petitioner *and* the agency as "trade secrets." Finally, and in line with emergent judicial philosophy, the administration bill allowed for "citizens' suits," granting standing during cancellation to "anyone adversely affected." This provision was added to FIFRA in 1964, but all that time pertained largely to farmers. By 1971, however, the same phrase was interpreted

Table 7. Two FEPCA Versions Compared

Provisions	*H.R. 1425 Administration Bill*	*H.R. 10729 House Committee Bill*
Classification Scheme	"General," "restricted," and "by permit only"	"General" and "restricted" use
Essentiality as Basis for registration	Yes	No
Data disclosure	Open, save for "trade secrets"	Highly restricted
Basis for cancellation	"Unreasonable" environmental effects	"Substantial" environmental effects
NAS advisory panel	At discretion of EPA	Mandatory
Public hearing format	At discretion of EPA	Mandatory format
Judicial standing to	"Anyone adversely affected"	"Affected parties" only
Maximum civil pensalties for private users	$10,000	$1,000
Indemnity Payments	No	Yes
Applicable to Pesticides exports	No	No, but desired
Federal preemption of stricter state standards	No	Yes, originally; changed on House floor

far differently by the courts, which increasingly expanded its meaning to include virtually any citizen and allowed environmentalists to influence policy.

All in all, FEPCA proposed to make registration more demanding, the cancellation process less cumbersome and byzantine. The dominant attitude within the EPA, and within the USDA for that matter, was to stiffen the criteria for registration to prevent recurrences of the frustrating DDT battle. Better a strong front door than a back door that made cancellation virtually impossible; an authoritative registration process would, it was hoped, make the need to cancel entirely unnecessary.

House consideration of H.R. 4152 began with hearings before the Committee on Agriculture in late February, and ran far into March. The contrast between this decisional context and its 1947 counterpart is compelling: the rather perfunctory subcommittee hearings of 1946 attracted relatively few participants, and predominantly those in agriculture and industry. In 1971, however, protracted committee sessions drew approximately seventy-seven separate witnesses representing a broad spectrum of agricultural, business, environmental, public health, scientific, and government interests. Second, media coverage enveloped the 1971 hearings, though perhaps not as thoroughly as other environmental debates, and illuminated the issue far more widely than was the case with the almost invisible hearings some twenty-five years earlier. Those supporting the widespread use of pesticides now found their views aired side by side with those of their critics, who more often than not enjoyed the more sympathetic coverage. The limelight of media attention on a traditionally secretive committee generated a degree of public interest that was uncomfortable for farm bloc representatives.

Finally, the issue under debate no longer was defined entirely by the agricultural community. If, in 1947, all involved had agreed that this inherently was an agricultural issue, no such unanimity now graced the committee room. One notes with particular interest the effort by Committee Chairman W. Robert Poage (D, Tex.) to set the agenda for debate: "So the question again arises: Should we accept some damage from the use of pesticides or should we let the pests take control and possibly cause even greater damage?"[43]

The committee's answer was clear. Most members, schooled as they were in the traditions of the pesticides paradigm, viewed any effort to tighten federal regulation with true concern, if not outright hostility. Stalwarts like George Goodling (R, Pa.) and Frank Stubblefield (D, Ky.) argued against *any* change, claiming instead that well-intentioned but misguided urbanites (and some not so well-intended ones) were robbing farmers of their very livelihood. Most members conceded the need for *some* changes in existing policy, but the wall of distrust built up over the years between environmentalists and the patrons of agriculture could not have been more imposing. One side feared chronic toxicity, carcinogenic effects, and food chain degradation; the other spoke entirely about pest eradication, or worried aloud about the future of the farm economy. Environmentalists tried to impress upon committee members that chronic toxicity posed a critical threat to all, but these less visible, less direct effects of the uncontrolled use of pesticides were scientific matters that failed to jibe with representatives' shared experiences. That scientists of equally impressive credentials disagreed vehemently about the side effects of pesticides probably did not help matters much, and very likely forced members to base their decisions more on initial biases than on the weight of scientific evidence. For them, as for Clarence Cannon some years earlier, out of sight essentially meant out of mind.

Some environmentalists no doubt complicated matters by coming before the committee with an air of moral superiority about the issue, a stridency probably cultivated during earlier air and water pollution battles. Representative John Kyl (R, Iowa), one of the less vitriolic defenders of agriculture, in fact complained openly about the tendency of some environmentalists to picture "the other side as amoral."[44] Environmentalists may in fact have occupied some moral high ground on the more compelling questions in this debate, but they would find their opponents unbowed. To deal with air or water pollution was one thing—since those battles pitted an aroused public against rather shaky opposition—but to deal with pesticides was to run into a highly complex, less public issue and against a well-organized, powerfully entrenched configuration of policy specialists. Veteran activists trod lightly before the committee, aware that environmental defenders alienated agriculture at their peril. Others, like Joel Pickelner of the National Wildlife Federation, were less careful, and less lucky.

Pickelner's primary target was a provision in the committee bill to indemnify farmers, vendors, and manufacturers for any products in their possession "suddenly suspended or canceled by the EPA. Indemnification, committee members explained, would protect innocent parties who produced or purchased products previously registered under FIFRA, only to have the EPA later revoke that registration. Those possessing banned products should not be left holding the bag, members argued, simply because the government changed its mind. A precedent set after the cranberry scare of 1959 would now be embodied in law, though it would be EPA, not the USDA, that would now pick up the tab. Pickelner, not content to argue against indemnification on its merits, attacked the provision as a blatant subsidy, arguing that the impact of pesticides on the environment "would be more disastrous than one man going bankrupt or one man being hurt financially."[45] Indemnification made dubious fiscal sense, and no other regulatory sphere allowed for such payments, but Pickelner's offhand dismissal of the economic impact on farmers raised the hackles of committee members. Worse—for environmentalists at least—Pickelner was only a law student, and his apparent lack of scientific credentials only added fuel to the fire. Committee members quickly let it be known that they wished to hear from the federation's real experts, not this amateur. Representative B. F. Sisk (D, Calif.) aptly summed up committee attitudes when he admonished Pickelner for his "brash statement, which, very frankly, is just another wild statement, like a lot of other things we have been hearing."[46] Pickelner, sufficiently chastened, promised a more even-handed showing in the future.

The committee mauled the administration bill, though not without sharp intramural debate. The once unanimous agricultural consensus on pesticides appeared to fracture into three viewpoints, with a majority caught between the most recalcitrant defenders of the status quo on one side and ardent reformers like John Dow (D, N.Y.) on the other. More than five months and three separate committee prints elapsed before that work group reported out its "clean" bill (H.R. 10729) in September, and only then by a 13–6 vote (with a bare majority of the committee even voting). So public was the internal strife over this half-agricultural, half-environmental bill that the committee took the highly unusual action of including three separate dissenting views in its final report.[47] Particulars of the committee bill are shown in table 7, above, with the major points of conflict discussed in brief.

Classification. The committee eliminated the "use by permit only" category, leaving the "restricted use" category to cover a broader range of products. Chemical makers and agricultural organizations both complained, and committee members agreed, that a "by permit only" proposal would "place on the Administrator [of EPA] and users an unworkable and costly burden far in excess of needs or benefits gained by such regulatory machinery."[48] Environmentalists argued that certain pesticides, like some drugs, required use by prescription, but the analogy failed to sway those more worried that administrative logjams might hinder prompt and timely application of needed chemicals. "Restricted use" proposals could be applied only by or under the direct supervision of "certified applicators," but the committee defined this so broadly that environmentalists despaired of any control in the fields.

Essentiality. EPA and its allies argued strenuously that a primary criterion for pesticide registration should be whether the product was essential, whether it fit a specific market niche. Committee members rejected this concept, which had been formulated earlier by the Mrak Commission and already informal EPA policy, noting that government nowhere else barred a product from the market simply because "it wasn't needed." And, they added, precedent for such government authority was not going to start with pesticides. The patrons of agriculture above all feared that strict application of an "essential use" doctrine eventually would winnow the number of available alternatives to a pitiful few, thus robbing farmers of their freedom to choose the product best suited for personal needs and pocketbooks. To the EPA, however, such a doctrine might eliminate market duplication, reducing the glut of available products to those proven most useful, cost-effective, and environmentally benign. Freedom of choice was more precious to most committee members, and they quickly eliminated the provision.

Science Advisory Panels. Fear that ungrounded "emotionalism" might influence regulatory decisions prompted committee members to reinstitute mandatory review of EPA cancellation orders by the National Academy of Sciences. EPA officials and environmentalists, no doubt recalling the academy's track record on pesticides, sought instead to make such referral discretionary. Mandatory referral would only delay action (the DDT case being fresh in mind), and EPA officials stressed the need for speedier appeals procedures. Committee members viewed such logic coldly, for they were more concerned about protecting manufacturers (and, by extension, farmers) from bureaucratic capriciousness than about speedy cancellation, and due process for appellants was a more prominent value than was regulatory power. The academy, much to the environmentalists' dismay, was still in the pesticides advisory business.

Data Submission. The Nixon adminstration dramatically broadened FIFRA's information and test data submission requirements, arguing that complete documentation would produce more authoritative registration. Chemical companies, however, complained that the requirements would make registration more expensive and time-consuming, thus putting a "chilling effect" on research and development. Committee members agreed, narrowing EPA discretion on data submission to such a point that, the critics charged, the new law might well be weaker than the old.

Data Disclosure. The logic applied by corporate representatives against broader data submission grew out of their fear of unlimited public access to registration materials. The administration provided that all save "trade secrets"—

as determined by the company *and* the EPA—would be made public at least thirty days prior to the effective date of registration. This provision sparked one of the sharpest disputes of the entire debate, since chemical makers considered *all* information submitted on behalf of registration to be "trade secrets," and claimed that public access simply would allow their competitors to copy data that had cost millions of dollars to generate. Such access might further erode corporate incentives to research and development. Worse yet, from the corporate perspective, public access simply would offer more ammunition to their critics. Dow Chemical Company official Robert Naegele noted, with some bitterness, that accusations come cheap: "You can accuse with no basis of fact and make any statements you want to, but to generate the data to back that up, again and again, or to counter other irrelevant assumptions or conclusions costs a fortune."[49] EPA officials regarded public access with somewhat greater equanimity, since it might give the agency additional perspectives critical to reasoned regulatory decisions. The agency in January 1972 in fact released data on the toxicity and efficacy of *all* registered pesticides products. This action, traditionally avoided by the USDA, infuriated committee members. For environmentalists, of course, access to such information was a valuable resource in their battle against the gratuitous use of pesticides. But committee members, worried far more about the "chilling effect" factor than about any "public right to know," tightened future access to almost minimal proportions.

Judicial Standing. The administration, in line with the 1964 amendment, granted standing in court during cancellation proceedings to "anyone adversely affected," which, by 1971, meant any interested citizen. It was only through such definition of standing that environmentalists gained access to the courts in the first place, and they fought hard to retain that standard. They were supported by the EPA, which regarded environmental litigants as useful, if occasionally irksome, allies. Farm groups, chemical companies, and the farm bloc in Congress viscerally opposed such access—opposing access in general as well as access to pesticides appeals—and the committee narrowed the applicable concept of standing to "affected parties." Only chemical companies appealing regulatory decisions would have standing in the courts; EPA would represent the public. Environmentalists, wary about relying on *any* bureaucrat to defend the public interest, attacked the change as a giant step backward.

State Authority. The defenders of pesticides acceded to regulatory change primarily because of their desire to reinstitute stable market and regulatory conditions after several years of upheaval. Chemical makers above all wanted uniform national pesticide standards, not a hodgepodge of state laws. The Nixon administration, however, provided for state preemption of federal standards on "general use" pesticides if the state standards were more stringent, a provision that sent industry lobbyists howling to the Agriculture Committee. Committee members, in an action that must have seemed paradoxical to their innate defense of state prerogatives, immediately imposed federal preemption on general use standards. The committee in fact forbade EPA delegation of *any* regulatory authority to the states, arguing that uniform and stable pesticides regulation required federal preeminence. "States' rights," it appears, did not extend to pesticides.

Dissent from within the committee came chiefly from John Dow of New York, who complained that the majority both undermined the administration's bill and, when all was said and done, substantially weakened existing law. Dow's most pointed critique was that H.R. 10729, when considered as a whole, in fact shifted the burden of proof from manufacturer to regulator. Nowhere in the committee bill did the manufacturer have to prove that a product submitted for registration posed no substantial environmental hazard, whereas the EPA had to prove that it did do so during the cancellation process. Restrictions on data submission effectively precluded adequate EPA review of a product's environmental effects if the applicant failed to include test data, while the appeals process essentially forced the agency to prove time and again that the product posed a hazard. Dow promised to carry his fight to the House floor, since, he concluded, the committee bill was "essentially weaker than the present law and not nearly as protective of our environment as such a bill ought to be in this enlightened year of 1971."[50]

EPA's own timidity on the legislative front at this time is instructive. William Ruckelshaus did move quickly in deploying his cadre of young lawyers against a wide array of polluters, but the agency as yet was too new and fragile to carry on the legislative battle without strong presidential backing. Nixon had offered hortatory support at the start, but evidence suggests that he wanted to avoid alienating powerful farm and industry constituencies. The EPA had its friends on Capitol Hill, but possessed little independent clout to counter a committee apparently hostile to the agency's mission. The agency was caught between the advocacy of its environmental allies and the realities of power, so it lay low and, much to environmentalists' dismay, accepted a committee bill that it really did not want.

The battle resumed on the House floor, but once there became a doomed cause for Dow and like-minded colleagues. Despite protests by liberal Democrats that the committee bill required extensive revision, a consistent coalition of Republicans and southern Democrats easily turned back all challenges made during the two days of often acrimonious debate. The conservative coalition coalesced around the pesticides issue more than any other "environmental" vote during the early 1970s. Their stance may be attributed to common perceptions about agriculture, commerce, and government regulation that conflicted sharply with those expressed by environmentalists and their allies. So crucial was this legislation to so many in the conservative coalition that its cohesion recalls the civil rights and social welfare battles of the 1960s. Dow's only success, in fact, was a concession by the agricultural bloc that allowed states to implement stricter standards on general use products. This was a mere sop compared to such major issues as essentiality, judicial standing, and data disclosure. H.R. 10729, amended but once, passed by an easy 288–91 vote.

Environmentalists, predictably, panned the bill, while agricultural groups and chemical interests called it responsible and reasonable public policy. The Sierra Club called H.R. 10729 "not a reform but a retreat from pesticides regulation" and "possibly the worst bill to be introduced this session."[51] The national Wildlife Federation's Joel Pickelner called the bill "a horrible goddam piece of legislation,"[52] while Rep. John Dow noted only that "manufacturers effectively carried the day."[53] The Agriculture Committee's majority defended the bill, arguing that H.R. 10729 was not a farmer's bill, not a manufacturer's bill, nor an environmentalists's bill, but "a mixture of each, a composite of all."[54] That logic, said Rep. Michael Harrington (D, Mass.), was an indictment, not a virtue: "The committee's bill, in attempting to be a composite . . . becomes, in effect, nobody's bill. The primary value that must be given priority is the health of all the American people, both now and for the future."[55] Rep. John Kyl (R, Iowa) said, however, that the bill was more than a composite, that it sought to give science the time and tools necessary "to find the truth" and thereby to allow government to avoid policymaking by emotionalism.[56] Conflicting views about "good" public policy remained poles apart, with the Agriculture Committee's still dominant.

Subgovernments change, but, in the words of Ripley and Franklin, they "are remarkably persistent and tenacious."[57] This phase of the battle of FIFRA, if nothing else shows just how tenacious a beleaguered subgovernment can be, how those entrenched in positions of policy leverage can overcome attacks by those claiming to represent majority opinion. The environmental wave in several ways crested in 1971, carrying along with it notable policy changes, but it broke apart upon the pesticides subgovernment shoals. The agriculture-agrichemical nexus embodied both in committee and in the conservative coalition succeeded in defending traditional prerogatives in the name of the public good against competing and apparently hostile values. The defenders of pesticides allowed for some change, but asserted their power on the most critical elements of overall policy direction and scope. That victory, however, came at a cost, since subgovernment actors more than ever before were forced to deploy all available weapons. Subgovernment power, like power in any form, is most effective in its threat, or reputation, than in its actual use, and the fact that the defenders of pesticides had to work hard for its House victory indicates that they clearly were on the defensive. Common values and shared perceptions about pesticides once guaranteed easy subgovernment dominance, but fractionation in values and perceptions ultimately led to an erosion of cohesion just as new actors sought access from the outside. Those central to subgovernment dominance must have realized at the battle's end that returning to the consensual politics of the past was impossible, and that each succeeding policy debate would be a real fight. The next battle, in fact, lay just on the other side of Capitol Hill.

Conflict and Accommodation in the Senate

Richard Fenno, in his study of congressional committees, makes the following distinctions between the chambers of our bicameral legislature:

> Since the House and the Senate are different institutions, it should not be surprising that their committees are also different. Senate committees are less important as a source of chamber influence, less preoccupied with success on the chamber floor, less autonomous within the chamber, less personally expert, less strongly led, and more individualistic in decision making than are House committees.[58]

The reader is advised to keep Fenno's comments well in mind as we move on to Senate deliberation of H.R. 10729. Until now, this study has devoted little time to the upper chamber, and only then in reference to "outsiders": Warren Magnuson expanded Department of the Interior pesticides research efforts in the late 1950s; Abraham Ribicoff convened his multiyear investigation into federal regulatory policies during the early 1960s. The history of federal pesticides policy before the 1970s is a history of House preeminence, if not outright control. Nowhere in the Senate do we find the kind of subgovernment maintenance and defense so commonly encountered in the House, which has the stronger, more expert committees and leadership in appropriations matters. The House, one may suggest without fear of hyperbole, has historically been the leader in pesticides (and many other agricultural) developments; the Senate has usually followed.

Structural differences of course account for much; but the reason also lies in the different conceptions held by the two chambers about the legislative role. Charles O. Jones argues, "The Senate is more oriented toward the institutional functions of debating issues and discovering the basis for compromise *across* issues," while the House debates far less and tends to emphasize compromise *within* issues.[59] House action on FEPCA showed internal subgovernment compromise, but compromise predicated on a shared definition of the issue as an agricultural matter. Those arguing other perspectives or pushing other issues, were left on the outside. The Senate, by contrast, was expected by those losing in the House to weigh environmental values more explicitly against agricultural ones. And, if the Senate Committee on Agriculture and Forestry failed to do so, a strong cadre of environmentally oriented senators were ready to slug it out on the floor.

The Committee on Agriculture and Forestry, like other committees in the Senate, historically relied on its House counterpart for the detailed information, analysis, and, often, policy leadership. In fact, both the authorizing and appropriations bodies in the Senate responsible for pesticides had tended to accede to House views and simply passed them on to the Senate floor. Few believed this time that conditions would differ markedly, and committee inaction on the administration proposal during 1971 bore out expectations. James Allen (D, Ala.)

Figure 3. Senate Action on FEPCA

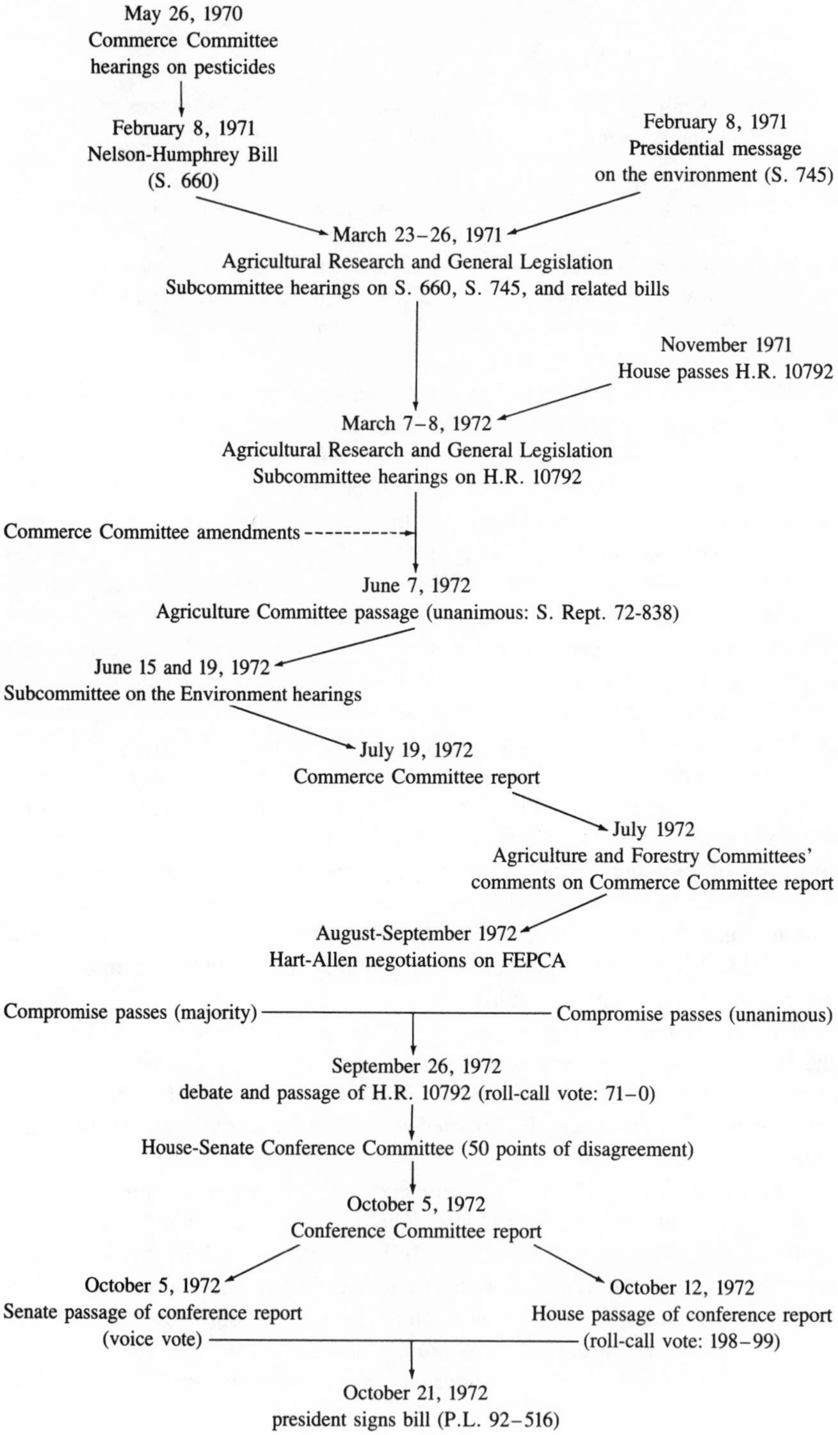

May 26, 1970
Commerce Committee
hearings on pesticides

February 8, 1971
Nelson-Humphrey Bill
(S. 660)

February 8, 1971
Presidential message
on the environment (S. 745)

March 23–26, 1971
Agricultural Research and General Legislation
Subcommittee hearings on S. 660, S. 745, and related bills

November 1971
House passes H.R. 10792

March 7–8, 1972
Agricultural Research and General Legislation
Subcommittee hearings on H.R. 10792

Commerce Committee amendments

June 7, 1972
Agriculture Committee passage (unanimous: S. Rept. 72-838)

June 15 and 19, 1972
Subcommittee on the Environment hearings

July 19, 1972
Commerce Committee report

July 1972
Agriculture and Forestry Committees' comments on Commerce Committee report

August-September 1972
Hart-Allen negotiations on FEPCA

Compromise passes (majority) — Compromise passes (unanimous)

September 26, 1972
debate and passage of H.R. 10792 (roll-call vote: 71–0)

House-Senate Conference Committee (50 points of disagreement)

October 5, 1972
Conference Committee report

October 5, 1972
Senate passage of conference report
(voice vote)

October 12, 1972
House passage of conference report
(roll-call vote: 198–99)

October 21, 1972
president signs bill (P.L. 92–516)

convened the Subcommittee on Agricultural Research and General Legislation for hearings on S. 745 during March 1972 (see figure 3), but, after a few days of rather desultory testimony, the subcommittee awaited final House action.

It is not necessary to describe in detail the subcommittee hearings on H.R. 10729, held in March 1972, since most of the discussion centered on the House bill and proposed amendments to the measure. Senator Allen may not even have bothered with the second set of hearings, which would have been consistent with committee behavior, were it not for the spate of reservations to the bill expressed by the EPA, environmentalists, and some chemical manufacturers. Allen, it appears, wanted to give all comers their day in court, revealing a view of the committee as an appellate body that runs consistent with senators' perceptions about their roles with respect to the other chamber. After all, and recalling George Washington's famous analogy, the Senate exists to "cool" the passions of its brethren in the larger body.

Several hundred amendments were proposed during those two days, with the most important changes offered by Senators Philip A. Hart (D, Mich.) and Gaylord Nelson (D, Wis.). As table 8 indicates, the Hart-Nelson amendments challenged the very core of the House bill since, as Hart testified before his Senate colleagues, "to report this bill as it has emerged from the House . . . would result in weakening existing law."[60] The amendments, in general, sought stricter registration criteria, more inclusive data submission, greater public access to data, revocation of the indemnity provision, and unlimited citizen access to the courts during EPA cancellation proceedings.

The EPA also returned to Capitol Hill to request changes, despite an admission by William Ruckelshaus that he would accept the House version if the only other choice were no bill at all. EPA Assistant Administrator David Dominick called the indemnity provision a bad precedent, arguing that to force the agency to reimburse holders of banned products would effectively chill EPA regulatory action. Dominick also opposed House provisions forbidding the agency to use data submitted by one company to judge applications for similar products submitted by others—a restriction that he warned would only foster oligopoly among those companies large enough to afford research and development.[61] The list of EPA suggestions overlapped those offered by Hart and Nelson in all but two cases. The agency, in a subtle narrowing of its earlier stance, opposed mandatory public disclosure of all test data immediately upon its receipt from the company; instead it wanted disclosure only *after* registration decisions were made. Hart and Nelson wanted predecision access to allow environmental groups to study health and safety data, while the House bill granted virtually no such access. The EPA also opposed citizens' suits against the agency, insisting that such suits be limited to actions against producers and users. For the most part, however, the agency sat resolutely on the fence.

Other interests and other issues were introduced formally for the first time during the subcommittee session. Senator Adlai E. Stevenson III (D, Ill.), noting that each year as many as 800 workers died and more than 80,000 were injured

Table 8. Competing Senate Versions of FEPCA

Provisions	*Agriculture Committee*	*Commerce Committee*
Classification Scheme	"General" and "restricted"	"General" and "restricted"
Essentiality as basis for registration?	No	Yes
Data disclosure	Highly restricted	Open, save for "trade secrets"
Basis for cancellation	"Substantial" adverse environmental effects	"Unreasonable" effects
NAS advisory panel	Required	At the EPA's discretion
Public hearing format	Mandatory format	At the EPA's discretion
Judicial standing to	"Affected parties"	"Anyone adversely affected"
Maximum civil penalties for private users	$1,000	$10,000
Indemnity payments	No	No
Applicable to pesticides exports	No	No, but stricter standards desired
Federal preemption of stricter state standards	No	No

by the production, use or misuse of agricultural chemicals, proposed to protect farm workers and chemical plant employees. Labor unions, because of these trends, for the first time took a keen interest in pesticides and their impacts on the workplace.[62] And, in a departure from the traditional agriculture community consensus, both the National Farmers Union (NFU) and the National Farmers Organization (NFO) endorsed *all* of the Hart-Nelson proposals. NFU official Weldon Burton argued that farmers were caught between chemical manufacturers and federal agencies on pesticides matters, and that they had a strong stake in establishing policy that was both environmentally sound and fair to farmers. "We're in favor of opening this whole thing up a little bit," said Charles Frazier of the NFO. "The test data provision is OK with us; we don't see anything wrong with the amendments on citizens suits and exports."[63] The American Farm Bureau Federation, the most conservative of the farm organizations, maintained its allegiance to the traditions of the pesticides subgovernment.

Senate Agriculture Committee members approved a number of changes in H.R. 10729 and reported out the bill in June. Key changes include deletion of indemnity payments, expansion of the definition of "misbranding" to include products having substantial environmental effects even if used "normally," and a "compromise" between the House bill and environmentalist demands on the issue of judicial review. In the latter case, the committee agreed to allow anyone adversely affected to obtain standing in court once the EPA began cancellation proceedings, but rejected the plea to allow citizens the right to sue both manufacturers and the EPA over other aspects of regulatory policy. The bulk of the Hart-Nelson amendments were rejected, as was the Stevenson proposal on worker safety, and the committee passed its version unanimously.

Then, in an unprecedented action that had a critical effect on pesticides policy, Agriculture and Forestry Chairman Herman Talmadge (D, Ga.) "re-referred" H.R. 10729 to Magnuson's Committee on Commerce. Magnuson, in turn, sent the bill to Philip Hart's Subcommittee on the Environment. Hart earlier had expressed disappointment about the Agriculture Committee's failure to accept most of his proposals, and also claimed partial jurisdiction over pesticides because of his subcommittee's environmental focus, but these two reasons fail to explain why Talmadge "gave" policy dominance to the "outsiders." Talmadge's action no doubt dismayed, if not infuriated, those in the agricultural community, so his reasons must have been compelling. Various public statements by the participants indicated a desire to move consideration along in order to obtain final Senate passage before the party convention recess, while administration officials expressed their desire for a law—any law—on pesticides before the election recess. Senators Hart and Nelson publicly had promised a tough floor fight if they did not get a second chance at the bill, a thought not at all welcome to agricultural and industry force fearing an election-year debacle on an important environmental bill. A floor fight would only attract unwanted attention, so both the administration and agricultural interests were eager to be accommodating, up to a point.

But the deal might be explained in far more prosaic and telling terms. Talmadge simply needed a favor from his Commerce Committee counterpart. According to Victor Cohn of the *Washington Post,* Talmadge agreed to grant Hart a second crack at H.R. 10729 if Magnuson agreed to delete from the Consumer Safety Act of 1972 a provision to transfer USDA meat, poultry, and egg inspection to a proposed consumer safety agency.[64] Agricultural interests naturally opposed a strong food inspection provision, particularly one implemented outside the USDA. Magnuson, for his part, was willing to go along because he was pessimistic that the consumer bill would pass the House with the provision included. Agriculture was caught this time between its traditional prerogatives on pesticides on the one side and by an assault on equally fundamental interests on the other, so Talmadge decided to swap. Environmentalists now had their first real crack at FEPCA—a clear opening made possible not by the compelling nature of their arguments but by the farm bloc's sudden need for temporary allies. Magnuson gave up the expendable, Talmadge the critical.

Hart held hearings in mid-June, and on June 23 the Commerce Committee reported out H.R. 10729 with fifteen amendments that embodied the entire Hart-Nelson slate plus Stevenson's farmworker proposal.[65] Agriculture Committee members objected, and floor consideration at this moment seemed remote. However, Talmadge and Magnuson agreed to one more deal: Philip Hart and James Allen (D, Ala.), aided by their staffs, would hammer out an intercommittee compromise, sort of a "mini-conference committee" whose report would require joint committee approval before going to the floor. This sort of formal intercommittee negotiating is rare in the Senate—and almost nonexistent in the House, where committee jurisdictions are defended jealously—but its occurrence marked a special turning point. For the first time "outsiders" were brought directly into policy decisions once dominated by agricultural and chemical company interests. Subgovernment *expansion* was temporary, but it occurred nonetheless, with telling impacts on the direction of policy.

Negotiations between Hart and Allen dragged on over two months, while House Agriculture Committee members watched from afar, many obviously angered by this intrusion into the inner sanctum. Leslie Sharp, general counsel to the House committee, argued that every Hart-Nelson amendment had been considered–and rejected—in 1971; The House, he protested, "is not going to go hook, line and sinker for something written by the Commerce Committee in two executive sessions after we spend months on the damn thing."[66] A tough conference battle lay ahead if the Commerce amendments were to make it through the Congress.

The final version of the compromise bill lay complete by mid-September, and was approved by both committees. The key compromises agreed to, shown in table 9.

Basis for cancellation. Commerce Committee participants substituted "unreasonable adverse effects on the environment" for the Agriculture Committee's "substantial adverse effects," a change in nuance that forbade adverse effects unless compelling benefits from a product's use dictated otherwise. House and Senate Agriculture versions allowed for stricter cost-versus-benefit considerations, where the latter clearly outweighed the former, but the negotiated version made costs less critical to registration decisions.

Data submission. Hart advocated more inclusive data submission, narrowed the definition of "trade secrets," and allowed the EPA to apply data received from one registration to other registrations involving similar or identical chemical bases. Agriculture Committee members, supported by the chemical industry, wanted narrower submission requirements and no cross-referencing. The final compromise incorporated the broader data submission requirements advocated by Commerce but allowed cross-referencing only when the "copycats" paid a "royalty" to the data developer.

Data disclosure. Commerce Committee views required the EPA to make all health and safety data supporting a registration accessible to the public upon receipt. EPA itself opposed predecision public access, arguing that it might complicate registration and thus should be allowed only after registration. Agriculture Committee members, worried about chilling effects on research, wanted no such access. The compromise allowed for public access to test data at least thirty days prior to the effective date of registration, after the agency's actual decision but before it went into effect.

Table 9. Senate Compromise and the Final Version of FEPCA

Provisions	*Senate Compromise*	*FEPCA OF 1972*
Classification scheme	"General" and "restricted" uses	"General" and "restricted" uses
Essentiality as basis for registration?	Yes	No
Data disclosure	More open; public access 30 days *before* decision; narrower trade secrets protection	Narrower; public access 30 days *after* decision; broader trade secrets protection
Basis for cancellation	"Unreasonable" effects	"Unreasonable" effects
NAS advisory panel	Required	Required
Public hearing format	Mandatory format	Mandatory format
Judicial standing to	Citizens, but who may only sue the EPA (not chemical users or makers)	"Affected parties" only; no citizens' suits
Maximum civil penalties for private users	$1,000	$1,000
Indemnity payments	No	Yes
Applicable to pesticides exports	No, with some clarifications	No
Federal preemption of stricter state standards	No	No

Science advisory panels. The Commerce Committee retained advisory panels for questions of scientific fact only, as distinct from the House's mandatory referral to panels during cancellation proceedings, and insisted that panel membership be strictly controlled so as to prevent conflicts of interest by scientists with public links to the chemical industry. The Agriculture Committee accepted the amendment, but did not think that it would prove effective.

Judicial standing. The committees agreed to allow citizens to sue the EPA for nonperformance, but disallowed suits against chemical makers, sellers, or users. The Commerce Committee advocated unlimited rights by citizens to sue, while Agriculture members wanted absolutely no right—fearing that "professional litigants" (i.e., environmental lawyers) would "interfere with the orderly administration of the law"—and the EPA wanted no suits against itself.[67] Agricultural interests later allowed for citizens' suits against the EPA alone, particularly after farm groups pointed out that farmers were citizens too—citizens who might need to sue the agency if it started obstructing the flow of pesticides to the fields. Commerce Committee members accepted this change, though not without reservations, while EPA officials were not at all pleased about being singled out as the object of suits.

The Committee on Agriculture and Forestry passed the compromise unanimously, Commerce by only a majority. Agriculture Committee members acceded, partly to move the deliberations along, partly because they knew that a House-Senate conference committee would likely restore some of the original provisions. Commerce Committee members were less enthusiastic about those possibilities, but approved the compromise knowing that key committee members would help represent the Senate in conference. The intercommittee compromise went to the floor on September 26, where it was adopted as a substitute for the House bill, and passed by a 71–0 roll-call vote.

Executive Retreat and Final Action

One cannot fully appreciate the activities packed into the month between Senate passage of H.R. 10729 and Richard Nixon's final approval of FEPCA without accepting a single powerful reality: 1972 was a presidential election year. This helps to explain why the Nixon administration, which between 1969 and 1972 played environmental one-up-manship with Senate Democrats, literally caved in on FEPCA as the bill went into conference. George McGovern was well on his way into the history books, Nixon was in no mood to alienate agricultural and corporate support, and the EPA was under orders to take what it could get and not to fight too hard about it. There is no other way to explain the wide gap between the agency's public support of the Hart-Nelson amendments and its clear retreat when push came to shove. Linda Billings of the Sierra Club, which was at the forefront of environmental pressure, commented angrily, "EPA completely sold us out."[68] Senate Commerce Committee staff counsel Leonard Bickwit's assessment is worth noting: "There were threats by the administration to come out in favor of the Agriculture Committee bill if we didn't give in. We were constantly threatened. I was yelled at a lot. . . . The industry set the rules; the administration agreed to play by them. They were so eager for the ends [passage of the bill] that they were willing to accept means that we found disgusting."[69]

House and Senate conferees met in late September and early October, filing the conference report on October 5.[70] Conferees disagreed on more than fifty separate items, and, according to Sen. James Allen, House conferees gave in on thirty-two occasions. Senate conferees gave in on seven, and the two sides compromised on the remainder.[71] One should note, however, that the final version of the bill reflected House views more than Allen's figures might indicate. The major accommodations reached in conference are shown in table 9, above.

Registration criteria. Conferees approved the Senate provision that allowed the EPA to reject a registration if the product posed "unreasonable adverse effects on the environment," as opposed to House insistence on the term "substantial." This dispute seems arcane, but in fact proved quite fractious, and the conference report's explanation that the change was merely a clarification, and not a change in substance, somehow does not jibe with the intense maneuvering that surrounded this issue.

Indemnification. The sharpest dispute among the competing interests centered not on procedure but on economics. Chemical companies lobbied hard within the administration and the congressional farm bloc (which, overall, didn't need much persuasion) to require that EPA indemnify all pesticides makers, sellers, and users possessing products affected by agency regulatory actions. The EPA naturally opposed this provision, fearing that it would harm the agency's ability and willingness to regulate pesticides, but, as Sierra Club's Billings points out, the agency "gave up on the indemnity provision in order to get a bill."[72] The Senate Commerce Committee went so far as to negotiate directly with the National Agricultural Chemicals Association on this and other key matters, apparently preferring to work directly with industry rather than with the Agriculture Committee. Committee staff counsel Bickwit noted that many Commerce members were keen on a floor fight, but that the industry, agriculture, and the administration were just as eager to avoid one.[73] Senate conferees, for the most part just eager to get a bill out before the end of the 92nd Congress, gave in on indemnification, much to the anger of environmentalists and their congressional supporters.

Data. House conferees prevailed on most matters pertaining to industry data. The conference report provided for a far narrower provision on data submission, one where registrants could designate some data as trade secrets. The EPA could reveal chemical formulas under extreme situations, particularly product suspensions, but such decisions would likely hit the courts. Senate conferees also gave in on the question of public access to health and safety data, which the Senate bill had required at least thirty days prior to the effective date of registration. The EPA opposed prior public access, arguing that it might muddy the decisional waters, and conferees agreed to require public access to such data thirty days *after* registration went into effect. This decision was essentially a compromise between Senate's prior disclosure and House's no disclosure, but the final outcome obviously favored the industry. Environmentalists could peek at industry data, but they would not be able to stop registration before it went into effect. They could only petition the EPA to begin cancellation proceedings shortly after registration occurred—a decidedly problematic undertaking.

Judicial standing. The Hart-Nelson amendment allowed for unrestricted citizen suits, while the House version forbade them across the board. The final Senate version allowed citizens to sue the EPA alone, arguing that the agency represented the public against manufacturers and users. The House once again prevailed; the interested citizen, who had access to the courts under other environmental statutes, found no such access under FEPCA.

Senate floor debate on the conference report centered primarily on the indemnity dispute. Senator Gaylord Nelson called indemnification "a very serious precedent," arguing heatedly, "I do not understand why the public should indemnify a manufacturer who puts a dangerous product onto the marketplace. When the public discovers it is dangerous, why should the public then have to pay for the losses?"[74] Robert Dole (R, Kan.) also expressed some dissatisfaction with the provision (though he supported financial relief for farmers), but, he added, "The question essentially becomes whether we would have a pesticide bill or break up in disagreement over one issue."[75] Philip Hart echoed Nelson's views, but agreed with Dole. FEPCA, Hart argued to his colleagues, was "a mixed bag" that contained "quite a few pluses as compared to existing law and one rather substantial minus."[76] Rather conspicuous in their silence during the debate were most farm bloc senators, since they got pretty much what they wanted. The Senate passed FEPCA by voice vote. The House, following even less debate, passed the conference report by a 198–99 margin, with the conservative coalition once again carrying the day. President Nixon signed the measure into law on October 21, just before the election.

Conclusion: The Path to Policy Change

Despite indemnification and the other controversial provisions in the law, FEPCA marked a new direction for federal pesticides policy. Registration requirements were tightened, and the EPA could determine much of the data to be submitted with each application. No longer could manufacturers produce, or users use, a product for any purpose not stated explicitly on the label, and the EPA had more streamlined procedures with which to cancel or suspend a product that posed unreasonable environmental effects. Those appealing a regulatory decision could no longer select appellate bodies at their convenience; both the public hearing and the advisory panel now would convene simultaneously, with the latter dealing solely with those scientific questions referred to it by the hearing examiner. "Anyone adversely affected" could obtain standing in court on cancellation or suspension matters, though citizens were not given the right to sue the EPA, manufacturers, or users in the absence of agency action.

Was FEPCA "good" public policy? Somehow the tired old saw about beauty being in the eye of the beholder comes to mind. However, as John Blodgett observes: "The Act which emerged from two years of concentrated legislative activity, including hearings, debates, votes, and conferences, was a collection of compromising among agricultural, health, and environmental proponents. Depending on his point of view, a person can hear contentions that any one side predominated."[77]

The point of view I take here is not that of the farmer, nor of the environmentalist, but of the political scientist. And, from this point of view, passage of FEPCA marked a critical change in the relationship among competing interests and between each interest and the government. FIFRA of 1947 was a compromise, but one hammered out among like-minded participants in an atmosphere of mutual trust, accommodation, and overlapping self-interest. Passage of FEPCA, by contrast, resembles a textbook description of the American legislative process, with its lively committee hearings, backroom maneuvers, debates (lofty and otherwise), and requisite bits of drama. However, this took a long time, not only in years, but also in the evolution and change in issues, actors, and structures. The pesticides subgovernment that so long had dominated the direction of policy was forced to compromise with those who did not share full faith in the pesticides paradigm. Congressional committees on agriculture (and their appropriations colleagues) still held formal jurisdiction over pesticides matters, but they no longer could operate unchallenged. Environmental and health views that once had been almost automatically screened out in committee were now integral to policy deliberations, whether the committee liked it or not. Whether future policy directions would reflect these insurgent views remained to be seen. That the committee on agriculture now dealt with the EPA, and not the USDA, was no small element in that calculation.

FEPCA now was law, and if environmentalists did not get all they desired, they at least got a law that by many standards was better than before.[78] But, as any student of the policy process quickly appreciates, policy formation is only the beginning; the politics of implementation, or of policy modification, had yet to take place.

Chapter Eight

The Policy Pendulum: Pesticides Politics into the Eighties

Hard Choices: Environmentalism and the Issue-Attention Cycle

American history normally marches through a succession of quadriennial presidential terms. But the rhythm of our history also can be divided into what we can call "issue eras," distinct periods in the national parade dominated by a single issue, or a cluster of related issues. We thus speak of the Depression, the civil rights era, or the "antiwar years," confident that such shorthand bespeaks a definite period as well as a dominant coloration in the national mood. To mark transitions between such eras clearly is more difficult than to separate presidential terms; the passage of personalities seems more certain than the passage of issues. Some issues accompany new leaders, while others erupt inescapably onto the national agenda through crisis or cataclysm—those are the easy ones to spot. But many, like the environmental era, often creep imperceptibly into the national psyche, building over the years until they reach a critical threshold, to spill over into our social and political consciousness.

To worry about "issue eras" is to accept an assumption about conflict and mass democratic politics. It is to assume, as Schattschneider clearly does, that "conflicts compete with each other," that the society and its government can only deal with so many issues at a time.[1] We therefore set priorities, a selection of conflicts that means all the world both to leaders and to the mass public. Each conflict has distinctive political alignments—those dominating the course of confict A may find themselves in the minority on conflict B—so that displacing B for A also fundamentally alters society's agenda for action and shifts overarching national priorities. An issue displaced is one apparently less critical to the polity (or to its opinion makers); it is on the political back burner. Once displaced, it is also, very often, then judged in terms set by its successors, held up to standards defined by other issues. This competition takes on additional importance when both issues vie for resources, budgetary or emotive. Resources are rarely meted out equally.

Anthony Downs advances this notion of "conflict displacement" with his "issue-attention cycle," a "systematic cycle of heightening public interest and then increasing boredom with major issues" that is "rooted both in the nature of certain domestic problems and in the way major communications media interact with the public."[2] The Downsian cycle has five distinct (though not easily distinguished) stages that, he argues, almost always occur in the same sequence. It is useful here to apply Downs's thesis to the pesticides case, if only to understand what happened to this policy area after 1972.

The "pre-problem" stage. The "objective problem" of pesticides existed many years before *Silent Spring* temporarily excited the mass public, and a decade more before policy change came about. In fact, in line with Downs's theory, the problem of pesticidal pollution probably became less acute the closer environmentalists came to achieving statutory revision. The most egregious examples of the abuse or misuse of pesticides occurred in the 1950s and early 1960s—long before most of the public became interested in the issue.

Alarmed discovery and euphoric enthusiasm. Rachel Carson alarmed the public about the pesticides problem, but, unlike what happened regarding "the environment" in general, no massive social attack on the problem emerged. The pesticides issue in 1972 did not lend itself to "crisis politics," did not allow for the quick and flashy answers typical of so many instances of "big issue" response. There was instead common pessimism about achieving any ideal solution, accompanied by a political atmosphere already vastly different from the halcyon days some two years earlier. Passage of FEPCA rode the fading environmental wave, and barely made it to the end.

If these first two stages of Downs's cycle apply to the pesticides case only partially, the next three provide key insights into the politics of pesticides far into the next decade.

Recognition of costs. Everyone attached to the pesticides issue knew that strict government regulation would be costly, but few expected the magnitude of financial commitment necessary for the EPA to fulfill its statutory goals. Within two years, as we will see, the EPA faced two alternatives: obtain massive injections of money in order to process some 50,000 products, or modify its goals. Recognition of a program's true costs often cools even the most ardent reformer.

Decline in intense public interest. Downs argues that as more people understand the difficulty and costs of promised solutions, they become discouraged, threatened, or bored with a problem. Mass attention wanes the longer an issue dominates the public agenda, probably because most people despair of finding a real solution. Problems seen as intractable move aside as new and perhaps more powerful issues climb into the public spotlight.

There is no doubt that the "environmental era" ended dramatically in 1973 after the OPEC nations abruptly introduced Americans to the "energy crisis." Environmental matters that once dominated headlines now ended up in the fine agate on the back pages, while television forgot them altogether. One era replaced another, with telling consequences for environmental policy considerations. The energy crisis soon meant more than a shutoff in the oil supply; it became a bundle of connected problems, a condition of perceived urgency that threatened the national fabric and brought into question the range and purpose of government action. Common perceptions about the need to realign what then passed for a national energy policy meant profound changes in national policies and priorities. Cries for a rational and comprehensive energy policy filled the air, in turn generating shifts in prevailing political norms and policy calculations. The policies of the earlier era, enacted under far different social and political parameters, were now compared almost dispassionately to national energy needs—and often found wanting. Cost/benefit analysis became the watchword, and many openly questioned whether the nation could in fact *afford* to clean the environment. Conscious tradeoffs had to be made; no longer was the environment a sacrosanct issue.[3]

Pesticides policy, like other environmental matters, became ensnarled in the new crosscurrents. Low commodity prices and crop surpluses had produced a soft agricultural economy years before OPEC turned the oil screw, and the subsequent escalation in energy prices squeezed farm incomes to the bone. Pesticides depend heavily on petroleum—as a chemical base or simply to fuel production—and prices rose sharply. Caught between destabilizing economics on one side, and more stringent government regulation on the other, farmers naturally sounded the alarm each time the EPA openly discussed new restrictions.

The "post-problem" stage. Finally, as issue eras switch places on the national agenda, the displaced issue moves into "a prolonged limbo—a twilight realm of lesser attention or spasmodic recurrences of interest."[4] Issues relegated to the back page may command center stage once again, should conditions change. This time, however, public reaction may be confused, since many no doubt thought the problem had been solved, or people may be angry, since the problem obviously lives on. Or a revived issue may simply produce apathy, particularly if the public is no longer in the mood to tackle major social problems.

The final stage in the Downsian cycle is a perfect description of pesticides politics in the decade following passage of FEPCA. The question of pesticides after 1972 world reappear haphazardly, usually after an accident or a major EPA action, but, for the most part, policy activities proceeded in their own limbo. The politics of policy formation gave way to the grind of policy implementation, interspersed every once in a while by the almost reluctant politics of oversight and policy modification. The issue once again enters the less public realm of the knowledgeable and the included—though the configuration of insiders is broader and far less exclusive than before—interrupted by occasional spasms of more visible policy activity. The modifications to come would in fact be rather narrow and often temporary, as slight changes in the political balance shifted the calculus of winners and losers. It is for this reason that I describe the issue during the ensuring period as on the "policy pendulum."

Bringing Power to Truth: Early Implementation and Its Problems

> It is hard to design public policies and programs that look good on paper. It is harder still to formulate them in words and slogans that resonate pleasingly in the ears of political leaders and the constituents to which they are responsive. And it is excruciatingly hard to implement them in a way that pleases anyone at all.[5]

Implementation, argue Pressman and Wildavsky, "may be viewed as a process of *interaction* between the setting of goals and the actions geared to achieving them."[6] It is more than "administration," or "getting a job done"; rather, it is a process heavily influenced both by initial goals and concurrent political, scientific, and economic realities. Implementation, properly understood, is a policy process in its own right.

Effective implementation, argues George C. Edwards, requires first of all that those who are to implement a decision *know what they are supposed to do:* "If policies are to be implemented properly, implementation directives must not only be received but they must also be clear. If they are not, implementors will be confused about what they should do, and they will have discretion to impose their own views on the implementation of policies, views that may be different from those of their superiors."[7] The problem, of course, is that legislative intentions, guidelines, or instructions often are imprecise, ambiguous, or at cross-purposes, resulting in the blurring of lines between legislator and administrator that confuses the citizen and infuriates the ideologue. "Modern law," charges Theodore Lowi, "has become a series of instructions to administrators rather than a series of commands to citizens."[8] Congress delegates excessive discretionary authority to bureaucrats, which would not be so bad were the instructions themselves clear and forceful. Alas, Lowi says, they are not. Congress—and liberal society in general—is capable only of ambiguity, of fuzzy legislative instructions to bureaucrats that, in turn, produce equally fuzzy administrative standards. Such delegation of legislative power is, to Lowi, "policy without law"; standards are ill substitutes for laws, and fuzzy standards are worse than none at all. It comes as no surprise that Lowi seeks legislative precision, minimal administrative latitude, and unambiguous lines of authority. To have none of these is to undermine justice.

Lowi is correct when he speaks of symptoms and effects, when he argues that imprecise legislative instructions create problems down the line. Whether the causes of this symptom lie inherently in "interest-group liberalism" may be a far different matter. It may very well be that we get fuzzy legislative instructions simply because of the *types* of problems Congress seeks to address, and because of unrealistic expectations. Richard R. Nelson, for example, ponders the "moon-ghetto metaphor," our propensity for comparing the success of the Apollo space program to our apparent and concurrent failure to "solve" the problems of poverty and other social ills. Nelson argues that one big difference between going to the moon and solving poverty lies in values: the moon program "had the advantage of not threatening significant interests and of promising something to several. Many of the proposed solutions to ghetto problems face quite the other way."[9] Problems like poverty, in this vein, simply may be insoluble. Pressman and Wildavsky state the dilemma in more prosaic, but equally powerful terms: "Expectations about new governmental programs violate common, everyday experience."[10]

"Successful" implementation ultimately hinges on the nature, depth, and speed of intended change. Going to the moon was, in this sense, "simple." Those who had to meet that national goal agreed with it, knew what to do, enjoyed considerable backing and resources, and could define "progress" in understandable and observable terms. Regulating chemical pesticides, by contrast, is far less simple. FEPCA contained multiple goals: The EPA was to ensure that chemical pesticides were safe, environmentally benign, *and* effective. Each goal is admirable but also laden with complex and problematic scientific, technical, and valuative dimensions. That all three goals were embodied in the same law only added to the dilemma, for they more often than not operated at cross-purposes—or, at least, they created cross-tensions—with no clear legislative indication of overarching priorities. FEPCA told the EPA to do several things, but because they were sometimes contradictory, the EPA was never really told exactly *what* to do.

Second, those who implement a policy must also understand *how* to do so. Lowi's answer to the problem of legislative ambiguity and administrative discretion is to instill strong juridical values into the system, to insist that Congress create no policy unless it is accompanied by precise operating instructions. Barring some transformation in our current method of governing, implementors at a minimum must know how they are to go about matching action to intentions. FEPCA was a sharp departure from previous regulatory operation, for it demanded that the EPA develop comprehensive and authoritative regulatory standards, uniform registration guidelines, and sophisticated testing protocols. The problem, according to one veteran agency official, was that "you didn't know what the regulations would be like when you passed the act. Before 1972 not many people really understood the implications of this."[11] Reaction among EPA careerists to the goals of FEPCA was, he aded, "disbelief": Everyone in the agency knew that it was "absolutely impossible" to design such standards where none existed previously, and nobody, including Congress and the EPA, really knew how to do so. Congress by necessity left most of the details to the EPA, where most careerists privately expressed uncertainty about what to do.

Finally, successful implementation depends on having the resources and external political support necessary *and* sufficient to meet both legislative intentions and, perhaps more important, deadlines. Congress not only sought broad, fundamental change in national pesticides policy, but also ordered the EPA to implement statutory goals within a rather short period. Uniform, comprehensive registration and classification guidelines were to be completed by late 1974, and these guidelines depended ultimately on sufficient and authoritative scientific data, organizational resources, and support by attendant interests. Even more daunting was the congressional edict that the EPA must *reregister* some 50,000 existing products and uses, following the new guidelines, by October 1976. Congress did boost program funding, and the Office of Pesticides Programs did add personnel, but the EPA nonetheless failed to meet either deadline. The agency successfully formulated its final registration guidelines a year late, but reregistration never even came close to meeting its deadline.

Subsequent congressional analysis of EPA's failure on both counts lays the blame on bureaucratic delay and ineptitude, not on the goals set initially by the legislature. The EPA in fact did have a rocky time in setting up the new program, problems born of the internal discord and organizational malfunctioning inherent in the agency's own birth—not to mention its multiplicity of new mandates. "The EPA," commented one veteran environmental attorney, "was formed out of the dregs of the USDA and the FDA, with a superstructure of bright but inexperienced lawyers and other people." Those "dregs," also known as the "old-timers" by EPA watchers, now staffed the Office of Pesticides Programs (OPP), a unit distrusted by environmentalists for its cautious and accommodative stance toward pesticides. The "bright but inexperienced" lawyers were those hired by Ruckelshaus to form the Office of the Counsel General (OCG), the agency's enforcement division and regulatory vanguard. OCG attorneys led the fights against DDT and other organochlorines, often to the dismay of OPP personnel long comfortable with, if not outright supportive of, those products. Ruckelshaus relied on the OCG because he wanted to quickly and dramatically establish the agency's "environmental credentials," almost regardless of the weight of opinion among OPP experts. One longtime OPP survivor describes EPA's early experience:

> The lawyers were running the place. They liked the adversarial style, armed with lots of "ammo" in the form of documentation. . . . The lawyers didn't trust the scientific people. . . . There *were* some "Neanderthals" in OPP, to be sure, but there were also some in the OCG. Some of the early registration decisions were strange; they were not made on a rational, scientific basis. . . . It was almost impossible to get USDA lawyers to do something, since they were almost completely out of the picture, which was not true in the EPA.

An environmental attorney puts matters differently, but comes to similar conclusions about internal EPA dissension:

> There was no question that the OCG had a very strong hand in policy decisions. They did not take the role of counsel but of policymaker. The OCG was at war with the OPP; . . . the OPP was the "fifth column" for industry. . . . But the OCG had the ear of the administration, and the oldtimers either left or retired.

One group sought dramatic and rapid impact; the other placed its faith in "scientific rationality." Yet the problems of internecine organizational warfare or widely disparate perceptions about policy direction and styles pale in comparison to the more fundamental problem of EPA's implementive capacity. That problem was the absence of data. Timely and authoritative registration depended ultimately on accurate and comprehensive test data and chemical information, but the ad hoc, case-by-case process of USDA registration before the 1970s left the EPA with very little uniform data in its files. Nor did the agency have comparable or uniform bases for judging registrations, since much data had been generated in studies performed ten to twenty years before. Scientific advances and the advent of new testing protocols and technologies made much of the data suspect, if not totally unreliable. FEPCA mandated that registrants submit data from both animal studies and environmental exposure tests—a requirement immediately applicable to new registrations but also *retroactive* to previously registered products after the agency finalized its guidelines and testing protocols. That finalization process alone took a year longer than expected, giving the EPA little more than a year for reregistration. EPA failed to ask Congress for an extension, though agency officials acknowledged privately that the mandated reregistration was impossible by the original deadline.

To speed up reregistration, the agency began to scan existing files for the mere *presence* of test data and other information. If "gaps" in a file came to light, the EPA required the registrant to provide relevant material, but the agency made little effort to verify its validity against the proposed (but as yet incomplete) guidelines. Experience with DDT and other cancellation actions made it painfully obvious that to verify test data for thousands of existing registrations was well-nigh impossible; better to have complete data at the present moment. But, as one OPP veteran later related, this whole process was based on faith: "There was a faith that the industry data and information were valid, since it would make no sense for industry to lie. Agency people *wanted* to believe in the veracity of the industry data and tests. *Not* to believe would've made life pretty hard."

That the agency made little effort to verify industry data suggests more than bureaucratic ineptitude or even duplicity as the problem. At its core lay contradictory—even mutually hostile—legislative goals, unrealistic expectations, and unexpected scientific and technical difficulties. Pressure came from everywhere: congressional farm bloc members and chemical interests wanted quick action on reregistration; environmentalists wanted zealous prosecution of suspect products. Whitten's subcommittee expected expeditious registration, with little or no "obstruction" to the industry or its customers. The defenders of pesticides in no way expected EPA to verify industry-supplied test data; their traditional faith in the industry led them to oppose efforts to establish large-scale EPA research capacities. Environmentalists, of course, expected precisely the opposite.

Implementation of EPA regulations also began coincidentally with dramatic changes in the national policy agenda and the overall political atmosphere. The energy crisis and its subsequent inflationary squeeze prompted a government-wide reassessment of program costs and benefits. The Ford administration by early 1975 demanded all federal expenditures to be framed in terms of their apparent inflationary impact—an angle of approach clearly not foreseen in the heady days of environmental policy formation. The bills were coming due precisely at the time when the EPA required massive injections of funds merely to carry out the initial intent of the program. Diffuse public support for the environment remained strong, but it now was tempered by growing awareness of economic costs and national tradeoffs. This perception *always* was strong respecting pesticides—costs to national food production had traditionally formed agriculture's strong opposition to stricter regulation—but those perceptions come to the fore dramatically as the EPA struggled to perform its mission. Good intentions clearly had run into hard realities.

Congress in 1972 ordered the executive establishment to seek three important, if contradictory, objectives. But, Charles O. Jones asks, "What if our solutions are acknowledged in the first place to be beyond the immediate capabilities of administrators and technicians to implement?"[12] Jones finds what he calls "policy beyond capability" in the EPA's attempt to implement the 1970 Clean Air Act; the same was true of FIFRA. The agency had to implement both laws, not to mention a spate of other new and important environmental statutes, *and* at a time of growing concerns about costs. This tells us quite a bit about the problematic intersection between intentions and capabilities.

Implementing beyond policy capability, Jones argues, "does not bring one directly to the problem at all."[13] One first must create and/or increase organizational capabilities—by reorganization, staffing, etc.—before tackling the problem. This dynamic held for EPA implementation of FIFRA, and early on the agency worked hardest to get its own house in order, with precious little observable progress made respecting policy goals. This lag in turn generated strong criticism from Congress, interest groups, and the media that forced the agency to "act tough" in a highly public manner. EPA officials concentrated their early efforts on canceling a few highly visible and widely used organochlorines (such as aldrin and chlordane) in order to establish the agency's public image as a no-nonsense defender of the public good. These efforts garnered both praise from environmental advocates and severe criticism from farm and industry interests. Congressional farm bloc members charged that the agency with precipitous and "unscientific" product suspensions, and efforts began soon thereafter on Capitol Hill to revive DDT. The agency's strategy might have been less risky to its congressional support if the issue had been more salient, but public concern with pesticides waned noticeably as the energy crisis dominated the headlines. By 1975 pesticides policy once again was the purview of the most knowledgeable and interested.

Finally, Jones notes, "The decisions themselves must include escape clauses for postponement and/or compromises unless and until the efforts to increase organizational and technological capabilities do in fact pay off."[14] Implementors must have some statutory safety valves to give them time, if need be, to prevent the *appearance* of policy failure. EPA's problem with FIFRA was, ironically, both ambiguity of direction and specificity of deadline. The agency found few "escape clauses" in the law behind which to hide while trying to get its act together. Failure to meet initial deadlines, when added to the dramatic cancellation of key pesticides, brought only grief to the agency in succeeding years, prompting attempts by all involved to "reform" both the law and the agency. EPA officials sought to achieve organizational and policy stability by negotiating with the regulated; however, these accommodations only infuriated environmentalists and—as a series of "scandals" came to light—the public, which of course assumed that the pesticides problem had been solved in 1972.

The result of all this would be years of incessant congressional scrutiny, sporadic refocusing of the issue, and endless, increasingly less productive, forays into policy modification by policy claimants unable or unwilling to let matters rest long enough to allow some regulatory consistency to be established. The pendulum of policymaking would begin soon after 1972, a dynamic powered in no small way by a fundamental reordering in the style and structure of congressional life already under way.

Oversight for Everybody

> Changes in political institutions always leave tracks. The problem for the analyst is that years may be required to distinguish which way they are headed, and sometimes the direction never becomes clear. The rash of changes in Congress is a case in point.[15]

A funny thing happened to the pesticides subgovernment on the way to the 94th Congress (1975–1976). Simply put, that long-entrenched configuration of policy insiders wandered into the path of the congressional reform movement—and was run over. History will record that the 94th—the "post-Watergate" Congress—perhaps reached its apogee in an era of governmentwide institutional and procedural "reform." These changes created a Congress scarcely recognizable to those with long tenure. Congress by the end of the 1970s would be more open and less centralized, its members better equipped with greater information and analytical capacities, and, perhaps, with greater pretentions toward program evaluation and purposeful oversight. Capitol Hill became a much busier place, with more actors having fingers in more problem areas, with greater fragmentation in policy jurisdiction, control, or even influence. As a result, there was far less predictability about either the process or its outcomes.

The pesticides subgovernment was "run over" by the reform movement because it was dislodged somewhat by forces it could not withstand. The same urges central to the social reform movements of a previous decade spurred fundamental alterations in the way government went about its business—chief among them a change in how power was allocated in the congressional committee system. These changes began in December 1974 when the House Democratic Caucus stripped the Ways and Means Committee of its traditional jurisdiction over committee chair selection. That power went to the caucus itself, and secret ballots thereafter decided who ruled the fiefdoms of the House. The caucus also formally recognized what all had known for decades—that appropriations subcommittee chairmen were equally, if not more, powerful as standing committee chairs—and ordered that they also be chosen secretly. The choice presented to incumbent chairmen was clear: demonstrate party loyalty or lose power.

The forces sparked in late 1974 coalesced dramatically in early 1975, as the massive and liberal (ninety-five new Democrats) "class of 1974" charged Capitol Hill and (at the risk of a little hyperbole) took a few victims. Among the more immediate casualties were three House committee chairmen, all southern conservatives, who were ousted by the new selection process. Removed were Wright Patman (Banking), F. Edward Hebert (Armed Services), and, in a result that shook the agricultural community, W. Robert Poage. The Texan's conservatism, the declining power of the rural southern bloc within the Democratic caucus, and his unabashed support for pesticides figured prominently in Poage's removal, and he was replaced soon thereafter by Rep. Thomas Foley of Washington, whom the young Turks felt would be less dictatorial and more amenable to environmental and consumer attitudes. House Majority Leader James Wright (D, Tex.) later summed up the impact of the ousters: "The day is ended when any committee chairman can run his domain like a feudal barony, oblivious to the wishes and sensitivities of other members. He can, however, lead with the cooperation of other members."[16]

One person who understood that point perfectly well and rather quickly was Jamie Whitten, who was apparently on his way to losing his appropriations subcommittee chair. Poage's ouster was ill news to the defenders of pesticides, since it marked a palpable shift in the power and prestige of the congressional farm bloc; however, Whitten was the real cornerstone of the pesticides subgovernment, for nobody else knew as much about or had as much leverage over federal pesticides policy as the "Permanent Secretary of Agriculture." House

Table 10: Congressional Hearings Related to Pesticides, 1973–1976

93rd Congress (1973–1974)

House of Representatives

1. Agriculture (FIFRA implementation; indemnities)
2. Agriculture, Subcommittee on Cotton (DDT)
3. Agriculture, Subcommittee on Dairy and Poultry (indemnities)
4. Agriculture, Subcommittee on Department Operations, Investigations, and Oversight (FIFRA; pest control)
5. Agriculture, Subcommittee on Forests (tussock moth control)
6. Appropriations, Subcommittee on Agriculture, Environmental and Consumer Protection (USDA, EPA funding)
7. Appropriations, Subcommittee on HUD and Independent Agencies (workplace exposure)
8. Appropriations, Subcommittee on Interior and Related Agencies (fish and wildlife programs)
9. Appropriations, Subcommittee on Labor, HEW, and Related Agencies (research, workplace exposure)
10. Interior and Insular Affairs, Subcommittee on Public Lands (predator control)
11. Merchant Marine and Fisheries, Subcommittee on Wildlife Conservation and the Environment (predator control)

Senate

1. Agriculture and Forestry (indemnities; weeds)
2. Agriculture and Forestry, Subcommittee on Agricultural Credit and Rural Electrification (pesticides needs and costs)
3. Agriculture and Forestry, Subcommittee on Agricultural Research and General Legislation (FIFRA; weeds)
4. Appropriations, Subcommittee on Agriculture, Environmental and Consumer Protection (USDA, EPA funding)
5. Appropriations, Subcommittee on Interior and Related Agencies (endangered species)
6. Appropriations, Subcommittee on Labor, HEW, and Related Agencies (worker exposure)
7. Commerce, Subcommittee on the Environment (FIFRA oversight; predator control)
8. Interior and Insular Affairs, Subcommittee on Public Lands (predator control)
9. Select Committee on Nutrition and Human Needs (effects of agricultural practices)

94th Congress (1975–1976)

House of Representatives

1. Agriculture (FIFRA implementation; extensions)
2. Agriculture, Subcommittee on Department Operations, Investigations, and Oversight (fire ant control)
3. Agriculture, Subcommittee on Forests (pest control)
4. Appropriations, Subcommittee on Agriculture and Related Agencies (USDA funding)
5. Appropriations, Subcommittee on HUD and Independent Agencies (EPA funding)
6. Appropriations, Subcommittee on Interior and Related Agencies (fish and wildlife programs)
7. Appropriations, Subcommittee on Labor, HEW, and Related Agencies (research, workplace exposure)
8. Education and Labor, Subcommittee on Manpower, Compensation, and Health and Safety (worker exposure)
9. Government Operations, Subcommittee on Conservation, Energy, and Natural Resources (FIFRA implementation)
10. Interior and Insular Affairs (general environment)

11. Interstate and Foreign Commerce, Subcommittee on Oversight and Investigations (FIFRA implementation)
12. Merchant Marine and Fisheries, Subcommittee on Fisheries, Wildlife Conservation and the Environment (effects on wildlife)
13. Select Subcommittee on Labor (worker safety)

Senate

1. Agriculture and Forestry (FIFRA extensions)
2. Agriculture and Forestry, Subcommittee on Agricultural Research and General Legislation (FIFRA; Kepone)
3. Appropriations, Subcommittee on Agriculture, Environmental and Consumer Protection (USDA funding)
4. Appropriations, Subcommittee on Interior and Related Agencies (fish and wildlife programs)
5. Appropriations, Subcommittee on HUD and Related Agencies (EPA funding)
6. Appropriations, Subcommitte on Labor, HEW, and Related Agencies (worker exposure)
7. Commerce, Subcommittee on the Environment (FIFRA oversight; Kepone)
8. Environment and Public Works (general overview)
9. Environment and Public Works, Subcommittee on the Environment (FIFRA)
10. Foreign Relations, Subcommittee on Foreign Assistance (agricultural assistance programs and pesticides)
11. Judiciary, Subcommittee on Administrative Practice and Procedure (FIFRA implementation)

liberals made no secret of their decision to add Whitten's scalp to their belts, and it looked for a while that the Mississippian would have a real fight on his hands.

Then, in an action that stunned the defenders of pesticides, Whitten suddenly surrendered jurisdiction over environmental and consumer protection appropriations, thus ending his de facto reign over pesticides. The reasons were clear: Whitten at all costs wanted to retain his role as overseer of the USDA, not to mention his being next in line for the Appropriations Committee chair after George Mahon left the scene. Whitten gave in to save his chair, and his influence over pesticides policy ebbed sharply thereafter. No longer would he control FIFRA appropriations, and his later assumption of the chair of the Appropriations Committee could not reinstate that power. Appropriations by the late 1970s was a pale imitation of its predecessors, a body led less by an autocratic chairman than by a college of cardinals (the subcommittee chairs), and with the chair acting as procedural broker. Whitten, after his rise to committee chair, would also become a team player, if only to keep his post. An Agriculture Committee long accustomed to Whitten's annual review of federal pesticides programs now would have to pick up the slack if the farm perspective was to be maintained.

Congress by the decade's end was a more open, more loosely structured, and, as a result, more complicated place. What is more, the issues Congress sought to address became less sectoral in nature, and ever more members of Congress began to comprehend the nuances of and their stakes in complex policy questions. As a result, everyone insisted on some sort of oversight, and pesticides policy, for one, became the purview of all. In the process, of course, it became the exclusive purview of none. Where once a single, sharply limited group of insiders controlled the shape and direction of policy, by the end of the 1970s that group would be far broader, far less exclusive, and far less authoritative in its ability to sustain policy closure. Policymaking as a result increasingly took place at the margins of the legislative process, with alterations sneaking in unpredictably and through indirection, and with no one willing to let the issue rest.

One trend that emerged rapidly was a surge in continual congressional scrutiny over EPA implementation. This scrutiny was sparked both by the evisceration of the pesticides subgovernment and by the institutionalization of rival perspectives in competing congressional work groups. Table 10 indicates just how extensive this scrutiny became; the EPA constantly found itself before one set of critics or another. The aforementioned problems generated both by the agency's initial strategy and its failure to meet deadlines left it in a veritable no-win situation: "Generally, those who benefit from the use of pesticides charged the Agency with unreasonably restricting pesticides. . . . Others charged the Agency with not being restrictive enough. . . . Both sides agreed that information was incomplete. . . . Caught in the middle, EPA frequently found that its decisions pleased no one."[17]

FEPCA's initial three-year funding authorization elapsed on June 30, 1975, which guaranteed that battles left unsettled in 1972 would be reopened. Reauthoriation had always been routine up until the 1970s, with the law's open-ended authorization and Whitten's stranglehold on appropriations guaranteeing rather colorless proceedings. No more: that battle was to be refought at more frequent intervals, at predictable decision and clearance points that would reignite smouldering conflicts and drag the agency and its program almost annually into the limelight. EPA officials were exasperated by this almost constant conflict and sought multiyear funding, but few on any side of the issue apparently sympathized. Everyone, it seemed, simply wanted "another crack" at the law and its implementors. Agricultural and industrial interests complained vociferously about agency attacks against persistent organochlorines and sought less zealous EPA "persecution" of all pesticides. Environmentalists charged the Nixon (and then the Ford) administration with welching on promises of agency support, since it appeared that budget austerity was undermining "active" regulation. They also complained that the agency itself was retreating from its original objectives. Nobody, given the myriad complaints, assumed that reauthorization would be easy.

It wasn't. The record of 1975 is one of protracted conflict, consecutive interim extensions, and pitched floor battles before Congress finally approved a compromise in late November (see figure 4). House Agriculture Committee deliberations dragged on for months, exposing an increasing fragmentation of

Figure 4. Action on FIFRA during the 94th Congress (1975)

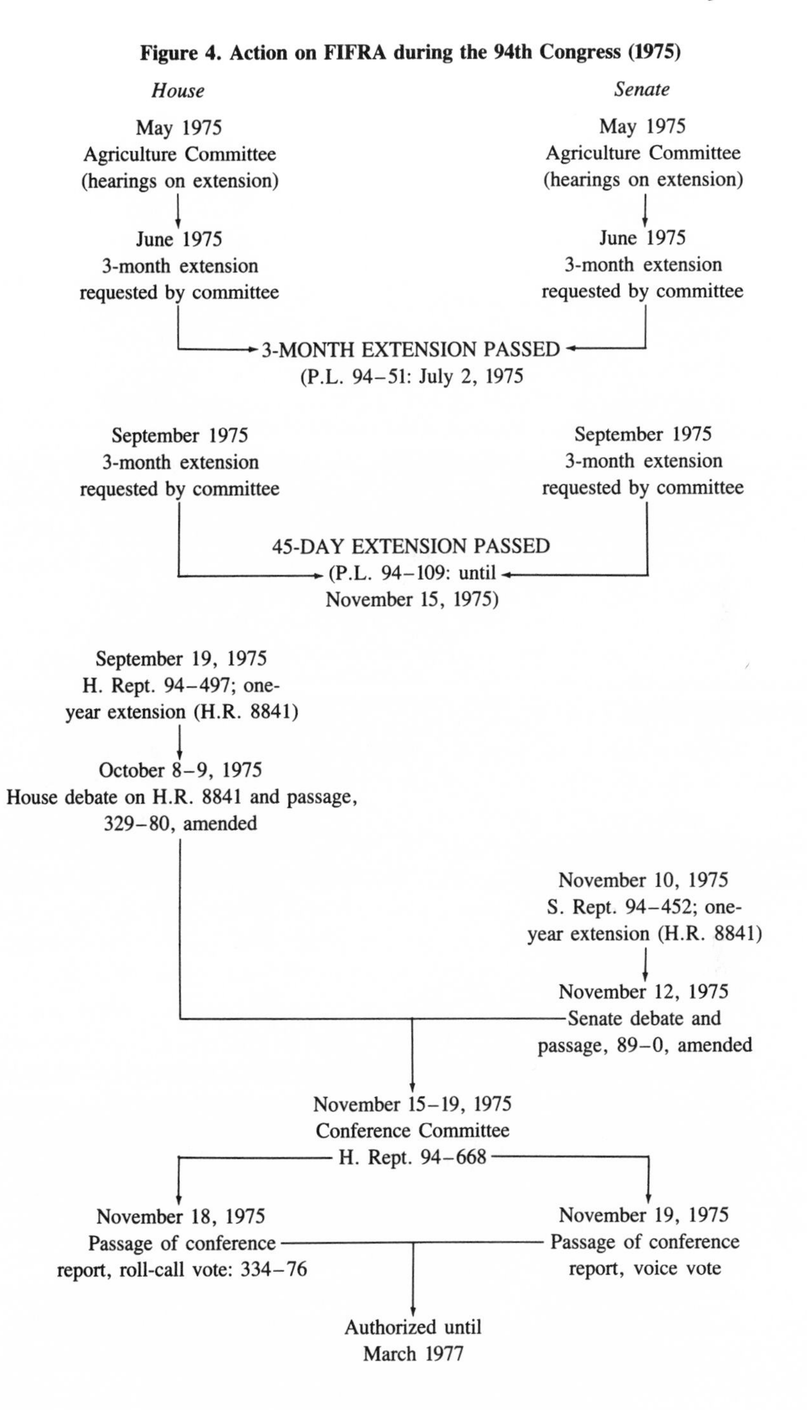

views even among the defenders of farm interests. The committee's inability to resolve major issues forced Congress to pass two consecutive interim funding extensions, creating a condition of uncertainty no doubt disconcerting to those mandated with implementation. Major issues concerned certification of pesticides users, EPA actions against organochlorines, and the agency's "hot line" that allowed citizens to report pesticides misuse. The EPA abolished the hot line soon after farm bloc legislators complained loudly that the program undermined farmers' civil liberties; the agency no doubt figured that it wasn't worth the headaches.

EPA actions against organochlorines soon surpassed other matters as the dominant issue. Farm bloc representatives, backed by the USDA and the industry, argued heatedly that the agency acted arbitrarily and with little or no scientific or economic rationale. These charges escalated dramatically after the EPA announced its intent to cancel most uses of chlordane, used heavily on corn, and mirex, the USDA's chief weapon against the fire ant. So high were feelings among farm bloc members that Representatives Poage and William Wampler (R, Va.) in late July proposed to give the USDA almost total veto power over EPA regulatory actions. The Poage-Wampler amendment was a bit too draconian even for many committee members, who in its stead required the EPA to notify the USDA prior to undertaking actions against any chemical or before proposing new regulations. The agency also would be required to submit its proposals to a new scientific advisory panel. A secondary, though no less emotional, issue lay in EPA proposals that farmers pass a certification test before being allowed to use restricted pesticides. Agriculture groups protested that this was a regulation written "by urbanites with the express purpose of reducing the use of pesticides and developing a federal pesticides record on every farmer in the U.S."[18] EPA officials, who privately advocated both of these objectives to some extent, argued that formal training and certification would prevent misuse, but the committee rejected this reasoning as a mere smokescreen for needless government intrusion. Farmers instead would be able to "certify" themselves by signing waivers indicating that they had read instructions for use and would abide by them.

The committee reauthorized FIFRA for one year, despite EPA's request for at least two—an action justified by the need for continued oversight. Two days of sharp floor debate preceded House passage, but not before EPA's harshest critics almost reinstated USDA veto authority. Agency supporters barely beat back the amendment by Representative Steve Symms (R, Idaho), a margin of victory so narrow (167 in favor versus 175 opposed) as to send a strong warning to the EPA that agriculture would brook no further "unscientific" adventurism.

Senate action closely resembled that in 1972, with Agriculture and Forestry Committee members standing alongside their House colleagues on almost every matter. And just as in 1972, Gaylord Nelson and Philip Hart pressured the committee to add changes advocated by environmentalists. EPA supporters could not rescind the USDA prior-notice provision, though they succeeded in getting simultaneous public notice through the *Federal Register* in hopes of offsetting any pressure to come from agricultural interests. The Senate committee similarly accepted the new science advisory panel, but agreed to Hart-Nelson proposals to include environmentalists and other "nonexperts," as well as rules governing conflicts of interest. Finally, with respect to farmer certification, liberals were able to delete language forbidding federal preemption of any state's desire to institute formal examinations. States' rights, at least in the convoluted world of pesticides regulation, had become an environmentalist banner, if only because several states in fact administered stricter pesticides regulations than those contained in FIFRA.

Senate floor consideration and passage proceeded rather quickly, but not before environmentalists were able to amend the bill in three key areas. EPA would not be required to notify the USDA prior to taking action against pesticides deemed to be an "imminent hazard" to the environment, while it would be required to expand its public notification and education programs. A third change, proposed by Jacob Javits (R, N.Y.), provided that only regulatory appeals preceded by public hearings and accompanied by a full record could proceed to the U.S. Court of Appeals. Javits's amendment aimed to curb what EPA officials and the courts considered "nuisance suits" by the industry, a flurry of which erupted soon after passage of FEPCA.

Senate passage came in November, three days before interim funding expired, sending the bill directly to conference committee. Conferees dropped the Javits amendment and some of the Nelson language expanding public access to information, while accepting states' rights on certification and the expanded advisory panel. Neither side claimed victory, though the requirement that the EPA consult with the USDA was regarded by environmentalists as a partial return to the bad old days. To agricultural interests the provision only made good sense—farmers, after all, still used the bulk of pesticides. Both chambers quickly passed the conference report, and FIFRA was authorized through March 1977.

Reauthorization for one year provided little breathing room, either for the EPA or for Congress, and the House Agriculture Committee was back at it the very next March. Committee members once again failed to agree on statutory charges—perhaps reflecting widespread uncertainty about future directions or even about what had been accomplished thus far—so the panel proposed a simple six-month extension until September 1977. Chairman Thomas Foley indicated that the committee wished to study the situation in greater detail, a task requiring more "breathing space." The irony could not have gone unnoticed by agency personnel.

The extension passed the House on August 5, 1976, but not before Rep. Dawson Mathis (D, Ga.) in a 36–7 floor vote attached a one-house legislative veto provision to the bill. The intent, as everybody recognized, was to give the House Agriculture Committee and its allies the opportunity to strike down "errant" EPA actions. Environmentalists, not to mention the Ford administration, were caught napping; the Senate passed the bill without debate the very same day. Ford opposed any such devices, either of the one-house or two-house variety, and he subsequently vetoed the extension, arguing that the legislative veto provision made the bill unconstitutional. Congress, hurrying to leave for the campaign recess, made no effort to override Ford's veto. The EPA already had its money through a separate appropriations bill, and legislators found themselves more preoccupied with other matters to bother any further.

Few disputes over FIFRA were put to rest during the 94th Congress. The three years of relative freedom enjoyed by the EPA gave way to short-term authorizations and constant congressional scrutiny, and this pattern would be repeated throughout the remainder of the decade. If the legislature proved unable or unwilling to grant the agency some sort of breathing space, the agency would have to get it on its own.

Narrowing Access: The EPA Builds an Escape Valve

I commented earlier, following Jones's notion of "policy beyond capability," that FEPCA provided the EPA with remarkably few "escape clauses" that allowed the agency to prevent the appearance of outright policy failure. Agency officials, aware that they could not meet statutory deadlines, desperately sought out administrative mechanisms through which they could reach authoritative registration decisions and work out some accommodation with the regulated. To fail was to give agency critics even more ammunition.

Environmentalists accepted FEPCA in 1972 despite its flaws, because the EPA, after all, was not the USDA. They hoped that the agency's own organizational mission to defend the environment, plus its early zealous advocacy, would ameliorate whatever shortcomings emerged from the legislative process. That faith appeared to pay off as Ruckelshaus's activist lawyers avidly prosecuted major organochlorines and seemed to direct the agency toward uncompromising defense of "the public interest" as defined by environmentalists. If the agency floundered a bit on registration and rule-making, these were not disasters, since agency allies also believed that constant public pressure and the justice of the environmental cause would sustain their momentum.

But, as we know, the momentum behind an issue, always susceptible to displacement by other, temporarily more compelling issues, can be fickle. Momentum on the environmental front, beset by new national priorities, implementation problems, recognition of costs, and a Congress apparently unable or unwilling to close the matter decisively, had definitely shifted by mid-decade. FIFRA, by the end of the Ford years, was an implementation quagmire, if not an outright regulatory nightmare. The earlier strategy of highly public assaults on organochlorines succeeded, perhaps too well for the agency's own good. Not only did the EPA restrict or cancel a number of persistent pesticides, but also it generated a severe backlash on Capitol Hill, badly strained its relations with farm and industry interests, and, most likely, undermined other FIFRA responsibilities through disproportionate allocations of funds. The Ford administration found this strategy counterproductive, particularly with respect to its relations with the agricultural community. It also appeared costly in economic terms; the energy crisis made these old organochlorines appear all the more acceptable to an administration intent on controlling inflation.

These pressures intensified in 1975 when the EPA, now headed by Russell Train, announced its intent to cancel major uses of chlordane and heptachlor, an action leading both to the Poage-Wampler amendment and high-level administration pressure to moderate EPA advocacy. The outcome was Train's decision to remove the Office of Counsel General from the regulatory vanguard; it would now serve other agency units. No longer would the agency be as zealous in its prosecution of allegedly hazardous pesticides, and it thereafter took a decidedly cautious—some would argue overcautious—approach to cancellations. Those in the Office of Pesticides Programs, meanwhile, set out to routinize the cancellation and registration processes in a way that would get the agency off the front pages and out of the courtrooms. Advocacy was fine for environmentalists and lawyers, but to OPP personnel it only caused major headaches in their relations with the regulated, not to mention their congressional overseers.

A major accommodative vehicle came in the form of the process known as RPAR—Rebuttable Presumption Against Registration—finalized by the OPP in mid-1975. The RPAR process was the agency's way of directing its limited resources against the most hazardous products out of the thousands awaiting registration—a procedure that agency officials hoped would eventually eliminate the reregistration backlog and produce more timely and authoritative cancellations. The philosophy behind the RPAR mechanism was superficially simple: the reregistration of a product would hinge on its meeting certain health and safety criteria. Issurance of an RPAR meant that "there are data indicating that one or more of certain criteria for determining a potential unreasonable effect, such as toxicity or persistence, has been met or exceeded, and that a formal review is therefore warranted."[19]

Industry and environmental groups alike criticized the new procedure—the former fearing arbitrary agency action, the latter initially questioning the stringency of the thresholds. EPA officials assured the industry that the RPAR process was as "objective" as possible, and that issuance of an RPAR notice was not tantamount to cancellation. It was, instead, an informal notice that some problem with the product or its accompanying data existed. Once notified, the registrant had a specified period within which to rebut the agency's presumption of guilt by showing that (1) the EPA was in factual error; (2) the pesticide did not pose a hazard; or (3) the pesticide's benefits outweighed its possible risks. This third provision infuriated environmentalists, since economic benefits accruing from the use of a specific chemical always seemed to outweigh far less tangible health or environmental costs, not to mention the economic costs associated with suspension. A registrant's inability (or unwillingness) to rebut the EPA's presumption automatically would prompt formal cancellation proceedings. Successful rebuttal, on the other hand, essentially ended the matter insofar as the agency was concerned; a pesticide that survived the RPAR process was assumed to be safe—or safe enough.

The agency set out on RPAR actions against some forty-five chemicals—an ambitious undertaking promised to be complete by early 1977. The process soon bogged down, however, as the EPA found itself unable to manage the massive volume of data and information attached to each case. Registrants demanded and received long extensions in response deadlines, which of course created a huge backlog in overall OPP activities. Agency personnel also soon found it far easier to negotiate directly with industry representatives on RPAR actions than to proceed unilaterally in the face of potential lawsuits and ever present political pressure. The agency did process RPAR "rush jobs" against a few chemicals, normally when the pesticide in question was under the spotlight of controversy, but for the most part the EPA's own action-forcing mechanism simply ground to a halt.

One major problem with the RPAR process—at least, according to some environmental activists—was that it too often excluded nonindustry representatives in resolving disputes. The procedure was designed initially to decide the guilt or innocence of a particular chemical in a less adversarial, more "rational" manner, but RPAR was not a formal procedure. It was, instead, a more informal *predecision* fact-finding process. Environmentalists could participate once a formal RPAR notice was issued, but, as Thomas McGarity argues, "For many pesticides the process was terminated prematurely when the registrant agreed to voluntary cancellation of all uses. For others the process was halted in midstream because the agency reached an accommodation with the registrant that was acceptable to both. In none of these premature "settlements" did outsiders other than the registrants play a significant role."[20]

A process designed to spur the EPA to registration decisions in a stable and "nonpolitical" manner thus had the result—foreseen or not—of excluding environmentalists and other critics from key regulatory actions. Environmentalists could participate in cancellation decisions after the EPA formally announced its intent to cancel, but often had no role in this precancellation process of discovery. Nor could environmentalists force the agency to act once an RPAR decision was reached. A U.S. appeals court in 1980 held that environmentalists could not appeal an EPA decision not to commence formal cancellation proceedings against chlorobenzilate following conclusion of the RPAR process.[21] Chemical makers since passage of the 1910 Federal Insecticide Act have had the right to appeal adverse regulatory actions, but, the court argued, a decision that satisfied both the EPA and the regulated was not open to appeal simply because it did not go "far enough." Environmentalists prior to 1972 used the courts to force cancellation under the ambiguous and broad provisions of the 1947 law, but FEPCA and the EPA's review procedures limited such action-forcing court challenges. RPAR was an informal process, but, insofar as the courts cared, its outcome, for all intents and purposes, closed matters. But, as one longtime environmental lawyer noted with dismay, this mechanism also narrowed the avenues for access into policymaking by those not tied directly to the question at hand:

> Environmental groups have left FIFRA nowadays largely because under the legal interpretation of the act they would be required to go to court against the EPA just to gain access. This is a legal restriction, a bias, that forces environmentalists to deploy resources elsewhere, toward states, for example. Everything is a cutback to the 1972 changes, either legislatively or administratively. There has never been less pesticides litigation than now.

The courts in no way closed the door to environmental litigants, but they did raise new barriers, hurdles that were based on the existence of regular administrative decision-making procedures. It is ironic that the vague and somewhat amorphous nature of the 1947 act gave environmentalists their initial point of access; the courts could force the USDA to carry out its mission under that law and the department had little legal basis on which to argue that it was already doing so. By the late 1970s, however, routinized and formal administrative procedures had replaced ad hoc regulation, at least on the surface, and the courts accepted these procedures as necessary and sufficient decision-making vehicles.

To closely monitor administrative rule making is a tedious, thankless, and decidedly unglamorous task. It is also critical to success. Administrative procedures are the guts of policymaking, the rules of the implementation game that decide who plays and how. Environmentalists helped to pass FEPCA, but most of them essentially left implementation up to the EPA; there were other, more glamorous, battles to fight. In doing so they also left the field wide open to the promoters of pesticides, who *were* willing to sit down with agency personnel and to reach accommodations to help all involved. Environmentalists learned the hard way that there is no such thing as a self-implementing policy. They also learned, much to their chagrin, that regulation is a two-edged sword: the rule that seems to structure someone else's behavior can just as easily structure yours.

Back to the Drawing Board: Retreating from Objectives?

The politics of pesticides faded in the public eye soon after the passage of FEPCA, but general somnolence about the issue was interspersed by occasional "spasms" of interest as the problem reemerged now and then and policy was reappraised. Pesticides policy went into semi-limbo during much of the late 1970s. Questions of policy shape and direction reappeared at intervals (often predictable ones), were—or were not—dealt with, then sank again into the purgatory of agency decision making. Unsettled questions occasionally rose again in the consciences of those who cared, yet, as we shall find, those who cared proved increasingly reluctant to bring up old and unresolved matters each time the problem returned, tenacious as ever. Policymakers hoped that the problem would just go away, resolve itself, or cause someone else to come to grief. By the late 1970s, we find "policy exhaustion"—not for the problem, but for policymakers.

Policy analysts and policymakers often talk past one another—illustrating the "two worlds" problem addressed by a number of scholars—but all agree that policymakers at a minimum should be willing to learn from their mistakes.[22] Call this dynamic what you like—"scanning," "feedback," or whatever—but it assumes some capacity to know what went wrong and why. More important, policymakers should try to learn how to correct their errors. Correction can take two routes: modify practice and realign it with intentions, or admit errors of intent and restructure objectives to fit prevailing realities. The first strategy proves difficult to achieve, but the second seems more painful. After all, to admit (at least tacitly) that one's intentions were ill-informed or inadequately conceived might also be to admit that policy directions were illegitimate. It seems far easier to blame the bureaucrats.

EPA's regulatory effort lay in a shambles as Jimmy Carter strolled down Pennsylvania Avenue to the White House. Recognition that FIFRA contained serious flaws accumulated slowly throughout the mid-1970s, particularly as successive congressional overseers discovered—or charged—numerous administrative false starts, questionable bureaucratic interpretations of legislative intent, and the agency's apparent accommodation to industry's needs. The RPAR process, for all its warts, at least produced closure on a number of major pesticides, though closure in such cases often infuriated environmentalists. The rest of the agency's effort—reregistration, classification, testing protocols—simply went unfulfilled, or failed to lighten EPA's regulatory burden. EPA officials blamed legislative ambiguity, implausible deadlines, and tight resources for their woes, but congressional overseers were reluctant to shift the blame to Capitol Hill. By 1977, after two particular incidents, almost everyone in Congress realized that substantial statutory changes lay in store.

The Allied Chemical Corporation, through an exclusive contract with several former employees, produced the chemical Kepone at the Life Science Products Company of Hopewell, Virginia. Life Sciences operated without apparent problem for several years, but in mid-1975 suddenly was shut down by state health officials after employees began increasingly to complain of respiratory and nervous disorders. Such ailments were traced eventually to Kepone, the apparent cause, and state officials charged Life Sciences management with sloppy production practices, including the absence of respirators and other safety equipment. Scores of former or current employees were found with high Kepone residues in their blood, and several were subsequently hospitalized for a range of illnesses. Worse, state investigators discovered a history of Kepone dumping into the

James River, which flows into the fertile Chesapeake Bay, and dramatically high levels of the chemical in regional water life. These discoveries promptly led to bans on seafood sales from the affected areas.[23]

The Kepone incident, and one similar to it involving the chemical leptophos, reawakened the public, or at least a good portion of it, to the dilemma of pesticides and their production. The Industrial Biotest (IBT) episode, however, was to prove more devastating to policymakers' faith in FIFRA, since it would reveal to all that pesticides regulation was a house of cards. That story began as Senate investigators, checking into allegations made by disgruntled former EPA lawyers (refugees from the truncated OCG), found that many of the agency's registration decisions were based solely on inadequate, if not obsolete, industry data. The EPA, inundated by its registration and reregistration burdens, long had abandoned any pretense of systematic data review, blaming their actions on tight resources and unrealistic statutory deadlines.

Such practices, brought to light by the Senate Subcommittee on Administrative Practices and Procedures, led to even more intense scrutiny of industry test data by the EPA, the FDA, and other congressional actors. Senator Edward Kennedy (D, Mass.) and his colleagues charged that "the EPA had no sound basis upon which to assume that data 15, 20, or 25 years old was generally good and reliable," and criticized the agency for unilaterally abandoning product-by-product registration in favor of "batches" of related products.[24] EPA officials countered that this approach allowed the agency to focus on the relatively few "active ingredients" that comprised most pesticides, as opposed to the thousands of separate products that hit the market each year, but the Kennedy subcommittee was not convinced by this reasoning: "Several years of regulatory effort will have to be completely reexamined, substantially redone, and fundamentally redirected if the Congress and the public are to have reasonable basis to conclude that today's pesticides do not pose a significant risk to human health and the environment."[25]

The FDA, pursuing Kennedy's investigations, soon uncovered disturbing patterns of questionable testing procedures by Industrial Biotest, the nation's largest private chemical test firm. Charges of spurious tests and data falsification eventually sent four company executives to jail. IBT performed tests for those firms unable or unwilling to do their own testing, and about 200 pesticides had been registered using data generated by the now discredited firm. Every one of those products now fell under suspicion, thus shattering the EPA's regulatory credibility. "It was a real shock to us," admitted one OPP official: "The IBT thing took us from a rather naive atmosphere into the real world, although there always were those, mostly on the outside, who asked if this or that was safe. There are more people now who are less credulous of the data, which may have made EPA stronger. No longer can industry say, 'trust us,' although it didn't really affect industry-EPA relations at our level." Environmentalists may have been tempted to say "we told you so," but they too were disturbed by the prospect that huge portions of industry data might prove invalid. The affair jarred congressional faith in both the EPA and the chemical industry, even among some of the industry's strongest supporters. Consecutive years of oversight paved the way for policy modification, but the Kepone and IBT episodes made policy refurbishing imperative. Nobody could argue credibly that matters were fine as they stood—though, not surprisingly, few agreed about future directions.

EPA's immediate response to the IBT scandal was to halt all registration actions. The agency publicly admitted its inability to meet congressional deadlines (now set for October 1977) and, in early 1977, proposed extensive statutory modifications. The Carter administration, with administrative streamlining a top priority, proposed to legitimate EPA's heavily criticized switch to "generic registration," an orientation that had been de facto agency practice since 1975. Industry data no longer proved trustworthy, agency officials argued, but to attempt to verify test data for thousands of separate formulations was to ask the impossible. The only way out was to focus on data submitted in support of the relatively few active ingredients upon which most products depended. Generic registration also might reduce the regulatory burden for many smaller chemical firms that could not afford to generate the required data, since most active ingredients already were registered and purchased largely from the larger corporations. The 1972 law provided that "copycats" would compensate those generating test data when such data were used to support succeeding registrations, but subsequent litigation by the larger firms—which regarded all data as "trade secrets"—effectively halted any move toward data sharing. Generic registration might help to alleviate that problem, since most active ingredients were beyond patent law protection.

Second, the agency proposed "conditional registration" to eliminate its crushing backlog and allow newer products to enter the market more quickly. The 1972 provisions mandated extensive data-generation and review procedures that apparently created a "double standard" among old and new products: older products, registered by the USDA under the 1947 law, remained on the market pending reregistration (which lay dead in the water), while newer products encountered more formidable regulatory barriers. FIFRA in effect was a market gatekeeper, granting old products—most of them owned by the large chemical firms—artificial market dominance while newer products languished within the EPA. Critics argued that the law stifled competition, erected discriminatory regulatory obstacles, raised prices, and reduced incentives for new product research. "Ironically," comments Francine Schulberg, "this disparity exists even if the Administrator [of EPA] has more information about the new product than the old one and has greater confidence in the safety of the new product."[26] Farmers complained bitterly that this double standard also robbed them of alternatives; insect resistance increasingly eroded the effectiveness of many an old standby, while EPA cancellations took others off the market. The administration's solution, aimed in no small way to placate both industry (at least smaller firms) and the farm community was to register products "conditionally," pending completion of testing and data review. Conditional registration (and reregistration), when married to generic registration, would improve EPA's ability to verify test data and to speed new products to market. Without such changes, agency officials warned, registration might drag on another fifteen years.

Both strategies depended ultimately on freer EPA use of existing health and safety data—information currently restricted by the failure of the FIFRA's data compensation scheme. The scheme had sought a compromise between unlimited information disclosure (which environmentalists advocated) and tight restrictions on access to "trade secrets." Applicants could rely on data submitted by others only if they paid "reasonable" compensation, which was intended to avoid unnecessary duplication of effort (that is, using identical active ingredients) and unfair "free-rider" problems (copycats gaining free access to expensive data). The scheme foundered on its key exception: the second registrant could not gain access to information deemed to be "trade secrets," a definition left vague by Congress and tossed squarely into the lap of the EPA. Major chemical companies of course argued that all information used in registration, unless voluntarily disclosed, fell under trade secret protection, since any information might be used for competitive advantage. The EPA in 1976 ruled that health and pesticide efficacy data no longer enjoyed protection, but an onslaught of litigation stalled application of that standard. Congress now would have to decide, else the registration provisions prove just as hollow as did their predecessors.

Finally, the EPA proposed to separate reregistration from pesticide classification, which was, in effect, a move to restructure the 1972 law. Classification was intended to follow registration, with the EPA sorting pesticides into either general or restricted-use categories based on the test data. Pesticides posing "unreasonable" health or environmental hazards would be restricted to use solely by certified applicators; Congress thereby hoped to eliminate the need for protracted cancellation proceedings. Classification ground to a halt because of the problem with test data, which, of course, further disrupted the reregistration process. Separating the two activities, EPA Administrator Douglas Costle argued, would speed up classification of the potentially more hazardous products.[27]

EPA's proposals encountered heavy criticism by environmentalists who were concerned less with administrative efficiency than with unimpeachable data review. Conditional registration, argued Shirley Briggs of the Rachel Carson Council, "essentially returns us to the pre-1972 system, where a product is put on the market until new evidence might show reason for cancellation or restriction of uses."[28] Environmentalists looked dubiously upon generic registration, and regarded separation of classification from registration as a priori categorization of pesticides, certain to demean the intent of premarket testing. Few environmental activists bought OPP's argument that these changes would alleviate a host of regulatory ills; for those most critical of pesticides use, regulatory streamlining meant expedient and insufficient consideration of all relevant data.

Congress favored streamlining over seeking perfection. FIFRA's regulatory disaster was a vexing embarrassment to many who had invested long hours in its development, but most on Capitol Hill dreaded any prospect of reopening Pandora's box too wide. To do so only would renew bitter disputes about the law's fundamental structure. Congress would accept most EPA proposals in the spirit of regulatory fine-tuning, though nearly all involved remained dissatisfied with the law as it stood.

Few in 1977 wanted lengthy debate over reauthorization, but debate ensued nonetheless. House and Senate consideration took nearly two years, requiring temporary funding through the appropriations committees and extensive negotiations among conferees (see figure 5). The Senate Agriculture Committee for the first time led congressional action, as its members accepted administration proposals with few changes and the upper chamber passed S. 1678 on July 7 by voice vote, with no discernible opposition. The Senate bill authorized FIFRA through 1979, provided for public access to health and safety data, and required that copycats compensate data generators for seven years after originally submitting information. The bill emerged far more unscathed than public statements might have led one to believe it could when deliberations began. Environmentalists, if satisfied with the expansion of public access to test data, were displeased with conditional registration, but found even their strongest Senate allies inclined toward regulatory simplification. Agriculture and industry interests, for their part, differed publicly over the issue of trade secrets and data compensation. Farm organizations sided with smaller chemical firms and the administration on the compensation scheme, concerned that the status quo meant oligopolistic pesticides markets. Such divisions ensured that the least common denominator was the only choice left.

The House, meanwhile, proved more fractious than ever, as Agriculture Committee members wrangled endlessly and, to their embarrassment, publicly. Agriculture Chairman Thomas Foley (D, Wash.) was far less hostile than his predecessors to environmental views during committee deliberations, while ranking Democrats like George Brown of California sided openly with the EPA against the stouter defenders of agricultural and industry interests. The committee no longer ruled a monolithic pesticides subgovernment, but now was a divided work group that proved increasingly incapable of reaching closure. Foley hoped to bring a bill to the floor before the August recess, but divisions within the committee delayed action until early October.

House action on FIFRA during 1977 in many ways resembled the Senate's experience in 1972, with other work groups openly challenging Agriculture's policy jurisdiction. Interstate and Foreign Commerce Committee chair Harley O. Staggers (D, W.Va.) and four of his subcommittee chairmen in fact petitioned Speaker Thomas P. O'Neill (D, Mass.) for sequential referral, claiming that matters of public health placed FIFRA squarely in their jurisdiction. Foley, unlike Talmadge in 1972, would not budge, while O'Neill proved reluctant to interfere with committee prerogatives. Staggers lost, but the battle had its impact as nonagricultural actors were able to gain concessions from an Agriculture Committee suddenly under siege.

Figure 5. Action on FIFRA during the 95th Congress, 1977–1978

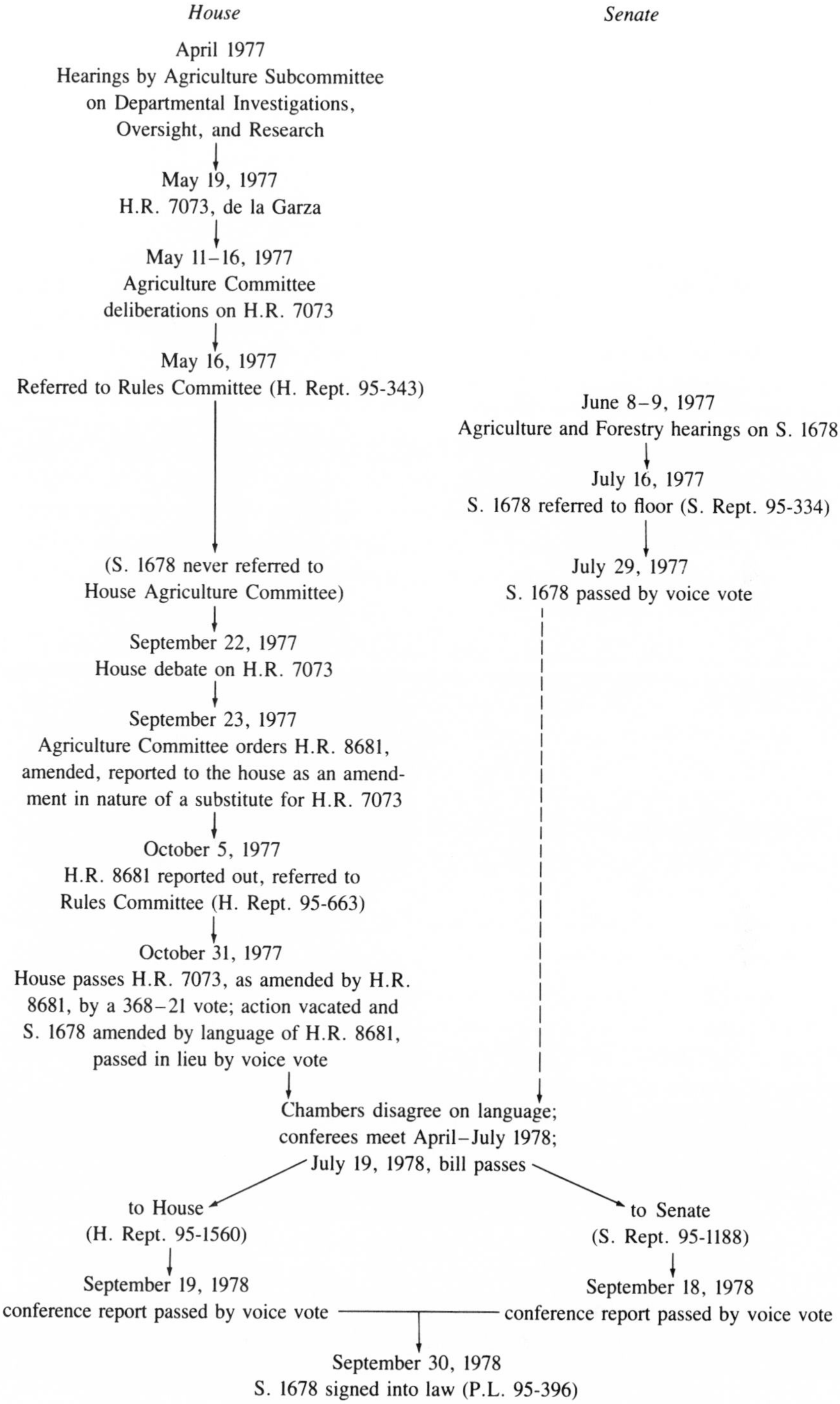

House

April 1977
Hearings by Agriculture Subcommittee
on Departmental Investigations,
Oversight, and Research

↓

May 19, 1977
H.R. 7073, de la Garza

↓

May 11–16, 1977
Agriculture Committee
deliberations on H.R. 7073

↓

May 16, 1977
Referred to Rules Committee (H. Rept. 95-343)

↓

(S. 1678 never referred to
House Agriculture Committee)

↓

September 22, 1977
House debate on H.R. 7073

↓

September 23, 1977
Agriculture Committee orders H.R. 8681,
amended, reported to the house as an amendment in nature of a substitute for H.R. 7073

↓

October 5, 1977
H.R. 8681 reported out, referred to
Rules Committee (H. Rept. 95-663)

↓

October 31, 1977
House passes H.R. 7073, as amended by H.R. 8681, by a 368–21 vote; action vacated and
S. 1678 amended by language of H.R. 8681,
passed in lieu by voice vote

Senate

June 8–9, 1977
Agriculture and Forestry hearings on S. 1678

↓

July 16, 1977
S. 1678 referred to floor (S. Rept. 95-334)

↓

July 29, 1977
S. 1678 passed by voice vote

↓

Chambers disagree on language;
conferees meet April–July 1978;
July 19, 1978, bill passes

to House
(H. Rept. 95-1560)

↓

September 19, 1978
conference report passed by voice vote

to Senate
(S. Rept. 95-1188)

↓

September 18, 1978
conference report passed by voice vote

↓

September 30, 1978
S. 1678 signed into law (P.L. 95-396)

Environmentalists managed to modify Agriculture Committee provisions on trade secrets and data compensation, though they once again were unable to block conditional registration. Committee member Charles Thone (R, Neb.) originally proposed to allow data generators ten years of "exclusive" use, forbidding any EPA application to other registrations. Supporters of this provision, particularly the major chemical corporations, argued that exclusive use would protect research investments, while the EPA and environmentalists—joined by some farm organizations and smaller chemical firms—opposed Thone's provision on grounds that the agency required all available information. Rep. Floyd Fithian (D, Ind.) countered with a ten-year mandatory compensation period, arguing that exclusive use robbed farmers of product choice and denied small firms the chance to compete. The Subcommittee on Investigations, headed by E. (Kika) de la Garza (D, Tex.), first passed Thone's proposal, but only by a 6–5 vote. Chances for success at the committee level, where Foley and other senior members were known to oppose exclusive use, appeared bleak. Rep. Jack Hightower (D, Tex.), who often sided with environmentalists, subsequently proposed to allow only five years of exclusive use, followed by another five years' mandatory data compensation. Any disputes would go to arbitration, similar to provisions in the Senate bill. Hightower's compromise passed 13–0, though, as de la Garza later commented, nobody was totally satisfied.[29] Everyone, it seemed, just wanted to finish before the session's end.

Final House passage came in late October, easing through on a 368–21 vote that masked sharp divisions among supporters. Rep. John Moss (D, Calif.) intended to offer a floor amendment deleting conditional registration, but was convinced that such a move would meet with defeat because it enjoyed support by the administration, farmers, and industry interests. Three amendments passed, including one by Elliott Levitas and Dawson Mathis, both Democrats from Georgia, that reinstituted the controversial one-house veto upon which reauthorization had faltered but a year before.

Differences between the chambers, not the least being that one-house veto, forced lengthy conference proceedings. Major disputes erupted over data compensation, with environmentalists and the EPA reluctantly supporting the Senate version—both ideally wanted free use of all information—while industry and some agricultural groups sided with the House. Conferees eventually agreed, Solomon-like, to give those developing *new* pesticides ten years of exclusive use, followed by five years of compensation, a period coinciding with established patent law. Chemicals registered before 1970 enjoyed no protection, while those registered between 1970 and 1978 enjoyed mandatory compensation for fifteen years after the date of original submission. House and Senate negotiators also sparred over how much authority states should wield in registration and enforcement matters. The Senate, seeking to ease EPA's burden, approved state primacy in enforcement when the state program complied with federal standards, while states operating without an EPA-approved plan would be subject to federal preemption. House members, however, advocated originally that states could register any pesticide for "special local needs," so long as the products in question were not banned by the EPA. EPA officials and environmentalists opposed this provision, arguing that it would end federal certification of state programs and balkanize regulation, while agricultural representatives argued strenuously that federal standards often failed to appreciate local needs, which were better understood by the states. Conferees eventually agreed to retain state enforcement, dropping the special needs provision in favor of one in which a state could tailor federal standards to local needs if the state's *overall* enforcement framework met with EPA approval. Environmentalists opposed even this provision, fearing that the special local needs angle might prove to be a gaping loophole in federal policy, but conferees opted for compromise.

Conference deliberations dragged on well into 1978, and Congress finally cleared the compromise reauthorization in September. FIFRA was extended for another year, indicating once again that many of those involved declined either to let matters rest or to give EPA any breathing room. Nobody, it turned out, really liked the data compensation scheme, but that dispute proved in some ways moot after Monsanto Chemical quickly filed suit in federal court to challenge the constitutionality of the plan.

The dispute hit the legislative agenda almost immediately into the 96th Congress. House Agriculture Committee members in May 1979 once again passed a one-year reauthorization, but one exhibiting even sharper committee divisions. What was intended to be a simple reauthorization quickly erupted into full-scale war on EPA enforcement activities, which the agency brought upon itself with a series of controversial actions. The administration in late 1978 suddenly suspended the use of Mirex against the fire ant, largely because it was discovered to include substantial amounts of the previously banned Kepone. The agency also moved against the important but long-suspect herbicide 2,4,5-T, known as Agent Orange. Suspension of Mirex infuriated southerners still seething over bans on DDT and chlordane, and prompted fractious disputes in the committee over the EPA's enforcement zeal. Rep. Dawson Mathis (D, Ga.) led a fight within the work group to approve emergency use of Mirex through 1980, which passed despite opposition by Chairman Foley and the administration. The committee also approved Mathis's proposal to institute the one-house veto over any EPA pesticide regulation.

Southern anger over Mirex did not, however, carry the full House. The Mirex use provision died, 167–224, with Southern Democrats and Republicans split and northern Democrats almost solidly with the EPA. The environmentalists' glee with that outcome soon ebbed as the House, by a somewhat impressive 278–121 margin, approved the one-house veto, with Southern Democrats and Republicans joining (209–20) to outnumber a somewhat divided Northern Democrat vote.

Senate action in 1979 focused exclusively on simple reauthorization, so the chambers once again squared off in a stalemate not to be broken until the waning days of the Carter presidency. Funding for fiscal 1980 passed via a separate appropriations bill, much to the dismay of the EPA's sharpest critics—Jamie Whitten no longer controlled the pesticides purse strings. House committee members once again found themselves divided, with Chairman Foley and subcommittee head de la Garza unable to push EPA's proposed two-year reauthorization through to the floor. Worse, the committee's own bill suffered defeat because it did not contain a one-house veto. De la Garza argued unsuccessfully that such a provision would kill the bill, since the Senate was known to oppose the measure, but Elliott Levitas prodded the chamber to stick to its guns.[30] Foley eventually returned with a two-house veto, which, he argued, the Senate might accept, and the House passed the one-year reauthorization in June.

Administration officials found opposition to a two-year extension even within the Senate, as Agricultural Research Subcommittee Chair Donald Stewart (D, Ala.) announced that he wanted to review the EPA's progress with the 1978 changes before approving long-term authorizations. EPA officials despaired over such a review, since corporate lawsuits continued to derail data compensation, which, in turn, sidetracked even generic registration. But this was an election year, particularly for many key Senate Democrats, and nobody was in the mood to dredge up the same arguments. The Senate passed the simple one-year reauthorization in late July, and conferees eventually agreed to the two-house veto provision, as well as a House amendment requiring the EPA to submit all data supporting suspensions to a scientific advisory board. Congress cleared the compromise in early December, in one of the last environmental actions of the Carter administration.

Conclusion

The seventies ended sourly for those concerned about the nation's reliance on pesticides, though those on the other side were no less disconcerted about the content and style of federal policy. The "environmental decade" that began with such fanfare and change ended, it seemed, with a whimper and policy retrenchment. Environmentalists complained that the 1978 amendments sent federal policy back to the pre-FEPCA days of *a priori* registration, while farm bloc representatives grumbled that the EPA still showed insensitivity to their needs. The EPA for its part, lay unhappily betwixt the rock and the hard place, immobile in its efforts to register pesticides and under attack for what actions it did take. Years of congressional scrutiny, prolonged debate, and statutory tinkering apparently failed to mollify anyone. Environmentalists found solace solely in their faith that the Carter administration largely sympathized with their values, a statement few of them would make about the administration to come.

Chapter Nine

The Endless Pesticides Campaign

The Reagan Revolution: Recalibrating the Federal Role

The law of politics, Walter Lippman once observed, is the law of the pendulum. Indeed, if the 1960s and early 1970s witnessed unparalleled spasms of social and political reform, "The political climate of the late 1970s was distinguished by the increasing ascendancy of conservative, antigovernment attitudes among the public at large and of conservative, antigovernment ideas among the intellectual elite."[1] Scholars spoke of policy failure, while politicians bemoaned the failure of massive sums of public monies to ameliorate some of our most vexing social and economic dilemmas. Failure seemed everywhere, since poverty and pollution obviously had not gone away.[2] This shift toward systemic skepticism about the role of government began in many respects with Jimmy Carter, who preached about limits to government's capacity to resolve social woes. However, skepticism blossomed fully with the ascendancy of an ethos of limited government and faith in the free market refracted through the election of Ronald Reagan.

To some, Reagan's victory was a "mandate" for a new brand of activist conservatism that would return America to its roots. Liberals who once warned that the structure of governance discriminated against the least affluent and most poorly organized now were challenged openly by those arguing that government had gone too far in the opposite direction. Not only had the bias of the previous twenty years been redressed, conservatives charged, but also a new bias now dominated, one in which those formerly on the outside—consumer groups, environmentalists, minorities, and so forth—now enjoyed preferential treatment by a system no longer sympathetic to the needs of "Middle America." A new disequilibrium reigned in which "the left" had institutionalized access into administrative structures, private foundations *and* government openly funded "anti-establishment" forces, a powerful media elite proved time and again its hostility to business, and above all, government no longer acted as the "neutral" arbiter. Government, conservatives said, must return again to its role as neutral referee if Americans once again were to enjoy unfettered equality of opportunity.

Conservatives argued that the market, not government, should determine national priorities. Government no longer could regulate simply for its own sake, but now must consciously weigh all relevant costs and benefits—social, fiscal, or political—before placing new burdens on society and its markets. The problem, it appeared, was that the once-limited nature of American government had become clogged by creeping statism, that the system threatened to fall into league with such bogeymen, to conservatives, as France or Sweden. To allow this was to hasten the decline of the United States, since such statism meant sclerotic social structures and overly powerful labor, educational, and—to some at least—business institutions. The welfare state threatened overcentralization and a systemic ethos of dependency, a condition to be averted solely by deregulation, decentralization, and rectification of the more egregious liberal follies.

EPA and Environmentalists at Loggerheads

The new conservative tide deeply troubled environmentalists, who saw Reagan as uninformed at best about environmental protection. Many in the movement admitted limits to regulation, acknowledging that other mechanisms *sometimes* provided credible incentives for resource management and pollution control. Few, however, probably agreed with Reagan's Council on Environmental Quality, which stated emphatically that "whenever possible, the achievement of environmental goals and the protection of environmental standards should be left to free market mechanisms."[3] Most environmental activists harbored deep doubts about the extent to which the market could protect the environment; government regulation was a counterpoise to a business sector that few in the movement trusted. Worse, for them at least, the business community had made a strong comeback from its 1970s doldrums. If the number and range of environmental and consumer groups had multiplied dramatically a decade before, by 1980 the rapid countermobilization by business lobbyists in Washington indicated to all that corporate America no longer was content to rely on its "privileged" position in national politics to get what it wanted.[4] Such trends, combined with a presidential administration avowedly oriented towards lightening the regulatory load, led the president of the National Audubon Society, Russell Peterson, to decry what he saw as a return to a mentality that the "business of America is business": "This is the mentality that regards wetlands as wastelands, that looks at a prime forest and sees only board feet of lumber, that looks at a wild river and sees only dam sites, that regards air pollution controls and stripmining restraints as a threat to free enterprise. This view regards environmental protection as un-American."[5]

Administration supporters of course saw things differently. The Republican party platform in 1980 endorsed environmental protection, but warned that it "must not become a cover for a 'no-growth' policy and a shrinking economy."[6] Many conservatives regarded the typical environmental activist as some twentieth-century Luddite, a hopeless utopian who, in a paradoxical way, was an elitist obstacle to economic opportunity and social mobility. Such was the antipathy between the two camps, and so sharp the ensuing acrimony, that Reagan later would moan, "I do not think they will be happy until the White House looks like a bird's nest."[7]

There is no need here to review Reagan's environmental record, nor the administration's relations with the environmental community.[8] Suffice it to note that the administration's visceral rejection of regulation and its preference for market mechanisms soon got it into very hot water with Congress—and not only with Democrats—and with an electorate that, despite Reagan's electoral success, apparently was still highly supportive of environmental protection. Perceptions, true or not, that the administration had "politicized" a traditionally "nonpartisan" issue area cost it dearly.

The heavily publicized tug-of-war between Congress and the executive over the EPA's direction obscured more mundane but no less critical regulatory matters. I recall attending two congressional oversight hearings on the same day in early 1983, one on EPA official Rita Lavelle and the Superfund "scandal," the other on reauthorization of FIFRA. The Lavelle hearing was classic public spectacle: television crews jockeyed for position among the "regulars" at the press table, lines of the curious wound their way down the hallway and around the corner, and every subcommittee member sat in public attention, surrounded by teams of doting staff members. The FIFRA hearing provided a notable contrast, attended only by a few members of the print media and industry trade journals, the "regular" farm, environmental, and industry representatives, plus the subcommittee chair, with an occasional appearance put in by other subcommittee members. The first hearing provided drama, while the second seemed more a case of going through the required motions. Everyone at the FIFRA hearing knew the stakes involved, and few surprises enlivened an otherwise pro forma exposition of competing arguments.

That hearing was but one small indicator, but it helped to suggest that pesticides regulation during the early 1980s was a backwater, an issue area on the lower end of the agency's budgetary and organizational priorities. It might not be too presumptuous to suggest that EPA became tired of FIFRA, tired of the regulatory morass and intractable disputes connected to pesticides matters. FIFRA sank back into oblivion during much of Reagan's first term, as toxic wastes and the travails of Ann Burford, head of the EPA, filled the headlines, and only those who truly cared took much notice about goings-on within the Office of Pesticides Programs. Congressional "oversight" followed each reauthorizing effort, but such efforts increasingly appeared desultory. Congress during the early 1980s, as we shall see, failed to settle many of the dilemmas stubbornly embedded into FIFRA—an absence of action that remanded policy making to administrative and judicial venues.

Reagan's new team at EPA lost no time in their efforts to reduce FIFRA's perceived regulatory burden. That zeal produced, critics later charged, an overemphasis on administrative speed over veracity of data, an overworked and progressively more understaffed OPP, and systematic shortcuts in regulatory procedures. Administrator Burford, acting on orders to reduce "unnecessary" advisory boards and—perhaps more important—to weed out any influence by Carter appointees, fired the entire FIFRA Science Advisory Panel, which had acted as a public hearing mechanism on cancellations and other regulatory actions. Congress's failure to reauthorize FIFRA on time in 1981 allowed the panel's mandate to expire, and Burford thereafter declined to name a new board. That action effectively eliminated a major, if little-publicized, avenue for access and information by nonindustrial interests. It was just part of what many charged was an administrationwide effort to restrict access to policymaking. This situation changed somewhat after Congress reauthorized the advisory panel and William Ruckelshaus returned to the EPA, replacing Burford, but environmentalists feared that conditions would deteriorate anew once Ruckelshaus left the scene.[9]

Environmentalists also argued that those entrenched within the OPP, freed at last from countervailing pressures by those higher up, became more sympathetic than ever to industry needs. One activist pointed to a 1982 speech by Assistant Administrator John Todhunter as evidence of the new and unwelcome orientation. Todhunter, in concluding his remarks before a pesticides trade group, made his views quite plain:

> Clearly the NEW DIRECTIONS which I have outlined WILL CHANGE the REGULATORY CLIMATE in the Office of Pesticides and Toxic Substances. I intend to REPLACE the generally ADVERSARIAL relationship between industry and government WITH one which is based on SCIENTIFIC DIALOGUE and REASONED REGULATION. I agree with a recent statement made by Dr. Alan Robertson of Imperial Chemicals Ltd that "BETTER AND MORE RAPID DECISIONS DERIVE FROM DECISIONS BETWEEN SCIENTIST AND SCIENTIST RATHER THAN BETWEEN LAWYER AND LAWYER." I also intend to ALLOW INDUSTRY every opportunity to effect meaningful and effective SELF REGULATION when problems are identified. . . . I believe that the end result will be BETTER PROTECTION at LESS COST.[10]

Environmentalists regarded such comments as evidence of collusion to come, but one OPP official discounted such charges as so much politics: "At the program level there was not a hell of a lot of difference between political administrations. A few years ago, people in the OPP were just plain discourteous to industry representatives, an animosity that probably grew out of EPA genesis. This animosity was tempered somewhat under the Carter administration because industry was disturbed by the way it was being treated. A majority of EPA people are not hostile to industry."[11]

A key congressional staff aide, regarded by others as the "gray eminence" behind FIFRA, commented in early 1983 that "EPA people at the lower levels, the program levels, are highly cooperative to both sides. These lower-level echelon EPA people are still open, though not as publicly as before. Environmentalists' beefs are primarily with the upper levels."

Perhaps one indication that not all of those in OPP were hostile to environmental concerns might be the number of internal EPA memoranda shown to me by critics of the agency to support their views that OPP was indeed cozying up to industry. These memoranda of course were leaked from within the EPA, and many environmentalists admitted that they still had friends in the agency.

There is little argument, however, that overall the EPA—at least prior to the second coming of Ruckelshaus—was less open than ever before to environmentalists and their demands. And the evidence corroborates charges that the agency through 1983 systematically sought to reduce, if not eliminate, whole sets of FIFRA provisions objected to by the chemical industry. The agency in 1982, for example, proposed broad changes in pesticides regulation, including almost total devolution of enforcement (but not standards) to the states, more lenient data requirements for registration, and voluntary compliance with EPA testing protocols. This latter proposal, commented one environmental activist, was analogous to making the fifty-five-miles-per-hour speed limit voluntary. The OPP in early 1983 also tightened public access to health and safety data—information technically unrestricted since the 1978 amendments. Environmentalists, labor unions, and many farm organizations complained bitterly that these restrictions severely undermined independent data verification and that, according to the NRDC, the agency was engaged in an "EPA-industry compact to prevent dissemination of pesticide data."[12] OPP chief Edwin Johnson, a career agency employee, countered that the restrictions simply aimed to prevent access to data by competitors, and that such a move was necessary until either the Congress or the courts resolved vexing matters concerning data use and compensation. Johnson may have had a point, but few environmentalists trusted his objectivity, and more than one complained that he was inordinately sympathetic to industry needs. That perception obviously tainted the relations between the two camps.

The question of data took a back seat to questions of EPA resources and overall agency interpretation of FIFRA provisions. A late 1982 staff report to a House Agriculture subcommittee noted with concern large cuts in OPP personnel during 1981–82, a greater diminution expected for 1983, and a concomitant 62 percent increase in the number of OPP actions. These figures, when taken together, indicated that fewer personnel were handling twice as many actions as in 1980.[13] Agency officials lauded these trends as movements toward greater bureaucractic efficiency, but critics retorted that much of this newfound streamlining in fact came through regulatory shortcuts and outright manipulation of statutory loopholes. FIFRA, for example, contains an "emergency use" provision designed to allow EPA to approve short-term and specific application of otherwise banned or restricted materials. Such flexibility was intended as a precaution against sudden pest outbreaks, where speedy control is a top priority. The 1978 amendments similarly empowered the agency to grant "special local use" authority to the states—a right designed to allow specific and limited use of a banned or restricted chemical in situations either where use is very minor (on certain specialty crops or for infrequent outbreaks) or where no suitable substitute yet exists. Neither provision was meant to bypass the regular registration process. However, the report noted, the number of "emergency use" exemptions skyrocketed between 1980 and 1982, from 198 to 505, while permits for "special local needs" climbed from 180 in 1980, to 750 in 1982.[14] Farm and industry representatives argued that the continued registration logjam (a snarl created in no small part by industry litigation over data use and compensation) made emergency and special use provisions the only solutions when farmers or state governments needed quick relief. Critics contended that extravagant use of such provisions effectively scuttled FIFRA's intent and effectiveness, and that the EPA simply was making ad hoc judgments about pesticide registration and use—a charge leveled at the USDA before the 1970s.

The report also charged that EPA had all but scuttled the RPAR process, disbanded the Office of Special Pesticides Programs, relaxed risk assessment standards, and put pressure on agency scientists to reduce the registration backlog at the expense of data veracity. The NRDC, in a 1983 lawsuit, charged the agency with hundreds of secret deals with industry, including those leading to the EPA's suspension of investigations into the controversial pesticides lindane and paraquat. The suit also alleged that high-level OPP officials allowed industry representatives to draft agency documents announcing important regulatory decisions, that they excluded the public from decision making, and that advance notice of key decisions went to industry representatives alone. Ruckelshaus, not long after returning to the EPA, settled with the NRDC out of court, agreeing to reopen PRAR reviews on thirteen products rather than face protracted court action.[15]

The administrative side of the ledger thus witnessed charges of program mismanagement, regulatory chicanery, and statutory shortcuts. The situation in Congress, by contrast, is summed up in one word: stalemate. While Congress fiddled during the 1980s, critics suggest unsympathetically, the administration, at least until mid-1983, allowed FIFRA to burn, and pesticides registration returned to its pre-1972 days.

The Pesticides "Subgovernment" in the 1980s: A Mirage in the Making

Mary Etta Cook and Roger Davidson observe that the 97th Congress (1981–1982) failed to enact a single new antipollution law. What was more critical, "Of the eight major environmental statutes on Congress' agenda for reauthorization, only two had been thoroughly debated and reauthorized when the 97th Congress adjourned sine die on December 29, 1982. The remaining six programs were prolonged by simple short-term extensions or continuing appropriations."[16]

Things did not improve markedly in the 98th Congress. Part of the problem of course lay in partisan conflict, not only between Congress and the White House but also between the Democratic House and a Senate controlled by Republicans for the first time in almost thirty years. But this legislative stasis on the environment also reflected deeper conditions, some of them characteristic of multidimensional, highly complex issues. Environmental issues no longer seemed so sacred to legislators bombarded by conflicting and equally compelling pressures, but instead carried with them multiple and competing costs and benefits. Interest-group coalitions no longer were so stable as before, but instead they fractionated and coalesced almost randomly as each segment in a complex issue area emerged. Budgetary pressures, and the growing dominance of budgetary politics on the congressional agenda, pushed most other matters to the margin. A noticeably disaggregated legislature, where political power after a decade of "reform" seemed more amorphous than ever, found itself increasingly unable or disinclined to force closure on more and more matters. Action on FIFRA through the early 1980s became a mirror image of that of the late 1970s, with recurring swings in the momentum of policy claimants, but with equally problematic outcomes. And, in fact, the experience of FIFRA during this time powerfully reveals a Congress in deadlock, much of it intentional.

A great deal of this deadlock grew out of the committees on agriculture, which no longer shared a unified vision about policy. The committees, once bastions where senior legislators jealously defended agricultural prerogatives on pesticides, became work groups whose membership more accurately reflected the massive changes in congressional priorities and the overall influence of agriculture on Capitol Hill. Tables 11 and 12 suggest one reason for this. The House committee, for example, now is a larger, less senior body of representatives, in which members remain less automatically and perhaps have less issue knowledge than during the 1950s and 1960s. Committee members also seem more willing to relinquish committee posts when opportunities beckon elsewhere. Senate committee membership displays similar trends.

Even more noticeable is the fragmentation of views within the committees. House committee chair Thomas Foley in 1981 ceded his position to E. (Kika) de la Garza (D, Tex.), a low-key legislator highly oriented toward consensus and

Table 11. House Agriculture Committee Tenure, 1945–1985

	No. of Members	*Mean Tenure (in years)*	*Adjusted Mean*[a] *(in years)*	*No. of Freshmen*
1945	28	5.21	4.74	6
1947[b]	27	5.74	5.19	7
1949[b]	27	5.87	4.58	8
1951	30	5.27	4.62	10
1953[b]	30	5.97	5.28	6
1955[6]	34	6.26	5.61	10
1957	34	7.03	7.96	4
1959	34	6.59	6.03	11
1961	35	7.54	6.97	6
1963	35	7.88	7.26	9
1965	35	6.97	6.26	13
1967	35	5.51	4.79	18
1969	32	6.06	5.22	7
1971	36	6.33	5.54	11
1973	36	4.61	3.71	11
1975	41	3.02	2.15	21
1977	46	3.47	2.66	11
1979	42	4.66	2.68	7
1981	43	4.37	4.19	14
1983	41	4.73	4.40	8
1985	43	5.96	5.61	6

a. Adjusted mean tenure = gross tenure minus the tenure of the highest-ranking member (in terms of years served on the committee) and divided by the remaining number of committee members. This was done to remove a notable outlier, particularly 1947–1977, when the committee chair tended to have at least twenty years of service. Removing this outlier gives us a better insight into the committee's overall experience.

b. Years of partisan *change* in committee control.

not viewed as a master of legislative detail. One House staffer who served another committee with interest in FIFRA, opined, "There is lots of internal dissent in the Agriculture Committee with de la Garza in control," though that turmoil was caused by differences in perspectives on issues and members' priorities more than by the chairman's managerial style. A chief source of trouble appeared to be subcommittee chair George Brown (D, Calif.), who gained jurisdiction over pesticides policy after de la Garza's move to the committee chair. Brown, long regarded by environmentalists as a good friend, often appeared at odds with his own subcommittee on major program matters. The same staffer commented, "Representative Brown is in the minority on the subcommittee. Most of the other members are more cautious than Brown with respect to the sensitivity of the issue and the nature of their constituents. Brown is more willing to challenge the farm groups and the industry." An aide on Brown's subcommittee countered that the real reason for the subcommittee's dissent over FIFRA is

Table 12. Senate Agriculture Committee Tenure, 1945–1985

	No. of Members	*Mean Tenure (in years)*	*Adjusted Mean[a] (in years)*	*No. of Freshmen*
1945	20	7.25	6.26	8
1947[b]	13	6.85	5.08	3
1949[b]	13	5.00	3.58	5
1951	13	4.08	3.16	3
1953[b]	15	4.73	3.86	4
1955[b]	15	5.93	5.00	2
1957	15	7.13	6.14	2
1959	17	7.12	6.12	4
1961	17	7.59	6.50	3
1963	17	8.41	7.25	3
1965	15	9.80	8.43	3
1967	15	9.93	8.43	4
1969	13	11.00	9.33	4
1971	13	11.30	9.50	3
1973	13	9.08	7.16	3
1975	14	7.86	6.15	2
1977	18	7.67	6.23	4
1979	18	5.89	4.23	6
1981[b]	17	3.53	3.00	4
1983	18	4.88	4.35	2
1985	17	6.23	5.62	2

a. See note to table 11. The all-time champion for tenure on the committee was Milton Young (R, N.D.) who, by 1979, had racked up thirty-four years of continuous service. Young unfortunately never was made chair, and left office before the Republicans took over Senate control in 1981.

b. Years of partism *change* in committee control.

that "Brown insists on an open door for *all* sides, so that peoples' positions would rise and fall with the merits." And, of course, nobody agreed on the merits. Whatever the case, and the truth appeared to lie somewhere in the middle, the subcommittee had tremendous difficulty in reaching a consensus, and, worse, frequently found itself at odds with the full committee. The committee, in turn, encountered stiff opposition on the House floor—a trend that had emerged previously but escalated during the 1980s.

The Senate committee during the early 1980s proved an even more dubious instrument for policy closure, and gave policy stakeholders true headaches. Gone was the genial and accommodative Herman Talmadge, replaced by Jesse Helms (R, N.C.), the abrasive conservative who often clashed publicly with members of his own party. Helms held few hearings on the program, and, according to observers, few on the panel really grasped the complexity of the issue. The situation for FIFRA on the Senate committee, concluded one staff member, was "horrible."

Fragmentation within Congress mirrored the divisions among interests concerned with the pesticides issue. Small chemical firms fought regularly with larger ones over data use and compensation, while the larger firms accused the smaller ones of seeking unfair market advantage. Farm groups, once comfortably aligned with the chemical industry, increasingly found themselves caught between the needs of small and large firms, not to mention between the interests of industry and environmentalists. Lost in the shuffle, one farm group official complained, were the needs of the farmer:

> Our coinciding interests with industry remain, but less often than in the past. Agrichemicals are bigger businesses than before, which leads them to defend their more central interests. Business is wrapped up in its own interests, which may or may not coincide with user interests. Those stronger patent and trade secret statutes ill-serve the farmer, since they prevent price competition. Farmers are sympathetic to business's interests, but not to the point where it includes a single company holding market dominance with regard to a certain type of chemical.

Farmers also no longer depended primarily on the USDA for information and guidance, but increasingly on chemical firms and, surprisingly, some environmental groups. Traditional linkages among the USDA, the land grant colleges, industry, and the farmers still existed, but grew weaker than ever. "USDA is behind the times," argued the farm group official. "It still does 'how to use' research, but the research now is done primarily in the industry, which has the best labs and scientific talent." And, even more startling, he was satisfied to keep FIFRA in the EPA: "There are public relations advantages to keeping it in EPA, no 'fox in the henhouse' perception, since EPA is mandated with protecting the public health and environment. Putting FIFRA back into the USDA would just bring out the old charges of 'capture.' I think that when the EPA gets its act together it is the right place for a broader perspective on environmental management than might be the case with the USDA."

Whether this view dominated the agricultural community at large is debatable, but there seems little doubt that the pesticides subgovernment experienced centrifugal pressures during the past decade. Environmentalists, meanwhile, saw renewed surges in public interest and organizational strength after the activist fatigue and general environmental-group doldrums of the late 1970s. The coming of the Reagan administration no doubt catalyzed environmental organizations, and the first-term appointments of Ann Burford to EPA and James Watt to the USDI proved two of the best shots in the arm for environmentalists since Earth Day. Organizational memberships skyrocketed, with the Sierra Club alone claiming a 60 percent increase (180,000 to 300,000) in the years 1980–1982.[17] Budgets grew, as those concerned about the Reagan agenda donated greater sums to those in opposition.

More critical, at least for their future political impact, environmentalists began to change their tactics to match the new realities. Groups that once relied totally on litigation sought out members, backed congressional campaigns, and hit the grass roots. One activist explained: "The picture is so bleak, as far as getting help from the federal government, that I think that we need to look at some other approaches. Unfortunately, the EPA was supposed to protect us nationally from pesticides, but history goes up and down, and we are sort of in a down period." A chemical association executive agreed: "Environmentalists have learned to operate within the system. the amount of money they donate to congressional races has grown fantastically, and this money is a pivotal point in some of the races." Indeed, industry representatives and environmentalists alike pointed to environmental group support for the 1982 reelection of Senator Robert Stafford (R, Vt.), a prominent player in later FIFRA action.

Environmentalists also worked harder at building coalitions—including such former foes as the AFL-CIO—on common problems like worker safety, and began to work harder in political venues outside Washington. Fear that a Reagan EPA might dismantle FIFRA administratively, if not through legislation, prompted dozens of environmental and public-interest law organizations in 1981 to create the National Coalition Against the Misuse of Pesticides (NCAMP), an umbrella organization dedicated to coordinated action at both the federal and state levels. "Environmentalists now are building up more permanent grassroots at the state and local levels," noted one participant, "because there is only so much you can do through litigation. Lawsuits seem to drag on forever." Lawsuits also became incredibly expensive. The fight over DDT, for example, cost the Environmental Defense Fund over $300,000, and that was over a decade ago. Environmentalists groused that FIFRA allowed for only narrow definitions of standing in court, but, they added, the increasing cost of litigation made this a moot point anyway. The judicial route simply was not as useful as before, so environmentalists turned their sights to other decision-making arenas. Industry representatives, when interviewed, say that they had not seen this level of environmental grass-roots activity since the early 1970s. Said one industry official, "[all this] has made my job a hell of a lot tougher," since industry too must now work at several different levels and in more decision arenas than ever before. The game itself has become fractionated, and the players involved find it far more complex.

Finally, the nature of the pesticides issue itself evolved. An issue once defined in stark food-versus-famine terms by the 1980s became multidimensional and complex, forcing all those involved to change their way of thinking, at least privately. Says one environmentalist: "These are the 'nuts and bolts' days, unlike the earlier days when all you had to do was say, 'trees,' and a lot of people would lend their support. You just can't say 'trees' like in the old days. Everyone now realizes the complexity of the issues involved, though everyone publicly may argue that it is a 'black-versus-white' issue."

These conditions—an administration with an agenda for policy redirection, an increasingly divided and diffuse policy community, and a general admission (at least in private) about the complexity of the problem—coalesced during the early 1980s to make issue closure difficult indeed. In fact, consensus under those conditions became well-nigh impossible.

FIFRA in the 1980s: Going Nowhere

To understand the course of congressional action on FIFRA during the 97th Congress is to acknowledge that the chemical industry perceived the coming of Reagan as a golden opportunity for favorable policy change. Environmentalists seemed in disarray, whipsawed between an EPA suddenly displaying suspicious orientations and a Congress of uncertain loyalties. The Reagan philosophy on deregulation and the wisdom of market forces seemed preeminent both in the executive and legislative branches, even among those traditionally regarded as friends of the environment. Industry officials were generally optimistic about their chances on Capitol Hill, particularly in a Senate that long had been a thorn in their sides. Such expectations fed industry proposals for statutory change to the degree that its alleged rapacity would provoke stiff opposition. Or, as one participant remarked, "Industry got too greedy."

FIFRA's authorization expired at the end of fiscal 1981 with neither consensus on the nature of the pesticides problem nor agreement about future directions. Both committees on agriculture reported out simple one-year reauthorizations in May, but further progress bogged down under industry protests that pet substantive alterations went unconsidered. Atop the industry's agenda for change were restrictions in data use and disclosure, public access to health and safety information, and state regulatory authority. So adamant were industry lobbyists on these changes, and the environmentalists were so equally stiff-necked against them, that both bills went right back to committee. To do otherwise would have guaranteed problems.

George Brown reconvened the Subcommittee on Department Organization, Research, and Foreign Agriculture for a series of hearings in summer 1981, and its sessions were attended by virtually anyone with an interest in FIFRA. Brown reopened the hearings to allow everyone a chance to speak, but also extended them because his subcommittee—and the Committee on Agriculture in general—lay terribly split on the major proposals under consideration. Brown also no doubt heard the covetous noises rising from other House committees with a keen interest in FIFRA, work groups that would be only too happy to take over jurisdiction should Agriculture fail yet again. One environmentalist later commented, "Brown offered us a very important opportunity to present our side of the story in a more comprehensive way than had ever been offered in the last ten years." Given the internal and external pressures on the subcommittee, Brown may have had little choice.

A fragile coalition of industry lobbyists, representing both large and small chemical makers, pressed hard to change the 1978 provisions governing data use and compensation. The larger firms wanted longer periods for exclusive use of data, fifteen years instead of ten, while the smaller producers sought relaxed registration standards for those merely formulating a product from purchased chemicals. The EPA regarded both as a bit excessive, and unnecessary, but the administration overall sat on the fence, allowing industry to take the lead. Environmentalists of course opposed both proposals, regarding the first as restraint of trade and the second as an erosion in registration standards.

Second, industry spokesmen sought to narrow public access to health and safety data—a right created by the 1978 amendments. Litigation immediately followed passage of that law, which effectively stalled EPA implementation, but the Supreme Court in 1981 *(Union Carbide* v. *Costle)* ruled that the public indeed had a right to access. Industry lobbyists soon clambered all over the Hill with proposals to either eliminate all public access to data or narrow it substantially. The Brown subcommittee, split on the matter and—desperate for a compromise—originally developed the idea of a FIFRA "reading room" where members of the public might examine industry data under EPA supervision. The agency would determine who could gain access to the files (to prevent access by competitors), as well as the materials allowed to be copied. This concept excited neither the industry nor the environmentalists—the one fearing that valuable data might fall into the wrong hands, the other accusing the subcommittee of impeding the public's right to know. The EPA, meanwhile, worked out an agreement with the industry to allow members of the public to look at selected data but forbade any copying. Brown and a few others on the subcommittee thought this idea overly restrictive.

Third, chemical firms sought to expand "trade secrets" to include all "innovative chemicals" or "innovative analytical methodology" that was used to determine toxicity. What would construe "innovation" was to be determined by the EPA—a prospect that sent environmentalists howling to the subcommittee, particularly since the agency had worked on the idea exclusively with the industry. Jay Feldman, coordinator of the National Coalition Against the Misuse of Pesticides, argued vehemently that such data protection would serve only to undermine independent verification, even more seriously than would the public access restrictions. Chairman Brown, who was willing to negotiate on most matters, chastised industry officials for suggesting what, in his mind, only "impede[d] scientific progress," and he warned chemical firms to tread carefully, else they find their intransigence reaping unexpected and unwanted rewards.[18]

Fourth, pesticides makers sought limits to state regulatory authority, arguing that "regulatory federalism" was fine for some issues, but not this one. Chaotic market conditions were the industry's public fear, but its rejection of state authority seemed based more on a recognition that many states by the 1980s were in fact tougher on pesticides than was the EPA. Chemical companies especially wanted to prohibit states from requesting data for registration beyond that required under FIFRA—a right previously upheld by the courts. Environmentalists, once passionately opposed to state prerogatives, now almost paradoxically promoted them, and not simply because of the Reagan EPA; a decade of environmental consciousness had prompted many states to pay greater attention to such concerns. And, in their opposition to narrowed state authority, environmentalists suddenly discovered an unlikely ally: state departments of agriculture. State officials, whether or not they sided with environmentalists on any other facet of FIFRA, agreed wholeheartedly that state prerogatives were sacred, and industry lobbyists soon encountered wellsprings of opposition that they probably never anticipated.

Finally, pesticides industry lobbyists pushed to restructure FIFRA's legal framework, seeking both to expand *their* judicial rights and to narrow rights of access by private citizens. Companies generating registration data wanted the right to sue those using their data without permission, and such suits could produce treble damages for the plaintiffs. Only the specter of stiff penalties would preserve their proprietary rights, industry officials claimed, and the courts were the most effective guarantors of compliance. What was good for the goose did not appear to be good for the gander, however, as the industry and some agricultural groups proposed to deny private citizens *any* right to sue for use or misuse of pesticides. The EPA adequately defended the public good, they argued, and farmers needed to be shielded from "nuisance" suits by environmental and farm worker organizations. The agency sided quietly with industry on both counts.

Public sparring over these industry proposals, and a subsequent deadlock within the Brown subcommittee, drew legislators from other House work groups into the fray. Environmentalists, increasingly worried about the direction of policy debate and impatient with Brown's efforts to negotiate compromises, asked allies on other committees to intervene. "The Agriculture Committee is not an appropriate body to discuss scientific issues or health issues," charged Jay Feldman, suggesting that only drastic measures could prevent industry success.[19] House Energy and Commerce chair John Dingell (D, Mich.), a longtime industry critic, contemplated what should be his own action should Brown fail, and in May formally requested sequential referral of H.R. 5203. Dingell also lobbied for exclusive future jurisdiction over FIFRA, since his committee already dealt with a wide range of environmental and food safety issues. To make matters more difficult for Brown and company, House Science and Technology chair Don Fuqua (D, Fla.) also threatened to intervene, but later was persuaded to drop his plans after de la Garza promised quick and fair consideration of the bill. Dingell's request for referral, meanwhile, was rejected by a cautious Speaker O'Neill.

Such threats to Agriculture hegemony over FIFRA but did not cease, and they no doubt spurred Brown and his colleagues toward decisions. The intensity of the conflict in 1982 prompted Brown to warn his panel that things might get out of hand unless they acted. "If they get much worse," he cautioned, "it endangers passage of the bill."[20] Environmentalists promised a bloody floor fight should the subcommittee not tone down or delete some of the more controversial industry proposals, while industry lobbyists increased their own pressures. "Both sides played their cards heavy and hard," recalls an aide to Henry Waxman (D, Calif.), who chaired an Energy and Commerce subcommittee. "The lobbying got very heavy handed, the worst I have ever seen. It was definitely hard ball."

House Bill 5203 eventually surfaced in April 1982 with most of the changes suited to industry views, though subcommittee members unanimously registered their disapproval with industry pressure by allowing private citizens the right to sue pesticides makers and users (though only for injunctive relief). Environmentalists had argued throughout mark-up that the industry's provision was unequal justice—an argument apparently seductive to a subcommittee willing to rebuke industry in at least one case. One subcommittee aide later complained that industry lobbyists "got increasingly stubborn" and were "acting in a paranoid fashion." The private lawsuit measure was industry's only real subcommittee defeat, but that it occurred at all was seen as a message to industry lobbyists to lay off.

Even that setback to industry was redressed by the Agriculture Committee, which overturned the private lawsuits provision by a 21–19 vote. The committee also endorsed, by similarly close margins, industry proposals on exclusive data use, restricted public access to data, expanded trade secrets protection, and truncated state authority. Farm groups complained about the data provisions, arguing that they unnecessarily hindered free market dynamics, but they reluctantly supported the overall package. Labor, including the AFL-CIO, opposed the bill as a whole, primarily because of the private lawsuit provision and because of the committee's failure to address the bill's impact on farm and chemical workers. Environmentalists vowed to carry the fight to the floor.

The administration, meanwhile, vacillated publicly, first opposing the restriction on state authority and then, after "persuasion" by chemical lobbyists, switched to support federal preemption. Office of Management and Budget (OMB) officials steadfastly opposed the bill on fiscal grounds, since the Agriculture Committee provided for higher funding levels than desired by the White House. The EPA in the end officially sided with the OMB, but everyone on the Hill knew that Burford and Todhunter sided with the industry.

If industry lobbyists congratulated themselves on the committee vote, their merriment quickly dissipated when the bill hit the House floor. The lower chamber would pass H.R. 5203 in August, but not before it shredded the committee bill and tacked on a few amendments opposed both by industry and the EPA. The chemical makers' reach for comprehensive statutory revision eluded their grasp, as a broad coalition of environmentalists, labor, state officials, and segments of the scientific community persuaded legislators to take a careful look at the committee package. That coalition turned out to be more potent than even its partners envisioned. Industry lobbyists were prepared for an amendment by Rep. Abraham Rosenthal (D, N.Y.) to expand citizen standing in court on regulatory matters—an amendment that was defeated—but were caught completely off guard by a spate of reversals that sent them scurrying out of the galleries. Rep. Thomas Harkin (D, Iowa), a member of the Agriculture Committee and notable dissenter from the committee bill, reinstituted state regulatory authority by a surprising 95-vote margin (250–154) that split Republicans and southern Democrats. The House also deleted industry's trade secrets proposal, reinstated private lawsuits, and renewed the 1978 provision granting public access to health and safety data. Neither industry nor concerned Agriculture Committee members could staunch the tide.[21]

Nor were they prepared for the final amendment, one that coalesced concerns about the impacts of pesticides held by so many House members. In an address to his colleagues that one industry lobbyist later called "the most emotional thing I've ever seen," Rep. Sydney Yates (D, Ill.) pleaded for a ban on toxaphene, once one of the nation's most widely used pesticides and an alleged cause of his wife's cancer. The Carter administration had judged toxaphene to be carcinogenic, Yates said, but the current EPA dragged its feet on further action. Despite pleas by Brown that the EPA, not Congress, was better equipped to ban a pesticide, the House overwhelmingly supported Yates.[22]

The House did leave most industry proposals intact, but the reversals served notice to the Agriculture Committee that it might be out of step with majority opinion. One pesticides industry association official called the House votes "emotional" and "a repudiation of the Agriculture Committee," while an environmental lobbyist said simply, "We turned the House of Representatives into a health and environmental committee of the Congress. It should have sent a message to the leadership of the House that the committee structure and the authority that has been given to the Agriculture Committee are not appropriate." A subcommittee aide, however, retorted that "environmentalists cannot accuse the Agriculture Committee of ignoring public health interests," adding that things could have been a lot worse without Brown in charge. The floor votes really were no surprise, he said, though he was amazed by the scope and diversity of the anti-industry coalition. Perhaps one environmental activist best explained the results: "There is hardly a congressman today who doesn't have a pesticides problem in his district." That reality, one perhaps not appreciated by an industry convinced of its momentum, would powerfully influence future policy deliberations.

The two-year reauthorization now lay in the hands of the Senate Committee on Agriculture, which quickly reinstated the measures killed on the House floor. Committee members restricted state authority, eliminated private lawsuits, and generally accepted provisions on data exclusivity and access first approved by their House counterparts. The only notable industry defeat in committee came when it rejected a Helms proposal for expanded trade secrets protection—a vote prompted apparently by threats of a floor fight coming from Environment and Public Works chair Robert Stafford (R, Vt.). Industry lobbyists nonetheless claimed victory, while environmentalists said that no bill at all was preferable to this one. Subcommittee chair Richard Lugar (R, Ind.) subsequently hoped aloud that the warring factions might keep their differences "down to a low roar" because, he concluded, "it would take no skill at all to kill this bill."[23]

Lugar turned out to be far more prescient than he might have wished. Environmentalists quickly convinced Stafford, Howard Metzenbaum (D, Ohio) and William Proxmire (D, Wis.)—people who needed little convincing—to issue "holds" on the bill. A hold is a bit of senatorial courtesy that says that no bill thus held will proceed further without the approval of the concerned senator. Stafford's hold was particularly influential among his Republican colleagues. Said one Senate aide: "Stafford wanted to play a role in FIFRA this time, and it was pretty well known, although unannounced, that he was prepared to offer amendments on the floor if it came up. This factor led to widespread realization that there was no quick solution in sight." Farm groups, concerned that Stafford and other environmental sympathizers might push through a private lawsuit provision as part of a compromise, persuaded Orrin Hatch (R, Utah) to issue his own hold. Not that it mattered: environmentalists by autumn were resigned to no bill at all. The 1982 congressional elections loomed, and most on Capitol Hill were in no mood for a costly floor battle. "It was easier to stop the whole process than to fight it out," commented one staffer, "which is why we did it."

Slim as hopes for compromise were, they were dashed completely during an unusual lame-duck session that convened after the elections. Senator Helms held up all action in the chamber with his filibuster against the administration's tax bill—a tactic that proved costly to his own role in FIFRA matters. One industry lobbyist later sighed, "[Helms] is not much of an asset to us anymore," while an aide to a senator opposed to the industry provisions exulted, "We didn't think we could win anyway, but Helms stopped it for us." FIFRA reauthorization, after two years of tough, protracted conflict, died quietly at the session's end. Program funding continued under separate appropriations measures, just as it did in 1981.

The Brown Subcommittee reconvened early in 1983, but amid dramatically altered political circumstances. If pesticides makers were wounded by the subcommittee's behavior in the 97th Congress, the public controversy over the EPA and its handling of toxic wastes programs would send industry lobbyists running for cover. EPA Administrator Burford, Assistant Administrator Todhunter, and a score of other high agency officials soon were gone, and environmentalists suddenly had the initiative. The change in tone was stark. Environmentalists, who in 1982 fought hard to stall progress on a bill, now pushed equally hard for thorough substantive change, attempting to exploit their perceived opportunities just as industry had in 1981. Albert Meyerhoff, testifying before the subcommittee on behalf of a coalition of environmental organizations, said, "The time is ripe, finally, for a fundamental reform in this critically important statute."[24] But, as environmentalists quickly discovered, nobody else wanted to play. "Everyone is sick and tired of it" groused one subcommittee member. "Industry came through the briar patch in 1982 and they don't want to get into it again, with the political situation as it is." An aide to Waxman's subcommittee commented, "It seems reasonable to just postpone the debate for a while. If you open up FIFRA right now you just open up a Pandora's box of problems. . . . It was a very bitter struggle in 1982, and nobody came out of it feeling very good."

Industry representatives acknowledged their sudden reticence. "There's a whole new ball game at EPA right now," said one lobbyist, "and we have a terrible image in the eyes of the public." Several interviewees also admitted that the chemical industry itself was split over future directions, and industry lobbyists pointed fingers at one another over who was to blame for the debacles of the previous session. "Industry has gone from leading the charge to being discreet," commented one Senate Agriculture Committee staffer, "and now they're in the trenches trying to hold their own." Industry unity had dissipated, he added, and it might be a long time before industry can present a common front on policy directions.

Congress during 1983 eventually reauthorized FIFRA for one year, but even that decision took considerable effort. Industry lobbyists and the new EPA leadership sought a simple two-year extension, which would have carried FIFRA beyond the 1984 presidential election and given Ruckelshaus and his people time to evaluate the program. Environmentalists and their congressional allies retorted that the sudden changes at EPA made short-term authorization all the more necessary; to go further would only support the status quo. Besides, environmentalists wanted to strike while industry and the EPA were in the congressional doghouse.

Brown's subcommittee nonetheless reported out a two-year authorization, with panel members hoping aloud that it would give them the breathing space to study the program in peace. A simple reauthorization also seemed the path of least resistance. "We'd like to do this without raising a big stink," Brown said after the vote, but he soon was embarrassed in committee, which accepted Tom Harkin's amendment for a one-year reauthorization.[25] Environmentalists threatened a reprise of the 1982 floor fight if they didn't get the one-year extension and promise for committee attention in 1984—and everyone this time around knew that they could succeed. Several influential House members, including Waxman, James Florio (D, N.Y.), Elliott Levitas (D, Ga.) and Albert Gore (D, Tenn.), warned de la Garza that a two-year extension "would forestall timely attention to urgent legislative concerns that may underlie severe problems at EPA. . . . The public needs to know that Congress is reviewing programs and taking actions."[26] De la Garza reluctantly acquiesced, and Harkin's amendment passed. The House quietly approved the one-year extension on May 17.

Matters went pretty much the same way on the Senate side, though Helms and Lugar apparently stalled action through the year until the Senate in November pulled the bill out of committee and put it on the calendar. The chamber then passed the extension, which was approved by the White House without comment. The outcome in 1983 pleased no one, and particularly those members of the House Agriculture Committee for whom FIFRA fast was becoming a dirty word. "We went through this process during the past two years with a great deal of heartburn," moaned one member. "Why has this thing dragged on so much?"

The stalemate continued through 1984. EPA administrator Ruckelshaus, under tremendous pressures from both ends of Pennsylvania Avenue, pleaded for more time so that he could appoint an advisory committee and review the pesticides program thoroughly. House Agriculture Committee members were only too willing to go along, expressing their desire to devote their attentions to a high-priority farm bill. Besides, argued de la Garza, his committee was not going to touch FIFRA during an election year unless he saw signals from Helms and the Republican leadership that the Senate would act first. The Senate committee had held but two days of hearings in three years on the matter, and it appeared obvious that things weren't going to change now. Senate committee Staff Director George Dunlop said simply, "Without something concrete that the EPA or the administration could advocate, we have no leverage."[27] Helms himself was in a very tough reelection fight, and the administration displayed virtually no interest, so it soon was apparent that nobody save environmental groups wanted to bother with the program.

Election year finger-pointing and buck-passing infuriated environmentalists, who complained bitterly that they had supported the simple authorization in 1983 only because of promises for action in 1984. But with neither Congress nor the administration in any mood for a fight, and, with industry quietly happy to let matters rest, environmentalists were stuck with a simple extension. They didn't even get that. FIFRA, despite a spasm of public alarm when ethylene dibromide (EDB) residues were found in grain, simply went unauthorized after September, with program funding maintained through appropriations legislation. It was, one Agriculture Committee member commented, "a hell of a way to run a Congress."

Transforming Adversaries into Collaborators

The organizers of Earth Day—to say nothing of their critics—would have been dumbfounded. In early March 1986, following almost a year of private, face-to-face negotiations, representatives of ninety-two chemical manufacturers and a coalition of forty-one environmental, labor, and consumer groups presented to Congress a compromise proposal that all involved hoped would finally lay to rest the acrid and lengthy battle over FIFRA. The compromise, hammered out independently of the EPA and farm organizations, called for a five-year reauthorization that would mandate new schedules for data submission and product registration, boost program funding through fees levied against registrants, expand citizen access to health and safety data, and authorize EPA action against any pesticides previously registered on the basis of false or invalid data. The proposal met with immediate promises for consideration by both committees on agriculture.[28] That something novel was afoot was acknowledged by all involved. "It is rare that such traditional opponents agree on anything," said Nancy Drabble, director of Public Citizen's Congress Watch, "let alone a major piece of environmental legislation which breaks a fourteen-year impasse."[29] Carl Kensil, speaking for the Ciba-Geigy Corporation, said simply, "Times change. It's time for the agrichemicals industry to change."[30] Both Drabble and Kensil could be accused of severe understatement.

What brought about this apparent dénouement? The main reason, though there appear to be a number of secondary incentives, was the sheer frustration felt by all with an interest in the shape and direction of federal pesticides regulation. Two things had become clear to all by early 1985. First, the evidence was compelling that FIFRA was bankrupt as a policy. Only six of some six hundred active ingredients mandated for reregistration under the 1972 amendments had been fully tested, and most of the more than 40,000 products on the market were unknown quantities in terms of their health and safety implications.[31] Chemicals banned in the United States continued to filter into the nation's food via imported fruits and vegetables, since FIFRA did not extend to chemicals exported for use elsewhere. Highly public controversies over residues of the fungicide EDB in grain and of the herbicide Temic in watermelons raised a furor regarding EPA enforcement activities. It had become more obvious than ever that both the law and the agency had little but post facto impact on pesticides and their effects, and a law that never really worked as envisioned now appeared not to work at all. FIFRA, argued Drabble, simply did not protect the public: "Most Americans probably assume that all pesticides have been carefully examined by the EPA and have passed rigorous health and safety tests. This is simply not true."[32] The EPA privately agreed; many of the same conclusions had been reached by the agency's own special advisory panel during its two years of study of the program.

Second, and probably more critical, the events of the first half of the decade finally had convinced everyone that no breakthrough could occur without some kind of bargain struck among those who had slugged it out in the trenches so long. There simply was no way that the committees on agriculture, much less an entire Congress, could pass a worthwhile pesticides bill so long as the acrimony among policy claimants continued unabated. Congress as an institution had become so permeable to outside pressures on this issue that the apparent equilibrium among those claimants had translated into policy stasis, and only some shift in the relations among those claimants could break the deadlock.

A broken policy, a Congress unable or unwilling to settle the matter, plus an administration with little or no interest in pesticides policy, finally added up to frustration for those who cared. Environmentalists by 1985 were ready to take half a loaf in order to get something, *anything,* while chemical makers had come to the realization that FIFRA no longer served them well, if it ever had. Those on Capitol Hill admitted their inability to hammer out a policy, while EPA personnel expressed their dismay at being caught in the middle, for what seemed like forever, between combatants.

Nobody involved would have predicted the events that would take place in 1985. FIFRA had gone unauthorized after October of the previous year, and those who cared were fully prepared for yet another year of long and probably inconclusive policy conflict. Environmentalists and industry lobbyists once again were ready to promote widely divergent proposals for policy reform. The EPA, following its two-year study of FIFRA, now wanted to move forward with statutory reform, but found itself hamstrung by an Office of Management and Budget that showed no inclination for action. The House Agriculture Committee was in flux, with George Brown moving on to another post and ceding his subcommittee chair to Berkely Bedell (D, Iowa). On the Senate side, Jesse Helms indicated that he had no intention of fitting FIFRA into his committee's agenda. Other committees in both chambers declared again their intentions to move against the committees on agriculture. The whole situation seemed liked a likely replay of the previous several years.

Yet, paradoxically, these very conditions may have eased the way for negotiations. Industry lobbyists were nervous about the continuing stalemate in the wake of the late 1984 tragedy in Bhopal, India, and the controversy over the fungicide EDB. These events had reignited public questions about the impact of pesticides, and raised the specter of a new wave of litigation relating to groundwater contamination and manufacturing plant siting. Many also were worried that the continuing stalemate, the growing resentment among members of Congress about industry's past behavior, and their increasingly dominant image as nay-sayers might finally tip the balance and propel policymaking into directions not to their liking. Chemical lobbyists also still smarter from a disaster in the previous session, when a cherished patent term extension bill went down to defeat. Chemical makers wanted such extensions to compensate them for the time and profits lost while the EPA determined whether their pesticides were safe. The bill had been sponsored by Rep. Dan Glickman (D. Kans.), and prospects for passage looked good since such relief had been granted previously to drug makers. Glickman, however, soon was embarrassed to find that the bill, which had been authored by the National Agricultural Chemicals Association, also would eliminate public access to pesticide health and safety data. That attempt to change FIFRA through indirection proved costly, since both Glickman and Sen. Charles McC. Mathias, Jr. (R, Md.) indicated no support for the patent extension *until* progress was made on FIFRA. "I'm being held hostage," said NACA president Jack D. Early, "but that's the way things work in this town."[33]

House Agriculture Committee members were more worried than ever about losing their cherished jurisdiction over FIFRA should the stalemate continue.

Bedell, despite the administration's and the Senate's opposition to any change, pushed ahead in the spring of 1985, hoping that the events of the previous fall and winter might create opportunities for statutory revision. Bedell's efforts no doubt were spurred on by yet another challenge from Henry Waxman, who was seeking to overhaul pesticides policy through changes in the Food, Drug, and Cosmetic Act. On the Senate side, Robert Stafford's Environment and Public Works Committee displayed increased impatience with the pace of policy consideration, and Stafford announced his intent to co-sponsor a FIFRA bill with William Proxmire. The pressure once again was on the embattled defenders of agriculture.

Yet it became clear, as Bedell convened his subcommittee for hearings in late April, that chances of a breakthrough were as problematic as ever. A coalition of environmental organizations, calling itself the Campaign for Pesticides Reform (CPR) authored one bill—sponsored by George Brown—that would tighten registration standards and deadlines, mandate training for those applying restricted pesticides, grant greater public access to health and safety data, allow for citizens' suits against all violators, and set strict timetables for EPA reregistration efforts. Industry lobbyists, backed by the White House, advocated a simple reauthorization, suggesting that chemical makers, not sure anymore about their leverage with the subcommittee, were ready to maintain a largely defensive position. The EPA originally had intended its own bill, which would end the indemnity payment program, streamline procedures, and grant the agency more authority to ban products registered on false or invalid data. Overall, the agency bill in fact coincided well with the concerns expressed by many environmentalists and subcommittee members. Agency officials, however, were overruled by the OMB—which increasingly dominated agency proposals. This action clearly infuriated Bedell and Pat Roberts (R, Kans.), who proceeded to incorporate the EPA proposals into their own bill.

Industry representatives appeared in a quandary. They opposed the changes proposed in *both* bills, but to lobby hard against them was sure to destroy whatever good will toward them remained among subcommittee members. One subcommittee aide commented, "The Agriculture Committee is less friendly these days toward agrichemical interests;" many industry lobbyists still were tainted by the debacles of previous years. Industry representatives probably could forestall movement, since both the administration and the Senate Agriculture Committee opposed any revisions at this time, but the costs to industry in the future might be too high. Besides, chemical makers desperately wanted the sidetracked patent extension, and lack of movement on FIFRA guaranteed trouble on that front.

Environmentalists, for their part, apparently had begun to tire from the incessant battles and simply wanted to put the issue to rest. Stalemate only maintained the unacceptable *status quo,* and it seemed obvious that prospects for any bill were as dubious as ever. More than a decade of activism had reaped considerable benefits, but, for many in the environmental movement, the game had to change somewhat if gains already made were to be solidified and future ones made possible. They no longer could count on the EPA, despite its apparent willingness to cooperate, if only because the White House stood in its path. Environmentalists had shown their capacity to withstand industry's efforts to make statutory change, but the costs both to their resources and to overall policy had been high. The only option left might be to talk directly with their opponents. Bedell, Roberts, and Brown, worried about mounting attacks on the Agriculture Committee and concerned that yet another session might pass without progress, publicly urged just this sort of direct negotiation. Besides, the congruence between the CPR and EPA bills offered fertile ground for some sort of compromise.

And so it went. Representatives of the Campaign for Pesticides Reform—led by Nancy Drabble of Congress Watch and Albert Meyerhoff of the Natural Resources Defense Council—sat down over the summer of 1985 in private discussions with David Wathke of Elanco Products and, later, Scott Ferguson, vice president and general counsel of NACA. The CPR represented some forty environmental groups, while NACA spoke for almost a hundred major agricultural chemical producers. Each side began negotiations by staking out areas for possible agreement, a process that of course left some issues—and some interests—excluded. Nowhere to be seen, for example, were the major farm groups, which over the years had played a declining role in pesticides policy as the issues involved became ever more urban and environmental in nature. Also conspicuous in their absence were representatives from the smaller chemical firms, led by the Chemical Specialties Manufacturers Association (CSMA), which during the 1980s clashed repeatedly with NACA over data use and compensation issues. Finally, several environmental organizations, including the National Audubon Society, the Environmental Defense Fund, and the Izaac Walton League, chose not to be included in the negotiations, largely because NACA from the start refused to budge on the indemnity question.

Most negotiating sessions were one-on-one, between Drabble and Wathke—while Meyerhoff and Lawrie Mott (also of the NRDC) joined in on occasion. Some environmental lobbyists not involved in the effort criticized the CPR representatives, suggesting that Drabble's relative newness to the pesticides debate (Congress Watch had not played a major role in FIFRA deliberations prior to this time) and Meyerhoff's lack of continuous participation (he is based in San Francisco) hurt the environmental side. Many worried that the CPR was letting the chemical industry off the hook just as public opinion and political momentum had shifted against it. One House aide, however, discounted such criticism, arguing that Drabble and Meyerhoff were on the phone to each other continuously. Both sides also had to get clearance from their respective groups before they could agree to anything, so the style of negotiating was not all that critical. Besides, these negotiations were the only hope for a breakthrough in what otherwise promised to be another inconclusive run at FIFRA reform.

The negotiators eventually agreed to a legislative package, first revealed in September 1985. NACA president Early called the agreement "a very important first step" that "will be in the best interest of the public and the pesticide industry." The NRDC's Meyerhoff said simply, "While no panacea, . . . it's an important breakthrough toward major reform."[34] The CPR-NACA agreement called for the EPA to review all existing health and safety data and identify gaps, which pesticides makers would have four to five years to fill or face suspension of their products from the market. The agency also would be required to re-register all products within one year of completing data files, and would have the authority to cancel all registrations for pesticides registered previously under invalid data. The agreement also specified quicker and more streamlined registration and cancellation procedures, allowed for greater public access to health and safety data, and mandated the public's "right to know" with respect to chemical manufacturing facilities (a growing concern after Bhopal). Finally, in a bid to woo agricultural groups, the compromise called for a ban on the importation of all commodities containing detectable levels of residues from pesticides that had been suspended or otherwise pulled off of the market in the United States.

Negotiations on technical languange continued on into 1986, and the coalition submitted its legislative package in March. Bedell, Roberts, and Brown—three who didn't exactly agree on FIFRA provisions in previous years—introduced the measure in the House, while Richard Lugar did so in the Senate. Lugar, obviously pleased by the compromise, also promised to convene his subcommittee and consider the package sometime in the summer, a marked departure from the lack of Senate committee action on FIFRA during most of the decade. Legislators in both houses hailed the agreement as a historic breakthrough in the interminable logjam respecting pesticides policy reform. The compromise, announced George Brown, "gives hope to all of us who have tried to reconcile the ideals of participatory democracy with the realities of trench warfare of legislative deliberation.[35] William Proxmire called the package a "stunning achievement and one which should be imitated again and again. I can think of no other example of historic adversaries coming together and drafting a strong consensus bill which protects human health while considering effects on the regulated industry."[36]

The compromise clearly had buoyed a legislature long frustrated at the deadlock in FIFRA reform. "The agreement is pretty strong," commented one House committee aide, "and everyone around here is encouraged." But, he added cautiously, prospects for passage in 1986 were not exactly preordained:

> Everything depends on the House. If we get bogged down this year the Senate will never get to it. But, if we get moving soon, Helms and Lugar might be willing to push things along. The OMB is giving the EPA a longer leash than last year, but we're concerned that the White House still isn't behind the agreement. Things have improved, but not totally.

The warning was appropriate, since the CPR-NACA package did not address concerns expressed by some who did not take part in the negotiations—for whatever reasons. EPA officials, though pleased at the progress made by those long on opposite sides of the regulatory fence, expressed doubt that the agency could reregister thousands of pesticide products within the six years targeted by the bill without sufficient funds. Such monies, if the OMB had anything to say about it, were not to come out of general appropriations. Many environmental lobbyists criticized the lack of "sufficient" action on groundwater and indemnities issues, which the negotiators had avoided for the sake of action on other matters. Strong opposition to the package came from farm groups, concerned that farmers might be harmed by lawsuits unless protected from liability for groundwater contamination caused by even proper use of pesticides. Finally, and perhaps more important, the chemical firms not represented by NACA complained loudly that the compromise did not address the critical data use and compensation question that still divided smaller and larger chemical makers. NACA and the CPR did not debate this matter during negotiations, since environmentalists considered this a purely business issue and NACA was divided internally on the matter. Some NACA members also belong to the CSMA, or to the Pesticides Producers Association (PPA), and the trade associations long had split on how those making "copycat" products were to reimburse those generating the data used in the registering the original chemical formulation.[37]

Marked improvement in the coalition's chances came during House subcommittee deliberations in April, which, when compared to previous years, moved along rather quickly. The American Farm Bureau Federation—not known traditionally as an ally of environmentalists on pesticide matters—signed on with the CPR-NACA bill after getting some relief for farmers from liability questions. Legislators meanwhile strongly discouraged attempts by the CSMA and its allies among the smaller chemical firms to broaden their focus and include the data compensation question. The lateness of the session and the already crowded nature of the legislative calendar—dominated by tax and budget matters—spurred subcommittee members to narrow their considerations to the ten issues central to the CPR-NACA package. That vehicle proved decisive, when compared to the half-dozen previous years, and the subcommittee unanimously reported out the bill in May. Lugar, meanwhile, began deliberations on the Senate side.[38]

House Agriculture Committee consideration moved with equal alacrity. The committee, in marked contrast to previous battles, accepted the bulk of the subcommittee package and reported out a bill in mid-June by a 42–1 vote. Few on the committee could recall such unanimity on FIFRA, and what attempts to alter the CPR-NACA package emerged largely were rebuffed by committee members as the coalition held together. The CSMA, for example, failed to gain sharply limited compensation rates for products registered using previously generated data. The committee also approved in principle an amendment to prohibit states from setting tolerances on pesticides residues that varied from federal standards, but deleted the provision from the bill after both NACA and CPR representatives warned that the amendment would jeopardize their agreement and provoke a turf fight with the Energy and Commerce Committee. Ground-water contamination was not addressed, as committee members and lobbyists decided to defer the matter in favor of progress on the basic compromise package.[39]

The CPR-NACA sat in limbo through early August, partly because of forces beyond the control of those central to the compromise and partly because of issues those who negotiated the package deferred or ignored. On the House side, FIFRA awaited the good graces of a Rules Committee already burdened by a crowded legislative agenda and facing a calendar short on working days and long on recesses and election-year concerns. One House subcommittee aide said simply that if FIFRA does not get to the floor by early September it may run out of time as tax, budget, and other "high priority" matters take precedence. In the Senate, the Agriculture Committee, long unused to debating FIFRA anyway, became locked in debate on data compensation, an issue deferred or glossed over previously. "Things can still split apart," argued a House subcommittee aide. "The Senate is the key right now, and things are 50–50 right now for passage before the year's end." But, he added, if the Congress should succeed and send a bill to the White House, the breadth of the coalition behind the bill should be enough to reap presidential approval; chemical companies and farmers hardly represent the sort of interest that a Republican president can ignore in an election year.

Conclusion

The long-sought reform of FIFRA at this writing (mid-August 1986) stands at the mercy of the congressional calendar. The Senate Agriculture Committee on August 13 unanimously approved an amended version of the CPR-NACA package. The differences are important, however, since they involve questions of state regulatory authority and registration data use opposed or deferred by House committee members. The House Rules Committee, meanwhile, has included FIFRA into the floor schedule, with debate set for early September. Those involved can only speculate on the probabilities for success. Whatever the eventual outcome, there seems little doubt that matters would never have progressed as far as they did without the year of negotiations among many long opposed to each other.

What are we to make of the events of 1985–86? More than thirty years ago, Earl Latham made his oft-cited (and much criticized) statement that the legislature acts as a "referee," ratifying the outcomes of battles among interest groups.[40] As E. E. Schattschneider rejoined a few years later, "It is hard to imagine a more effective way of saying that Congress has no mind or force of its own or that Congress is unable to invoke new forces that might alter the equation."[41] Subsequent studies of course showed numerous cases where interest-group strength is not the controlling variable when compared to committee dominance, the power of a chairman, or ideology. Congress, or at least many of its parts, certainly were not mere "referees."

But, in reviewing pesticides policy conflict during the 1980s, and recalling similar battles on a wide number of policy fronts, one cannot help but remember Latham's characterization and wonder whether the image is not more apt today than it was then. The Congress of 1986 certainly is a far more open and permeable place, and its members certainly seem far more responsive to outside forces (i.e., pressures) than when Schattschneider penned his retort. Congress *is* a more conflict-ridden institution these days, in no small part because the issues it deals with are more national—and more divisive—in scope. Congress never was meant to force closure before a consensus had formed, but the current condition of decentralized decision-making power and entrepeneurial memberships certainly makes the task all the more difficult.

To belabor that point is unnecessary. But, if Congress is as permeable and as incapable of authoritative policy closure as we say it is, then the task is to look at how interest groups—and, increasingly, coalitions of groups—strive to work it so that Congress *can* resolve conflicts. Schlozman and Tierney observe the trend toward coalition building among interest groups and comment, "Since the organizations involved in coalitions work hard to iron out any differences they might have so as to present Congress with a united front, they reduce the level of conflict that legislators see and thus make it more likely that Congress will act favorably on these 'predigested policies.'"[42] The willingness of legislators to accept the bulk of the CPR-NACA package in this case certainly lowered the level of conflict on pesticides policy reform, and chances for policy success were the strongest in many years. That a "predigested" policy paved the way for legislative action should cause us to look more at the changing nature of the relationship between Congress and the interest-group community, and among interest groups themselves.

Chapter Ten

The Pesticides Perspective Revisited

Thinking about Change

There is nothing permanent except change.
—Heraclitus

Looking back over this study dredges up memories of my undergraduate geology course. We would slog through fields and quarries, returning to the lab with a panoply of pebbles, crystals, or whatever looked interesting. Then we would study our finds, often for days, turning each over under ultraviolet light, dripping tears of acid on each face, or scraping off shavings for further tests. Each session ended by locking up our materials, to which we would return the next day, ready to pick up where we left off. The rocks were still there, unchanged.

Change. I am reminded of how confidently I studied those rocks as I struggled here to understand political change. The problem with studying change is that, well, most things *do* change, and a lot more rapidly and less predictably than do rocks. Understanding that any study of change is by nature always incomplete, there comes a time when we must stop and reflect on where we've been, where we are now, and, one dares to presume, where we are going. The present is always a convenient stopping point. If the past is prologue, this must be epilogue.

This study began rather simply. I wanted to select some case, given certain theoretical concerns, and observe how it and its attendant policymaking configuration weathered the test of time. Why, I wanted to know, were some actors and interests attracted to this issue in the first place? Why not others? What roles do the structures and processes embedded in the policymaking system play in the relation between issues and interests? To what effect? And—no doubt the most presumptuous question of all—how did all this change over time? These questions guided this study. They also frame this final (though hardly conclusive) discussion. I structure this discussion according to the framework first shown in chapter 1. This study arbitrarily ends in August 1986, although things will continue to change.

Problems, Perceptions, and the Age of Enlightened Uncertainty

Problems are artificialities, defined through decidedly subjective interpretations of real-world conditions. A condition does not produce "a problem" but a whole host of *possibilities* awaiting someone's perceptual linkages. Each phase of this study began with conditions that I thought relevant to the pesticides case on the assumption that context does matter. Problems are contextual; so must be policy responses.

One dynamic stands out after forty years of national concern regarding synthetic chemical pesticides: the objective conditions have evolved to higher orders of complexity, but the fundamental relationships paradoxically remain pretty much the same. That is, the conditions, in a strict sense, are far different today than in the 1940s, but their relationships to one another, and—more critical—to those perceiving problems, almost defy the test of time. I get a strong sense of déjà vu as I observe contemporary events; I get a feeling that I've seen all this before.

U.S. agriculture again is at a crossroads. The farm community faces fiscal pressures perhaps unparalleled since the 1930s. The abject poverty characteristic of the 1930s certainly seems a bad memory to most (though certainly not all) in the farm belt, but farm incomes once again lag badly behind total farm costs. Imagine that the year 1977 is our base (that is, 1977 = 100) in reckoning farm income and expenditures. On this scale, farm income in 1984 was 134, indicating some growth, but the prices farmers paid for goods and services equaled 161, indicating outflow exceeding income that accelerated dramatically after 1980.[1] The nation has enjoyed an economic boom (at least in the aggregate sense) since 1983, with healthy growth in gross national product and an even healthier decline in inflation. Consumers were happy, but agriculture appeared to suffer from triple blows of low commodity prices (due in no small part to the usual surpluses), high interest rates, and a strong dollar (which devastated exports). Many farmers who expanded their acreage during the speculative seventies, aided and abetted by the federal government, also found their mortgage payments outpacing the current value of their land. The result was a aggregate farm debt that grew by almost $60 billion between 1980 and 1985.[2]

The aggregate figure of course hid some notable trends, which came to my attention in the form of a front-page headline: "Rural protests grow 'like a prairie fire.'" The accompanying story detailed the surge in midwestern farm community unrest that has reached proportions not seen since the days before the New

Deal.[3] Farmers unable to repay loans are losing their farms at record rates. Banks, particularly the smaller rural outfits, are finding it hard to recoup their losses, and the number of bank failures has jumped dramatically. The federal farm credit system, the traditional lender of last resort for many farmers, itself is in a shambles. A farm community accustomed since the 1950s to a middle-class life style now finds that standard seriously threatened by macroeconomic forces beyond its control, and the answer increasingly is to take to the streets of the state capital. Smaller farms, always the most vulnerable to macroeconomic swings, are facing the auctioneer's gavel at the highest rates since the Great Depression, and the concentration in agriculture roars ahead at a faster clip. Only one percent of landowners today hold one-third of all ranch and farm land and account for nearly 65 percent of all farm income.[4] Critics on all sides agree that traditional federal farm policies, heavily based on acreage restrictions and subsidies—all costing some $35 billion in 1986 alone—paradoxically aid overproduction and hurt the family farm.[5] Low commodity prices, high capitalization demands, and the export market make economy of scale an overriding objective. Only the largest and best capitalized survive. "If the trend continues," warns former USDA Secretary Bob Bergland, "family farming will disappear and, in its place, there will be enormous economic power concentrated in huge corporate operations.[6] The Office of Technology Assessment calculates that about 1 million family farms will fold by the century's end, and estimates that some 50,000 large agricultural operations will produce 75 percent of our food by that time.[7] Readers who get the feeling that they've heard this before are not alone.

This newspaper, no doubt unwittingly, juxtaposed the story with another that, on the surface, had no bearing on the farm problem. That headline—"28 Chemical Leaks from W. Va. Plant"—no doubt made some farmers, not to mention most chemical industry executives, wince, because the story had direct implications for the agricultural dilemma.[8] The Union Carbide Plant at Institute, West Virginia, was but one of two plants in the world to make methyl isocyanate, a highly toxic chemical used to produce the popular pesticide carbaryl. The Institute plant in January 1985 came under acute public scrutiny precisely because its sister plant in Bhopal, India, had recently caused the greatest single pesticides disaster in history. More than 2,000 people in Bhopal died in December 1984 following a massive leak of methyl isocyanate—an incident that shocked the world, rocked the agrichemical industry, and sent hordes of lawyers scurrying for their liability casebooks. The Bhopal plant and its products inarguably aided India's "green revolution," but the accident tarnished both the company and its product.

The events at Bhopal directly affect American agriculture. Farmers today rely more heavily on chemical pesticides than they did in the 1940s. Small operators depend on chemicals to cut costs and stay afloat; larger operations are even more chemical-intensive as they seek to maximize efficiency and profits. American agricultural policy continues unwittingly to encourage chemical dependency—the pesticides treadmill wields its perverse magic—and policies directed toward laudable goals often only increase that dependence. The trend toward "no-till" farming, for example, seeks to reduce the soil erosion that threatens to return much of the Great Plains to Steinbeck's Dust Bowl, but no-till operations depend heavily on herbicides to control weeds and make the practice economically feasible. Seemingly discrete policies in fact are parts of a seamless web of interrelated problems, which makes answering one problem depend powerfully on solving a good many others.

Total pesticides production has doubled since Rachel Carson unleashed her attack on organochlorines, rising from 730 million pounds in 1962 to more than 1.5 billion pounds in 1980.[9] Two trends hide behind this aggregate figure. First, use of persistent organochlorines like DDT and chlordane has virtually ended in the agricultural community, the result both of increased insect resistance and the bans put in place since 1970. "The EPA took 40% of total pesticides volume out during the 1970s in its bans on organochlorines," claims one agricultural representative. "That is success if you measure it that way." Indeed, use of organochlorines dropped by more than 80 percent since the late 1960s, which, in turn, apparently improved prospects for some endangered species and reduced persistent residues in the soil and the human body.[10]

Second, there is a greater reliance today on less persistent, more specialized chemicals, which by definition must be used in greater volume and at more frequent intervals. The old organochlorines were popular precisely because of their persistence and broad-spectrum capabilities. Newer products are less persistent, but also more highly toxic, which raises concerns about the total volume being applied and about the safety of those who make and use the chemicals (for example, migrant farm workers). The massive growth in total pesticides use, much of it comprised of herbicides, worries many environmentalists and health officials. Pesticides users, on the other hand, complain that the newer formulations are more expensive, require greater amounts to do the job, are less convenient, and more temporary. Farmers who once used a single product for several purposes now complain that they must use several products for a single purpose. They worry both about their increasing inability to procure effective, affordable pest-control technologies and that chemicals in use today may not be available tomorrow. Such concerns are but new versions of those raised decades ago.

If objective conditions seem essentially unchanged, popular perceptions about those conditions decidedly are not. If we know little else, we certainly understand that problems beset us from all sides—or so it seems. We have observed the expansion in public awareness of problems. A question that once concerned relatively few has become over time a problem to an ever expanding array of worriers, many of whom have no direct attachment to the condition but who profess stakes in its effects. Why did this occur?

The first answer must be *knowledge,* both in its absolute and relative forms. Knowledge is the great leveler, and the more people who comprehend the consequences of some condition into means a broader range of possible policy claimants. It also guarantees a fragmentation in perceptual consensus. Knowledge, after all, neither assures acceptance of the condition nor certainty about effects. If anything, knowledge assures uncertainty; the more we know about a problem, and the more who know it, the less it seems that we understand, or agree upon. To understand is to question, and more people today are asking more questions about more things.

We not only know more about pesticides as a discrete problem, but—and this is even more important—we now perceive ever greater relationships between pesticides and *other* problems. A limited number of policy claimants were able to formulate FIFRA in 1947 simply because the remainder of society did not fully understand (if they understood at all) how DDT and its kin might affect the environment and human health; nor did they perceive that the issue bore consequences beyond the agricultural community. Time and "progress" bring knowledge, as well as new conditions, so that with time more people can and will make the perceptual connections among superficially discrete problems. The result, of course, is an even greater expansion in the community of claimants.

Today we are confronted by vast knowledge, sophisticated technology, forests of data, but precious little agreement about answers—or even about the nature of the problem. Suspending disbelief is problematic in a society where knowledge abounds and where so many profess equally plausible interpretations of the same referent. Policy closure seems more possible when all or most who are attracted to a problem share the same angle of vision, when perceptual consensus smooths the edges of conflict and melds all into apparently seamless support. But time and knowledge erode that consensus, slowly at first, and then in a tumble as consensus gives way to a perceptual free-for-all. New knowledge and new angles of vision introduce all involved to perceptual pluralism—most disconcerting to those seeking policy closure.

A brief case might better illustrate my point. Farmers since the 1950s have used ethylene dibromide (EDB) as a fumigant to prevent spoilage. In 1956 the FDA judged the chemical safe and therefore exempt from tolerance standards. Almost twenty years passed before the EDB matter came up again, this time as the center of scientific doubt over its safety. Greater knowledge, new testing protocols, and more sophisticated measuring equipment now allowed scientists to measure residues in the parts per *billion,* and the National Cancer Institute in 1974 judged EDB to be an acute carcinogen. "When you look at this data, it's exceedingly good," said Dr. Richard Hill, an advisor to the EPA. "It's just about as good a case as one can have."[11] Other scientists were not so sure, pointing out that EDB posed far fewer risks than did antifloxin, a natural carcinogen that occurs in peanuts and other crops. All agree that EDB causes cancer in laboratory rats, but disagree about its implications for human health. This battle went on within governmental and scientific circles, largely out of the public eye.

The question of EDB suddenly flowered in late 1983, when Florida health officials unilaterally recalled dozens of grain-based products (such as cake mixes) found to contain more than one part per billion of the chemical—the lowest detectable level. Discovery of EDB residues in drinking water, due apparently from its use as a soil fumigant, prompted the action, which sparked tremendous turmoil within agricultural circles and in the halls of government. California immediately banned importation of all citrus fruit fumigated with EDB; the EPA publicly weighed a decision to ban most uses for a chemical with no previously established residue tolerance; USDA officials pondered the possible impacts for both domestic agriculture and the grain export market. Food industry executives warned darkly of commodity shortages should other states emulate Florida's action, while Secretary of Agriculture John Block called the whole matter "ridiculous" and one that "could destroy the entire economic system." Only 5 percent of the nation's grain inventory was fumigated with EDB, but USDA officials—who for years had blocked a proposed EPA ban on the chemical—worried that the public furor might stigmatize all sales. Congressional actors called for investigations, while environmentalists cried for a total ban on EDB use. "The federal law is clear," said the NRDC's Al Meyerhoff; "it does not allow cancer-causing agents to be in our food."[12] Meyerhoff was referring to the Delaney Clause, which disallowed *any* level of a carcinogen in food. But this was a law born in 1958, before we even could measure in the parts per billion. We now are more precise and, not surprisingly, also encounter criticism that "zero tolerance" is both unrealistic and potentially disruptive to the nation's economy.

During all this time the average citizen must have asked: "But is this stuff dangerous?" Experts of all stripes agreed that EDB was bad, but nobody agreed about how bad. Said one biologist: "It is even more dangerous than Aldrin and Dieldrin banned by the EPA several years ago. Residues are showing up in drinking water, in fruits, carrots, and wheat products. It's also a gasoline additive, so it's in the fumes—the total consumer exposure is horrendous."[13] Other scientists however discounted the total risk, arguing that EDB's short-term risks were nothing compared to other hazards (for example, cigarettes) and that regulators and politicians were unnecessarily scaring the public.

Environmentalists charged that any risk was too much, that EDB merely piled yet another potential hazard onto the growing mound of suspect products and practices. Industry and agricultural experts countered that to desire a risk-free society is utopian, and that EDB posed far fewer risks than many "natural" carcinogens. The antifoxin example arose numerous times, and the author recalls with some amusement the industry "expert" who claimed on national television that the risk from a peanut butter sandwich was greater than of eating EDB-contaminated foodstuffs. Maybe he's right, but how is the citizen to know?

The EPA in 1984 set voluntary interim standards of 900 parts per billion for raw grain intended for human consumption and 150 parts per billion for cake mixes and other products. Massachusetts officials, however, banned all products exceeding 10 parts per billion, while an Ohio official declared, "When you're talking about parts per billion you're talking about a spit in the ocean."[14] EPA head Ruckelshaus later commented, "We have to admit our uncertainties and confront the public with the complex nature of the decisions about risk."[15]

And then what? The EDB case shows that every thesis prompts its antithesis. Betwixt the pesticides Panglosses and the environmental Cassandras we find a vast gray area of uncertainty that makes the hope of consensus laughable. We need not suspend disbelief because we know enough not to accept someone else's word. Our knowledge allows us to reopen disputes or carry them ever onward in hopes of achieving greater precision, perfect solutions, or, maybe, capitulation of our foes. The latter seems the most likely, but that doesn't mean that we've solved our problems. Knowledge is our weapon, and the exchange of scientific opinion an intramural activity among the knowledgeable, each claiming what is wrong but not agreeing on what is right. We seek answers, but cannot even agree on the questions.

The public, meanwhile, only knows, or perceives, that more problems seem to exist and that fewer remedies appear to work. Our technology brings us vast informational resources, cadres of experts, policies, but startingly few answers. John Dewey more than fifty years ago noted, "The machine age has so enormously expanded, multiplied, intensified, and complicated the scope of indirect consequences . . . that the resultant public cannot identify and distinguish itself."[16] Dewey might return today and despair. He advocated freedom of inquiry and full publicity about consequences as the answers to the problem of the public in modern society, but one must wonder if the babble of voices belonging to those claiming to know in fact confuses matters even more. Such communication may fail to assuage public concerns precisely because the signals themselves are both contradictory and equally persuasive. Such policy debates among the knowledgeable, Hugh Heclo argues, "yield more experts making more sophisticated claims and counterclaims to the point that the non-specialist becomes inclined to concede everything and believe nothing he hears. The ongoing debates on energy policy, health crises, or arms limitation are rich in examples of public skepticism about what 'they,' the abstruse policy experts, are doing and saying."[17]

Such squabbles among "policy experts" not only create public skepticism, but also undermine the value of those experts' own knowledge to policymaking. Nelkin and Pollack, after examining the roles played by scientists in French and German nuclear policy disputes, conclude, "When scientists expose the conflicting technical views within the scientific community, they raise public doubts about the neutrality and independence of science; yet by engaging in a debate, they too seek credibility as a source of expertise."[18]

Rachel Carson's 1962 attack on "subjective science" provoked such strong reactions precisely because those charges hit too close to home for too many. The rise of "public-interest science" during the past twenty years has balanced the scales of expertise, but, ironically, probably has added to the dilemma. Average citizens, whipsawed between contradictory yet equally plausible constructions of reality, befuddled by arcane scientific jargon, and increasingly uncertain about just who is right, react predictably. They accept whichever view fits common sense or their own biases. Such reactions seem rational to them but may fluster those dedicated to educating the public.

A society bombarded by allegations of harmful pesticidal effects and counter-charges of hysteria-mongering can only deduce that *something* must be wrong—and it no doubt must be a worst-case scenario. Perhaps much of this pehnomenon can be blamed on the television medium, which relies heavily on images. We see tragedy in Bhopal, switch by satellite to Institute, West Virginia, watch industry representatives grope for public answers, hear environmentalists warn of the apocalypse, and assume the worst. A society cultivated over the decades into believing that "everything causes cancer" must only regard any new concerns raised—no matter how legitimate—as the scientific version of crying wolf. Or, paradoxically, we react in ways feared by policymakers, filling town halls with screaming residents, crying children, and glaring television lights. Apathy and cynicism somehow complement spasms of hysteria, lost, it seems, is the middle ground of constructive reason—if there is such a thing. John Dewey no doubt would respond that we need to communicate with and educate the public about its problems, which probably would not hurt, but one begins to wonder just how much citizens can or will absorb before it all runs off their backs like water off a duck.

Interest group theorists might rejoin that we should not worry, that the public interest is served better than ever before by the myriad organizations now attached to the pesticides case. But I wonder about that. We find numerous "publics" dancing the policy minuet, but it is less evident that "the public good" is being served. Chemical makers assure the citizen that pesticides are safe and effective, but their reluctance to reveal test information, interspersed by alarm caused by the occasional and public pesticides accident only seems to breed public suspicion about them and their products. Some environmentalists, on the other hand, talk about pesticides and their effects in apocalyptic terms so often that the citizen must wonder whether mysticism is not to be found only in churches. Organized interests stake their claims, and propound their views, but the mass public may be growing increasingly diffuse, confused, and alienated. A law that supposedly protects the public from harmful effects is on the books, but citizens are forgiven if they wonder whether the law is more symbolic than real. No education campaign by *any* organized interests will dispel these doubts, since the sources themselves are tainted. All the public knows is that promise once again crumbles when held up to performance, even if for very good reasons.

The citizen in a free society bears a tremendous load, particularly when the issues being debated do not fall into easily understood partisan or social categories. Contemporary society is made up of many "publics"—those perceiving public consequences or some condition—but no "public" in a unified sense. More problems *do* exist, and we do have greater knowledge about them, but we appear increasingly incapable of reaching solid answers. Under such conditions the public becomes skeptical. Such is the state of enlightened uncertainty.

Interests, Incentives, and the Structure of Bias

A "participation revolution" is occurring in the country as large numbers of citizens are becoming active in an ever-increasing number of protest groups, citizens' organizations, and special interest groups. These groups often are composed of issue-oriented activists or individuals who seek collective material benefits. The free-rider problem has proven not to be an insurmountable barrier to group formation, and many new interest groups do not use selective material benefits to gain support.[19]

If the pesticides case shows us anything, it is the emergence of an interest-group population dramatically different from that studied by pluralists in the 1950s. The pesticides community today is broader, more variegated, and more fractious, with organizations claiming to represent almost every facet of the pesticides dilemma. There is, simply put, far greater *competition* among policy claimants. We know the general reasons why there are more interest groups today in American society, but what factors help to explain the gradual and then accelerated expansion in the pesticides policy community? What forces have emerged to expand that configuration of policy claimants?

The first dynamic must be the nature of the issue itself, tied to the growth in knowledge and problem perceptions discussed earlier. More people today perceive broader arrays of claims in the pesticides problem. These include economic interests (agriculture and industry groups), concern about the undesirable effects of exposure to pesticides (environmental and public health groups), equity (public-interest law firms), and the distribution of political power (associations of state regulatory officials). Increasingly multifaceted problems can only produce equally complex communities of stakeholders. We have also seen that such dynamics build slowly at first, as knowledge and awareness comulate incrementally, and then accelerate noticeably as those initially drawn to the problem spread the word to others. The nature of the problem itself, and how it has evolved, must be a prime factor in generating expanded attention to the issue.

A second and far less concrete point is that the nature of political participation in American politics has undergone significant reordering during the past two decades. The "reformism" of the past twenty years, tied to increasing public perceptions about the problematic link between citizens and their elected representatives, has produced two trends. Traditional forms of participation, such as voting and party politics, have ebbed, while more direct forms of political action have skyrocketed. Environmentalists interviewed for this study all point to a marked increase in grass-roots activity aimed at specific pesticides problems, and many environmental groups once concerned solely about Washington politics now work to mine opportunities at the local level. Industry representatives see similar trends, and they too work harder to educate and mobilize the grass roots. Both worry about FIFRA, but both also seem to put less emphasis on a national policy and more on local action, where problems have their most direct impacts.

The public may be uncertain about who is right or wrong on the pesticides question, but they also do not appear to trust government to represent their views. If it is true today that there isn't a representative who doesn't have a pesticides problem in his or her district, it is also true that each representative has constituents who are waiting less for action from above and are more likely to take matters into their own hands. Reluctance to rely on federal initiative equals local activism—the sort of activity that makes the hope of a coherent national policy all the more ephemeral. Citizens may display impatience with elections, parties, and elected officials, but show equally impressive beliefs in their own ability to take action. Such perceptions of political efficacy, while essential for a democratic citizenry, only further fragment and expand the policy community, and make the tasks of consensus-building even more difficult. One might add, however, that democratic rule never was meant to be efficient or bloodless.

Third, the pesticides community expanded in no small part because more patrons in more venues emerged during the past two decades to lend a hand or to open doors. Private foundations, and government itself, supported many previously excluded from the debate, while those in government—no doubt aware of broad public concern about the problem—have granted access to those who previously did not enjoy such privileges. Congress appears to reflect the competing views on pesticides more accurately than ever before, which, of course, may mean a diminished institutional capacity for reaching authoritative policy closure. Creation of the EPA catalyzed the legitimacy of the environmental perspective within government, and even if a political administration seeks to narrow access it cannot be eliminated; the agency would lose its own legitimacy. Ann Burford and the toxic waste program is a case in point, and only the coming of Ruckelshaus and his detente with the environmentalists saved the Reagan administration considerable grief over agency management. Those dedicated to shutting down certain bureaucratic units understand a key fact of organization life: any agency or office serves those seeking some benefice (if only in the form of access), and to eliminate that unit is to undercut those claimants. Agency supporters logically fight bureaucractic triage. They may not get the policies they want, but they certainly keep their patrons alive and (to some degree) available.

Fourth, laws—and changes in the laws—powerfully affect who gets to play the game, and if a law allows you to play, you certainly have a great incentive to do so. The National Environmental Policy Act of 1969 gave the still embryonic environmental movement a tremendous participatory incentive, and the particulars of that law no doubt eased environmental advocates' access to and influence in the game. The Freedom of Information Act—despite its flaws and attempts by some to stonewall—acts as a statutory guardian over public access to government information. The courts certainly seem willing to use the law to grant citizens access where some prefer they didn't go. The tax code still discriminates unfairly, but not as powerfully as before, and more people today contribute to more groups, both out of altruism or belief in a cause and because the tax code makes giving far less painful. FIFRA, despite its problems, does at least give statutory affirmation to the legitimacy of interest-group claims. We simply have more laws today that grant more people access to policymaking. Laws of course can be changed, narrowed, or eliminated, but we find that, like bureaucracies, laws generate clienteles and protectors. We also find, in fact, that some of the sharpest disputes over statutory change involve not substantive intent, but procedural guidelines. One can lose on substance, but still win if the procedures favor access and influence. The Reagan administration, in its efforts to convince the courts to narrow judicial standing, certainly understands that point.

Finally, government and society themselves are more open, less cohesive, and far more fractionated than even James Madison might have imagined. The diffusion of government power matches, if it does not condition, the expansion in the interest-group population. A single subcommittee no longer dominates FIFRA—implying of course that the avenues for access available to claimants are more numerous and varied. The appearance today that nobody dominates FIFRA suggests either that government itself has become so fragmented that nobody can control the direction of policy or that the plethora of interests attached to the issue cancel one another out. That both possibilities seem certain exacerbates the problems of reaching a policy closure. Closure may in fact come only when one side can authoriatively discriminate against the other, when it can dominate the parameters of debate and manipulate the mechanisms of decision making. That no such dominance currently exists—at least not to the extent necessary to keep the opposition quiescent—suggests why the interests involved became so fed up with the situation that they sat down and negotiated directly. This development, in retrospect, seems more logical all the time, since it is only through such efforts that policy claimants may be able to construct an atmosphere conducive to policy agreement. Nobody involved at this time knows whether the compromise of 1986 will survive, but that it occurred at all is worth more study.

All these factors coalesce today to create a far different "structure of bias"—the bundle of laws, structures of decision making, values, and power resources that screen out "illegitimate" claimants—from what existed prior to the 1970s. We may in fact suggest that the structure of bias today is in many ways far more permeable than ever before. The political system increasingly grants legitimacy to virtually any claim: agricultural groups once were confident that dominant social values and the structure of political power would shield their claims against nonagricultural ones, but no more. The agenda of government today looks more crowded because *it is,* because more of us today perceive that more problems deserve their moment in the sun. That line once separating "private" from "public" spheres of life now seems trampled under by a society that perceives *all* problems to be public—or, at least, meriting some consideration by government. The trumpeted genius of the American political model was that it erected strict walls between private (that is, commercial) and public concerns, thus insulating capitalist economic forces from populist pressures. One is no longer sure, in these days of government attempts to fine-tune the economy, television-dominated politics, and a general perception that all problems are political, that this wall even exists. Notions about limited government depend ultimately on this bifurcation between the private and the public, and, while a good many still speak wistfully of limited government, almost everyone wants his or her interests considered by the state. Industries demands government protection of their trade secrets, but at the same time claim their business to be private. They find increasingly that they cannot have it both ways, that far fewer among us believe that any problem can remain exclusively private when its effects spill so often into the public. We perceive far more public consequences of private action, and expect far greater public response. All this prompts greater interest-group activity.

Moreover, such activity now takes on entirely different forms. Pluralists once spoke of the role played by disturbances in mobilizing interests, but in the 1980s few interests seem to need disturbances to *keep* them active. The pluralist thesis applies to the pesticides case in the sense that most of today's interest groups were organized during the last twenty years, under conditions of tremendous social upheaval. However, they have unexpectedly matured. They did not fade away once they achieved their results; rather, they have stayed on, either because they failed to get all they wanted initially or, more likely, because they must guard their victories and look for new causes.

One rather superficial indication of this phenomenon is the number and range of organizations and interests formally represented during FIFRA hearings. Prior to the 1970s, there was rather limited participation, with industry and agricultural groups virtually the only claimants in attendance. By the mid-1980s, however, we find a diversity in representation that, on the surface at least, gives pluralists some vindication. Environmentalists, labor officials, public-interest lawyers, and state government lobbyists now crowd committee rooms that were once the exclusive preserves of those expressing the most direct economic stakes in the issue. But there is a difference here: all those present are regulars, part of an apparently "permanent" (at least for the forseeable future) policy community whose representatives express far less momentary stakes. These are not temporary policy actors whose interest in the issue rises and falls over time, but instead are legions of policy activists who show up regularly and predictably each time the issue comes up for discussion. Year after year the same organizations, and, often, the same faces, appear. Everyone in attendance knows one another, and, more important, one another's views. They may not like each other, and they certainly don't agree, but all seem to accept each other as permanent and legitimate players in the game—more so than one could have suggested two decades ago. All maintain offices in Washington, recognizing at least tacitly that "presence" at all times is far better than temporary bursts of policy interest. The flashes in the pan have fleeting and spasmodic impact; the regulars maintain constant vigil over their prerogatives. Surprises are fewer, but so are uncertainties about where any of those involved stand on the issues. Interest-group activity, at least up until the attempt to hammer out a compromise, increasingly displays all the dynamics of intramural sport. The players are known, their positions staked out, and everyone knows the rules. Everyone is in the game for the long haul. Each player naturally seeks victory, but the others are equally adept at playing the game, in recruiting allies, and in staving off defeat long enough to gain the advantage, however temporary.

This condition of representative stability in no way presumes, despite the efforts to negotiate compromises, any real capacity to resolve differences nor to reach policy closure. Roger Davidson points out that increasingly "policy concerns are effectively articulated by a multiplicity of groups, but little aggregation takes place."[20] Indeed, the pesticides policy community is so fractionated and broad as to make coalition building highly problematic, and that the events of 1985–86 occurred at all suggests the degree of frustration felt by all involved. "Presence politics" may lead logically to such efforts, since those involved eventually want to settle the matter, but it also in many ways may doom that effort once the compromise leaves their hands.

The public, meanwhile, sits on the sidelines, witness to a policy minuet that it neither comprehends nor accepts as legitimate. The cadres of policy activists may achieve permanent mobilization, and may even succeed, despite the odds, in aggregating interests, but they cannot endow their tenuous agreements with the mantle of legitimacy. Only Congress can.

Fractionation and Permeability in the Decisional Structure

> The unusual number of blockage points in the American legislative process compared with that in other Western democracies reflects both fears of concentration of leader power among the founders of the Republic and the self-assertion of large numbers of participants in the legislative process.[21]

The 99th Congress at this writing faces the expiration of many of the nation's key environmental statutes. These choices about future policy were deferred by its predecessor, which instead simply appropriated needed funds and convened yet another round of oversight hearings. Not that the 98th Congress is alone to blame; the 97th, or, for that matter, the 96th, fared little better. A half-dozen years of constant wrangling produced precious little movement, and few expect significant action on many of these laws. FIFRA may turn out to be the exception, if only because the way was paved by direct negotiations among key interests, but, as of August 1986, nobody knows how the agreement will fare on Capitol Hill.

We know by this time the shape and style of Congress today. Because of the massive changes of the past two decades, power seems more disaggregated internally and members openly brandish their independence. Members who worry ever more about reelection (despite assurances that they need not be) increasingly resemble case workers more than legislators, even in the Senate. Committee dominance has been eroded by subcommittee warfare, which in turn has undermined any pretense of respect for jurisdictional boundaries.

The result, argue Dodd and Schott, is a "vast proliferation in the number of congressional actors who could claim some authority with regard to agencies and programs of the bureaucracy."[22] More actors in more venues claim some right over policy, thus complicating opportunities for accommodation. This disaggregation of policy authority, Roger Davidson argues, "confronts outside agencies and groups with a bewildering array of access points. Not that access is difficult to achieve; far from it. What is difficult is deciding which access points are worth exploiting."[23] A veteran lobbyist puts the matter in infinitely practical terms: "There used to be two to five guys on each side who had absolute control over any category of bills you might want. All you had to do was get to them. Now getting to the top guys is no guarantee. You have to lobby every member on every relevant subcommittee and even the membership at large."[24]

The evolution of pesticides policy confirms these generalizations. Formerly subgovernment dominance over policy depended on the capacity of a few congressional actors to control the parameters of debate, discriminate among policy claimants, and protect the subgovernment's flanks from attack by other congressional actors—those who could not claim "legitimate" stakes in the issue. The House Agriculture Committee took care of the legislation, while Jamie Whitten handled annual appropriations and oversight. Subgovernment actors in industry and agriculture knew where to go and whom to talk to. The cozy world of pesticides policymaking revolved around a few key members of the legislature.

No more. Today a beleagured Agriculture Committee struggles to maintain its jurisdiction, while other committees look on like hungry wolves. George Brown opened his subcommittee to all comers not simply because he was less antagonistic to the environment but because he had no choice. To give the impression of exclusion on an issue that is important to so many is to risk legitimacy, and quite a few other House actors would relish the opportunity to kidnap what for decades has been guarded in that committee. But the very need to open the doors to all comers was not Brown's only problem; his own subcommittee rarely exhibited unanimity, or even authoritatively supported, the same issues. The subcommittee suffered from internal disarray not because Brown was unfair or weak, but simply because its members didn't agree on matters so central to agriculture. Brown's successor, Berkely Bedell (D, Iowa), fared little better.

Even when the subcommittee could pull itself together and report a bill, there were no guarantees, unlike the old days, that its measure went very far. The subcommittee increasingly found itself at odds with the full committee, while the House as a whole proved to be an obstacle to any bill favoring agriculture and industry too extensively. Gone are the days when committee bills slid through on the floor simply on the assurances of a committee. Instead policies are increasingly crafted on the House floor, a decidedly less predictable process.

The Senate displays similar tendencies, and more. Senators of even the slightest seniority now assert the equality given them by the Constitution, and senatorial courtesies now extend to the pervasive (and maybe perverse) use of "holds" on any and all bills disliked by even a single member. Indeed, the hold today seems more powerful than a filibuster, and the leadership in the 1980s thus far seems incapable or unwilling to stop being so courteous. When Robert Dole (R, Kans.) succeeded Howard Baker (R, Tenn.) as majority leader in early 1985, he promised to change matters, but one wonders whether he can overcome what apparently is a collection of princes enthroned amid an atmosphere of sharpened partisanship. The gentlemens' club seems decidedly less gentlemanly, and thus many incentives for policy accommodation have been lost.

Both chambers devote greater chunks of the legislative calendar to a budget process once thought (or hoped) to be a more rational and efficient way to allocate resources. No mechanism, however, streamlined, makes austerity any less divisive or easy. Passage of the Gramm-Rudman Act in late 1985, regardless of that law's constitutional or practical future, must be seen as a cumulative congressional reaction to budget process fatigue. Members on both sides complain increasingly that life on Capitol Hill lacks its old allure (though most continue to stay on), that legislating no longer seems like fun.

In some ways all of this is a lobbyist's dream. Environmental groups, once faced with a monolithic agricultural defense system, now play committee against committee, subcommittee against subcommittee, hoping either to pressure the agriculture committees to yield on important points or (and more desirable) to provoke a palace coup by those members frustrated by inaction. Industry groups know that they can lose in subcommittee and still win in committee—bills no longer seem so sacred at any level. Subcommittee bills are overturned in committee, committee reports are reversed on the floor, one chamber is countered by the other. The veto points are everywhere—myriad roadblocks that neither the nation's founders nor even Judge Smith and his Rules Committee of old could ever have imagined. Compromise, at least so long as the battle remains on the Hill, no longer seems necessary; there is always the game. It is a lobbyist's playground.

It is also, paradoxically, a lobbyist's nightmare. If everyone claims a stake in making pesticides policy, who then is important? The route to policy once went right through W. R. Poage and Jamie Whitten; now it seems that you may as well send out a mass mailing. Industry lobbyists discover to their dismay that victory in committee guarantees nothing, while environmentalists realize that they must do more than put pressure on a subcommittee. Both sides must lobby an entire Congress. Nobody relishes that prospect.

Policy thus drifted along, fed by annual reauthorizations (if luck held) and almost endless hearings. Jamie Whitten once performed both of these tasks quietly for the pesticides subgovernment, his subcommittee acting as the permanent overseer and appropriator, thus freeing the Agriculture Committee to work on more pressing matters. Today, by contrast, no single work group performs these tasks: several appropriations subcommittees fund different slices of federal pesticides policy (for example, research, enforcement, control), while any number of actors struggle to reauthorize the statute itself. Pieces of the EPA and its policies are parcelled out to different overseers, while funding often sits in the no-man's-land between appropriators and the increasingly dominant budget committees. All this begins to resemble the legislative version of "death by a thousand cuts."

These dual trends of continuous oversight and short-term authorizing, according to Dodd and Schott, may work "to inhibit authorizing committees from doing the one thing they are best prepared to do—write legislation."[25] Everyone who cared about pesticides policy agreed on one thing alone: changes were needed. The compromise hammered out among the relevant groups may prove decisive, but few on Capitol Hill assumed that it would hold up, or even that it would get through both committees on agriculture. Continued inaction might prompt jurisdictional interloping by other committees, but the others are too busy trying to reauthorize the host of other environmental programs within *their* jurisdictions. The picture is of a foreshortened legislative agenda, dominated powerfully by the budget process. One also sees policy fatigue, and incrementalism without direction, since lack of legislation in no way means lack of action. There is always the EPA.

And what about the EPA? Faced with multiple and competing social interests, unclear and often contradictory legislative instructions, court orders, and demands by a public aware only that problems with pesticides occasionally recur visibly, the agency falls back on its own predispositions. Steve Kelman argues that "as more interest groups become involved, . . . the leeway for decisions based predominantly on the values of agency officials increase again. As more interest groups organize, the accountability of agency officials may, paradoxically, decrease."[26] And, as Dodd and Schott suggest pointedly, Congress may be increasingly less suited to stop such trends. "Paradoxically, while power dispersion has increased the number of members active within subsystem politics, as well as the number of hearings held by Congress into agencies and their programs, it has at the same time weakened the ability of Congress to conduct oversight and administrative control."[27]

The USDA in its days of control over pesticides policy faced a single set of congressional actors over the long haul, as well as a limited and predictable range of policy claimants. This reality, ironically, kept the department on a rather short leash. USDA officials knew what was expected of them at all times, and Jamie Whitten was always there to remind them in no uncertain terms that he held the purse strings. Subgovernment politics, as constraining as it is cozy to those involved, forced its participants into regularized and predictable courses of action, with subgovernment maintenance a predominant goal.

The EPA in 1986, however, faces far less exclusive and far more inchoate sets of claimants, multiple congressional actors, and policy inclinations that may work at cross-purposes. None of this is conducive to a single and compelling course of action. The agency increasingly finds itself in a position to play one claimant off of another, to sidestep conflicting congressional instructions, and even to set its own course through the thicket of competing viewpoints and countervailing interests. Agency officials face far greater continuous scrutiny by far more actors, but, paradoxically, they may enjoy far greater leeway.

R. Shep Melnick, in studying the EPA and the Clean Air Act, argues that congressional oversight consistently reveals dual and conflicting purposes. "To many congressmen," he says, "controlling the bureaucracy means preventing administrators from taking actions that injure or anger their constituents."[28] Those not on the agriculture committees consistently seek influence over pesticides policy out of concern with a specific hazard in their districts or states, or because they see the agency making too many "deals" with the regulated. To these members, then, the EPA is another case of an agency "out of control," a bureaucratic unit apparently unaccountable to the needs and views of the public. We find a number of "outside" congressional actors appearing consistently before the committee beginning in the late 1970s, members expressing outrage over agency "inaction" regarding some specific pesticide or alleged hazard—complaints that essentially accuse the agency of not doing its job. But those sitting on the committees on agriculture profess far different notions of bureaucratic control, their views being thoroughly consistent with their own policy predispositions and their overall desire for policy influence. Says Melnick: "To those congressmen who sit on authorizing committees, controlling the EPA has a far different meaning. They expect the agency to obey not just the instructions contained in the statute, but also suggestions made in committee reports and even in committee hearings. . . . They object to the watering down of *their* statutes, and expect administrators to look to congressional committees, not the White House, for policy guidance."[29]

Indeed, we find throughout the history of pesticides regulation a tendency by authorizing *and* appropriations committees to sharpen, weaken, or otherwise modify the nature of congressional intent as chiseled into the law. Oversight hearings take on the dynamics of "nonlegislative legislating," where agency officials are expected to alter their actions more because of the demands of committee members than changes in the law itself. Jamie Whitten used his subcommittee hearings to send clear and compelling signals to USDA (and, later, EPA) officials about what he and the agricultural community wanted. Absent credible alternatives expressed by "outside" members, the bureaucratic unit has little alternative but to obey its masters. Money can be cut in places officials least want austerity, or put in places they least want to fund. Life can be made rather unpleasant for agency officials testifying on the Hill, or it can be rather perfunctory. But when the agency encounters equally strong and contradictory signals from multiple congressional actors, it is in both a no-win situation *and* a potentially liberating one. It all depends on the expertise and political savvy displayed by those at the top, as well as the ability of careerists to cultivate possible friends or soothe wounded sensibilities.

The White House, meanwhile, has its own priorities—even if, as in the case of the Reagan administration, they are expressed by simply doing nothing—aims inevitably contrary to some major portion of Congress. Executive priorities filter into the agency via political appointees, resulting often in the typical struggle between them and those they are supposed to "lead." Heclo argues that the typical relationship between appointee and careerist is conditioned by institutionalized conflict, in which those at the top have goals thoroughly inconsistent with the needs of the agency's career personnel. The result, he adds, often is a lack of trust on both sides and not a little "bureaucratic sabotage" of White House objectives by those below the waterline.[30] This duality of purpose is normal in a system where bureaucracies are permeable and where top appointees must respond to the White House. It is even more so in the always starkly "political" EPA, where the agency's overall direction appears more easily influenced by those at the top and less governed by bureaucratic inertia and entrepreneurship. Secretaries of Agriculture came and went, but the careerists down in the bowels of the Agricultural Research Service—held firmly in line by Jamie Whitten and his colleagues—pushed pesticides policy matters along a consistent and predictable path. Frustration about such bureaucratic policy dominance had as much to do with the revolt against the USDA over pesticides as the nature of that policy itself.

EPA, by contrast, seems far more vulnerable to pressure from above. Three factors produce this situation. First, the agency still is relatively young, not yet showing all the signs of mature (if not sclerotic) bureaucratic decisional styles. Second, EPA's genesis came not through any legislative consensus, but through an executive-initiated shotgun marriage of careerists yanked out of other units who openly displayed smoldering animosities toward one another and were unable to settle policy disputes that long had divided them. Agencies in which careerists are consistently at one another's throats are more likely to reflect the more unified goals of political cadres, whose objectives may be the only coherent force in an otherwise chaotic setting.

Finally, and probably most important, the EPA deals daily with problems and policies that simply may not lend themselves well to bureaucratic routinization and, in fact, prove far more consistently "political" (if not partisan) than programs administered by other units. Environmental policies, by their nature, prompt acrid disputes among equally determined and almost permanently mobilized sets of claimants because they exhibit structures of incentives more contagious to conflict than do agricultural subsidies or water projects. The American style of environmental regulation, heavily dependent as it is on legal enforcement actions, seems to lend itself more naturally to adversarial relationships than programs dependent on shared expertise or commonality of interest. Subgovernment politics, and the stability of bureaucratic routines, depends inordinately on exclusivity and commonality of purpose, but whenever EPA careerists (or their bosses) seek to establish such bonds with the regulated, they are quickly hauled into court or brought before Congress for not enforcing the law. The EPA, in this sense, is a poor incubator for consensual relationships. Environmental issues draw a broader and less homogeneous range of claimants, attract the attention of more actors in other governmental loci, and cause more consistently high-level political warfare—often under the glare of the television lights. Environmental policies *are* more political, a fact of life often bemoaned by industry and agricultural lobbyists who would rather return to the halcyon days of mutual accommodation. Such conditions tend to percolate down to the lowest levels of the agency.

Not that those at the bureau level don't try to blaze their own trails. The OPP during the past decade worked mightily to "rationalize" FIFRA according to *its* own needs and resources. Unilateral decisions on data inclusion, generic registration, and creation of the maligned RPAR process all exhibit EPA careerists' efforts to stabilize and optimize their administrative tasks. Such efforts succeed best when those on Capitol Hill both understand and agree with the rationale. USDA officials usually knew beforehand how far they could go; the EPA, on the other hand, faces so many diverse and contradictory opinions that it ends up groping gingerly, hoping not to step on any of the many landmines strewn about before it has time to build congressional approval for its actions. Such approval often comes, but only after considerable erosion in the agency's reputation (such as on the data veracity matter) and furious debate in a divided legislature.

One might chalk all this up to personalities, and suggest that more knowledgeable and charming political appointees might better steer the agency away from the rocks. But the EPA's problem, despite the Ann Burfords and Rita Lavelles of this world, is structural, not personal. Agency stability and effectiveness depend powerfully on time, undiluted commitment, and enduring political support. One exaggerates only a bit to suggest that the EPA has enjoyed little of any of these, particularly today. We find an agency torn between environmental advocacy and cautious weighing of all consequences by respective political leaders, strapped under statutory deadlines later admitted to be unrealistic, internally weakened by budget cuts and the hemorrhage of scientific talent under the current administration, and enjoying political support that is best (and charitably) described as flaccid. Agency careerists have every incentive to set their own course, but they are unlikely to stray far from the current policy muddle for fear of getting bushwhacked by any number of critics on virtually every side of the pesticides issue. Actions designed to enhance the agency's capacity and stability inevitably seem to contradict congressional intent—at least as defined by the courts, by interests of all sorts, and by later congressional overseers. EPA, after all, is required to enforce a law, even one that has gaping holes.

Which brings us back to where we began. A disaggregated and increasingly permeable legislature, surrounded by an apparent equilibrium of expert and permanent interests, seems incapable of fixing what all admit is broken. The courts decide that the law must be implemented according to "congressional intent," but the agency charged with that task possesses neither strong and consistent political support from the White House nor unambiguous instructions from Capitol Hill. Faced with judicial orders for action on one side and a welter of crossed signals on the other, the agency predictably drifts along, hoping for some break in the logjam.

But breakthroughs seem increasingly illusory in a system strongly tending toward policy stasis and in which the political culture nourishes participation. The American system, with its institutionalized fragmentation of authority, always seems to careen between sudden shifts in policy direction and long stretches of maddening stalemate. A structure of governance long on representation but short on steering capacity (at least when compared to other Western democracies) seems singularly inept at authoritative action except in a crisis. And the mechanisms available for conflict resolution, which are weak already, may not be sufficient for crisis management. This seems to be the case with pesticides policy.

Aberbach, Putnam, and Rockman compare the United States to other Western democracies and conclude, "In a country without a strong state tradition, lacking a central authority, possessed of a virile individualistic ideology, and lacking in well-developed parties, decision-making is highly political and depends on policy entrepreneurs."[31] Entrepreneurship is the great engine of the American system, but it is not governance. Entrepreneurs by nature are freelancers, concerned less about building coalitions than with technical perfection or intramural gamesmanship, at least until the game itself gets so frustrating as to require change. Heclo argues that such networks of policy entrepreneurs exacerbate structurally conditioned weaknesses in the American system. The greatest problem seems to lie in democratic legitimacy.

> The more closely political administrators become identified with the various specialized policy networks, the farther they become separated from the ordinary citizen. Political executives can maneuver among the already mobilized issue networks and may occasionally do a little mobilizing of their own. But this is not the same thing as creating a broad base of public understanding and support for national policies. . . . The trouble is that only a small minority of citizens, even of those who are seriously attentive to public affairs, are likely to be mobilized in the various networks. Those who are not policy activists depend on the ability of government institutions to act on their behalf.[32]

Exhaustive policy warfare among equally active entrepreneurs produces a stalemate, with all players retreating to lick their wounds before having at it again. Such a stalemate continues until some disequilibrating force dislodges things, be it a new set of policy claimants (as in the 1970s) or universally high frustration with politics-as-usual. Windows of opportunity prove fleeting, since this same participatory culture exhibits a tremendous capacity for countermobilization. Policy modification thus comes in spasms, often provoked by some perceived crisis, some event (such as the Bhopal disaster), or some fundamental shift in society.

Spasmodic policymaking has its values, dislodging stubborn inequities or entrenched policy defenders, but it also may produce promises that later fail to match expectations, changes without clear direction, or "reforms" that spin off unintended consequences. Spasmodic politics also may engender cynicism, both in those dedicated to political action and, more important, in the attentive (if not mass) public. Insiders get caught up in their policy games and either tire themselves out or give up their advocacy. Either way, they may find precious little reward for their efforts, and instead of wholehearted advocacy we might find cynical manipulation for the game's sake.

Witness to promises created during the spasm of action, the public is left in the dark while the activists thrust and parry, only to find that the problems thought to have been solved reemerge. Finger-pointing ensues, followed by a quick round of highly public (and, maybe, symbolic) policy tinkering, and the problem soon fades from the front pages. The public depends on the institutions of government to act on their behalf, but instead we might find that those very same institutions are caught up in the game of policy entrepreneurship, responding more to expert "publics" than to the public in general. The result often seems to be action that hits the margins of the problem but not its core—which, of course, just lies dormant until awakened once again. A problem reemerges, seemingly unsolved, and the mass public eventually becomes cynical. They had expected their leaders to tackle such matters, or at least to admit that they cannot. But, as Jimmy Carter discovered, to do the latter is to court defeat.

We are not speaking simply of the pesticides problem. Failure to resolve this dilemma, after all, in no way threatens the American system. But as Alfred Marcus concludes about policymaking regarding water pollution,

> Congress had ambitious goals when it gave the bureaucracy instructions to eliminate all discharges into the nation's waters by 1985, but numerous independent actors—among them the White House, the Justice Department, the media, industry, EPA's research office, outside contractors, and the courts—prevented this program from starting promptly and from being carried out without delay. Congress may authorize far-reaching goals, . . . but when autonomous actors have numerous opportunities to oppose these goals they are not likely to succeed.[33]

Add to this list any number of other environmental laws, not to mention many social programs, and one begins to see a pattern.

American politics always has been more sectoral than ideological, based more on issues than led by parties. This segmentation of political life has always made the system far more resilient than many others, far more capable of absorbing demands and tensions. It has always been a system in which interests legitimately pursue their particularistic goals and "the public interest" often seems little more than the cumulation of private wants. One must wonder, however, if there is in fact some threshold at which our inattention to the cumulative impact of sectoral (or segmental) policymaking returns to haunt us. We see some indications already in the current budget crisis, but we are less certain what happens when the agenda for action becomes overloaded, when sectoral politics produces unintended aggregate impacts on the capacity of the system, and subsequently on public trust. A system predicated on building consensus and refracting interests may prove painfully incapable of policymaking that transcends particularistic demands. A system that prefers issue politics may suddenly find that the whole really is greater than the sum of its parts. The public, after all, is that whole. While we can please many of the discrete publics, we may end up reinforcing the public perception that greed and self-interest transcend justice in the United States.

The Images of Policymaking Revisited

> At the nub of politics are, first, the way in which the public participate in the spread of conflict, and, second, the processes by which the unstable relation of the public to the conflict is controlled.[34]

The central purpose of this study was to examine change over time within the confines of a single, well-defined issue. I wanted to learn more about the dynamics of change and, more important, about what these dynamics imply for our understanding of policymaking in America. This section begins with an epigram from Schattschneider for a reason: it is, upon reflection, the thread that unites apparently discrete images of political life. I argued in chapter 1 that we owned many snapshots of policymaking but few documentaries, that we collected images but not dynamics. This study of pesticides leads me to one conclusion: the central factor joining our images together is the differential impact of conflict. The secondary factor, which powerfully influences the first, is that mechanisms for conflict management really do matter.

Table 13 shows our three major images in a slightly different form than was given in chapter 1. There I focused on characteristics; here I concentrate on dynamics and key transformation points. What is more, I replace the "issue networks" terminology advanced by Heclo with a term of my own—"presence politics." My overriding thesis is that, fundamentally, policymaking styles complement one another according to, first, the nature and duration of the public's overall role in policymaking and, second, the permeability of relevant structures and processes for conflict resolution. If politics involves the constant tension between the privatization and the socialization of conflict, then we need to focus far more on the devices we employ to achieve those ends.

Problems create politics, and information *about* conditions plays a powerful role in influencing policymaking styles. The more people who know about some condition, the more who wish to get involved. Subgovernments thrive under conditions of minimal information. Subgovernment actors are motivated less by truth than by commonality of interest—of the particularistic sort—so having total knowledge is less relevant than the need to control it. Existing information must be kept exclusive to those central to maintaining the subgovernment, lest unwanted actors—and, particularly, the "excitable masses"—become alarmed about some condition and demand inclusion. Pesticides manufacturers and defenders of the farm community alike wanted no government education campaigns that included warnings about the potential hazards of pesticides, arguing that the public would exaggerate the dangers and ignore the concrete benefits of pesticides.

More, less exclusive, information threatens subgovernments. Indeed, when information that discredited subgovernment views was thrust into public view, such as the airing of *Silent Spring* on television, new claimants suddenly emerged. Pluralist politics depends on "optimal" levels of information—enough, that is, to mobilize previously quiescent publics, which explains why those on the outside make good use of publicity. Subgovernment politics explodes into pluralist policymaking when enough people perceive their stake in the issue and where the nature of the issue itself no longer is defined solely by the subgovernment. Perhaps the most powerful change in the pesticides policy case was its being redefined as an environmental matter; loss of the power to define the issue was probably the most critical factor causing the decline of the pesticides subgovernment. Hence the recurring battles over information and public access.

Information needs are greatest in the world of "presence politics," what Heclo calls "issue networks." Because interests are more permanently involved, policymaking styles are far different from those of pluralistic politics, but the diversity and range of actors preclude the cozier dynamics found in subgovernment policymaking. What all share is expertise, not commonality of interest, and influence depends on knowledge rather than exclusivity of jurisdiction or organi-

Table 13. The Images of Policymaking Revisited

	Subgovernments	*Pluralist Competition*	*"Presence Politics"*[a]
1. Information needs and scope of information dissemination	minimal/exclusive (common interests provide bonds)	optimal, inclusive (key to interest mobilization/action)	maximal, varied (legitimacy based on expertise)
2. Goals of actors	"preferential policy"; defense of status quo	"victory"	"presence"
3. Key dynamic	privatization of conflict (exclusion)	socialization of conflict (inclusion)	varied use of conflict (expediency)
4. Results	mutual accommodation	majority coalitions	"equilibrium or stasis"
5. Role of mass public	virtually none	central, temporary	spasmodic, as needed by relevant actors
6. Conception of politics	"gatekeeping"	"conflict"	"intramural activity"
7. Requisites for "success"	closure-forcing mechanisms; hegemony over problem definition; lack of visibility	issue saliency; "open" processes for debate and conflict; disequilibrium in views, power	permeable structures; expertise as power resource; issues that *can* be publicized when needed
8. Transition point	loss of power over issue definition; procedural/structural change	issue expansion, visibility; social change	diminished saliency; expertise a requisite; implementation, modification needed
9. Nature of change	highly incremental	dramatic, deep	spasmodic, but usually incremental (policy margins)

[a]. "Presence Politics" refers to the activities of a rather stable, if somewhat inchoate, set of lobbyists and other policy activists. I use this term in lieu of the "issue network" concept proposed by Hugh Heclo in "Issue Networks and the Executive Establishment," in *The New American Political System,* ed. Anthony King (Washington, D.C.: American Enterprise Institute, 1977), pp. 87–124.

zational membership. Presence politics involves deep and expert knowledge of an issue, which excludes the public from policymaking. Pluralist politics hinges on breadth, not depth, while in the world of the policy expert, legitimacy depends on grasp of detail. Amateurs are not welcome.

Presence politics, ironically, proves potentially far *more* exclusive than even subgovernments, which can be destroyed through competition, by views that run counter to subgovernment interests. Presence politics, however, thrives on information and knowledge, not easily understood self-interest, and it is questionable whether the public can in fact discern which information is true or false, or even the most probable. The public may know enough to discern problems, since those playing the game of presence politics make little effort to hide the debate, but the arcane nature of that exchange may leave the public only with a shallow and incomplete understanding. Experts see complexities; the public sees only problems.

Subgovernment policymaking styles also thrive on a shared belief that their interests deserve support and should be given preference over others. Agriculture and industry representatives may disagree on substantive matters, such as trade secrets, but, like sibling rivals, they quickly patch up their differences and defend the family against outsiders. To do otherwise is to undermine common needs. Pluralist politics, by contrast, involves victory pure and simple, since coalitions created for the moment tend to dissipate with time. Windows of opportunity and advantageous coalitions are few and far between, and all involved are motivated primarily by thoughts of victory, not policy maintenance. Subgovernment policymaking in fact can turn pluralistic when one partner abandons the others for private victory. This normally occurs after new and powerful claimants seek change, but they can succeed best when the subgovernment is divided over an issue. Common interests remain powerful, but disagreements over facets of policy (such as trade secrets) provides strategic niches through which outsiders may join one side or another, and the other participants have no choice but to recruit their own allies.

Presence politics does not depend on maintaining preferential policy or temporary victory, but merely on being a long-term player in the game. All involved would like to win, which makes policymaking styles far different from the mutual accommodation common to subgovernments, but virtually none *can* win because the need to remain in the game permanently forces participants to alter their strategies. Presence is access: to emphasize policy stability while all others seek to win is to be overrun, while to seek victory might be to alienate those who consider all matters under debate as scientific, not political. Victory, in fact, might turn into defeat if what you win turns out to be wrong. Far better to nibble at the margins of policy over the long haul than to go for the big strike; you only gamble when conditions inarguably run in your favor, but even then you're never sure. The nature of the policy itself makes "victory" problematic. Better to maintain credible presence than to be a flash in the pan.

Different goals obviously create different conflict dynamics. Subgovernments seek to privatize conflict, while in pluralist politics the emphasis is on constant, rachet-like expansion. Each shift in the scope of conflict means exclusion or inclusion of actors, and in pluralist politics all involved have an incentive to seek more and more allies. Subgovernment politics involve "small" issues; pluralist styles concern "big" ones. But the pesticides case shows that the same issue can be "small" or "big" based entirely on the breadth of problem perception. Issues are not born big or small; they are made that way by actors seeking to keep out or bring in allies. Any a priori judgment about the "size" of an issue is to ignore the dynamics inherent in issue expansion.

Presence politics is a hybrid; the tensions of privatization and socialization compete as an ongoing dynamic among expert activists. Some among the expert always seek to privatize, some seek to expand, and each actor in presence politics consciously manipulates conflict—and, therefore, the public—according to the needs of the moment. Industry lobbyists argue that the pesticides issue is scientific, not political—a claim consciously intended to discredit those seeking to make the issue more public. Environmentalists, on the other hand, hold that the mantle of "science" too often is used by industry as a shield for the self-interested status quo. All sides play the game of conflict manipulation, and all play it well. The result, after the smoke clears, often is stalemate.

The public, meanwhile, plays the role of resource, ally, recruit. Subgovernment policymaking is by nature exclusive, and subgovernment actors make no bones about their view that only the most knowledgeable and, particularly, most directly tied to the issue should make decisions. Pluralist policymaking styles are by nature inclusive, with all involved seeking victory (particularly in elections); all sides woo recruits under the theory that the more who play the better. Presence politics, however, is not simply exclusive *or* inclusive, but a combination of both. Politics as intramural play among expert activists may in fact be the most cynical kind of all. All those involved rely on the public like chess players rely on their pieces, using each when neccessary, but *only* so long as necessary. To allow for uncontrolled access by the public is, after all, to disrupt the game. If the public's role in subgovernments is nonexistent, and in pluralist politics central, then in presence politics the public's role is spasmodic, depending ultimately on the tactical calculations made by the players than on any long-term interests expressed by the masses.

To what effect? Schattschneider defines "democracy" as a "competitive political system in which competing leaders and organizations define the alternatives of public policy in such a way that the public can participate in the decision-making process."[35] Subgovernment politics clearly does not fit Schattschneider's standard, while pluralist dynamics appear only to offer the occasional and highly public choices to the electorate or mass public. Presence politics on the surface seems more in line with Schattschneider's definition since it entails multiple actors with diverse views seeking to press many and varied policy alternatives. Such dynamics are more permeable than the closed world of the subgovernment, though far less so than the "wide-open" style of classical pluralism, but one must wonder what Schattschneider might say about the *types* of choices being offered. If alternatives structure the public role, the choices typically exhibited through presence politics seem equally complex and incomprehensible. Such complexity suits the strategic needs of those playing the game, but it might condition the public's role as a confused and only occasionally useful giant. The public responds to some highly visible and unambiguous cue—such as an accident or some cancer scare—that might work to the benefit of one side in the struggle among experts, but the essential intractability of problems such as pesticides soon betrays public expectations about success. The complexity surrounding such issues soon leads to policy stasis among the most knowledgeable and connected, leaving the public, back on the sidelines, perplexed and possibly more cynical about the whole matter.

We therefore need to know not only the scope of conflict, but also its intrinsic shape. *Whose* game we play is as important, if not more, than the scope of that game in terms of the public role. If the central players increasingly define the alternatives in complex ambiguities—because of the short-term strategic needs involved in presence politics—we should not then be surprised if the public seems confused and angry. The confusion stems from unabated issue complexity, where all views sound so plausible, while the anger emerges from a perception that experts and not the public control the future directions of policy.

This is not to argue that the current situation is antidemocratic. Schattschneider's standard for democracy increasingly looks demanding if only given the nature of many contemporary problems. If we instead measure democracy in terms of individual participation and capacities for access into the halls of power, the system does not seem so bad. Far from it; the twin trends of participatory democracy and greater structural permeability casts our politics into decidedly competitive, even individualistic, terms. Anthony King in fact goes so far as to argue that our politics today are "atomized," a condition of the body politic and its institutions that has profound implications for our capacity to create coherent and understandable coalitions. Says King: "American politicians continue to try to create *majorities;* they have no option. But they are no longer, or at least not very often, in the business of building *coalitions*. The materials out of which coalitions might be built simply do not exist. Building coalitions in the United States today is like trying to build coalitions out of sand. It cannot be done."[36] Why should this matter? King argues that atomized politics mean more problematic leadership capacities— nobody seems to want to follow—greater issue *and* political complexity, and spasmodic politics. We see all of this in the politics of pesticides, an issue area where policymaking becomes increasingly unpredictable, inchoate, and prone to spasms of action after periods of inaction. But it is King's point about the impact on the public that bears directly on this question of democratic choice. Atomized politics, he suggests, make us ever unsure about the efficacy of our choices: "More specifically, it becomes harder for the individual voter to predict what, in terms of public policy, will be the outcome of his or her electoral decision. There comes to be even less connection between what voters, even the majority of voters, want and what they get."[37]

The public today faces innumerable choices—perhaps too many—but it is not the quantity but the quality of choice that matters in the end. A public beset by a welter of complex alternatives may prove as ineffectual as one given no choice at all. Effective representative democracy may not be a matter of letting thousands of multifaceted choices bloom, since such complexity may pull us between simplistic answers on one hand and numbing inaction on the other. Representative democracy instead is about goverance *and* choice, about allowing those elected to govern a chance to do their job *and* granting to the public the *opportunity* to make reasoned and credible decisions about the issues before it. The public normally responds to the alternatives *structured by* those who govern, which should lead us to refocus our attention to the manner by which those who govern are linked to the public. Participation without effective and lasting impact may be just as bad as no participation at all.

In the eye of this systemic centrifuge lies a critical and singularly powerful centripetal force: the presidency. John Kingdon argues rightly that "no other single actor in the political system has quite the capability of the president to set agendas in given policy areas for all those who deal with those policies."[38] More to our point here, the presidency is the single most powerful institutional lever for policy breakthrough. It is also, not surprisingly, the chief bulwark of the preferential status quo. The presidency, in this sense, is the political system's thermostat, capable of heating up or cooling down the politics of any single issue or of an entire platter of issues. The literature on subgovernment politics tends to converge on one point: presidential intervention is a key variable.

Franklin Roosevelt, for example, successfully pushed through the 1938 changes in the FDCA despite concerted opposition by the farm bloc, thus interposing the FDA more squarely into the pesticides residues dilemma. Truman and Eisenhower, by contrast, pretty much left the pesticides subgovernment alone, and the latter in fact tended to support the defenders of pesticides at key junctures. The intersection of presidential and subgovernment views during the period may tell us a great deal about why the Fifties proved to be the glory years of pesticides.

John Kennedy helped to redefine the pesticides issue in his public support for Rachel Carson and her concerns, but his own problems with conservative congressional Democrats effectively blocked policy change. The pesticides subgovernment was hurt, but not badly, during the Kennedy years, while Lyndon Johnson obviously had so many other items on his list of priorities that pesticides policy questions drifted to the shadows. Such conditions support those defending the status quo, though those seeking change at least have the solace of knowing that the administration did not actively oppose their aims.

Richard Nixon proves critical to the direction of policy, for his ascendancy to the White House coincided with one of those rare "windows of opportunity" in American social history, the emergence of the environmental movement. And as we saw earlier, Nixon's active intervention on behalf of the environment (based on whatever motives) derailed the pesticides subgovernment and paved the way for substantive policy change. The opportunities presented to Nixon inarguably were more beneficial than those available to his predecessors, but he nonetheless had to take the lead if policy change was to come about. Leadership, after all, is that ambiguous intersection of context and individual—one must have the opportunities to lead in order to do so.

Neither Gerald Ford nor Jimmy Carter in many respects enjoyed such opportunities. Both of Nixon's successors followed paths of far more restrained executive intervention because they had little choice; the national priorities had changed. Both Ford and Carter stressed fiscal restraint and administrative streamling, priorities laudable in general but not the kind that lend themselves to coherent and undiluted policy direction just when such things are needed. Emphasis on efficiency and fiscal responsibility above all else can produce policy drift, since concerted policy movement can be pretty expensive. It is, one can argue, fiscally prudent to maintain the status quo.

And, finally, the Reagan administration. There is no doubt that Ronald Reagan has made much of his opportunities, particularly with respect to the overall direction of the federal government. There also is little doubt that the Reagan EPA, particularly during the first few years, sought to retrench in the environmental area. The administration failed on the legislative front, found itself in trouble over the scandals at EPA, and generally did not succeeded fully in reducing the scope of federal environmental regulation. Yet it succeeded in *narrowing* policy access and worked to privatize regulatory decisions. What the administration sought to narrow, however, seems to run counter to other powerful forces—those spawned two decades ago—and the result was policy stalemate. Not that such a condition is neutral; lack of change, after all, has its own array of winners and losers. Such a deadlock can continue for quite a while, to be shattered only under new disequilibria and new windows of opportunity. The attempt to hammer out a compromise on FIFRA is one such opportunity, but its success may hinge most on presidential support, or, at least, lack of opposition.

Trends, Structures, and the Policy Cycle

This study also shows two trends and a likely cycle. The trends lay at the micro and macro levels of politics, the cycle at the intermediate level. We find at the macro level irresistible urges toward greater societal involvement in our system of governance, a trend unlikely to be reversed in the forseeable future. Indeed, the history of American politics is one of gradual and relentless participatory growth, be it through the expansion of the franchise or in the scope of interest articulation. These trends are likely to continue, perhaps in quality if not in quantity. At the level of the individual we find concomitant growth in personal efficacy. We as individuals may not have the capacity to change the world, but *as citizens* we increasingly feel capable of governing our own affairs. Such attitudes no doubt reflect more widespread education, but they also reflect the pervasive American belief that individuals can and do matter. This too is a relentless theme throughout our history, one unfolding with each reform period and its subsequent consequences.

It is at the intermediate level, the level of institutions, structures, and rules, that a cycle emerges. The history of pesticides policy is a movement out of the cozy confines of subgovernment dominance into the more chaotic spasms of pluralistic conflict, followed by the apparent institutionalization of that chaos in our current condition. It seems unlikely that we will return to the days of subgovernment control—the trends previously mentioned probably mitigate against that—but we certainly can narrow what has been expanded by altering key processes and structures for policymaking. The expansion in the pesticides community was in no small way conditioned by changes in our structure of governance, changes that dramatically broadened the opportunities for access available to claimants. Change some of this again—e.g., reinterpret judicial standing or reinstitute stricter committee dominance—and you arguably might restructure those opportunities and influence future policy.

Structures, rules, and processes are all mechanisms for choice, devices we use to discriminate, and to change them is in many ways to redefine our politics. Formal structures and rules do matter, perhaps even as much as do perceptions, incentives, and motivations. Kenneth Dolbeare perhaps best states the case: "Neither scholars nor citizens invest much energy in serious consideration of ways to fit governmental forms to the functions we ask of them. The question of when or how the structure or practices of our national government institutions *change* is seldom noted.[39] This study was a modest attempt in that direction. After all, as Schattschneider suggests, "procedures for the control of the expansive power of conflict determine the shape of the political system."[40] We might wish, as the framers of the Constitution undoubtedly realized, to keep this in mind.

Notes

Index

Notes

Chapter One. Images of Policymaking

1. Floyd R. Hunter, *Community Power Structure* (Chapel Hill, N.C.: University of North Carolina Press, 1953); C. Wright Mills, *The Power Elite* (New York: Oxford University Press, 1954).

2. Robert Dahl, *Who Governs?* (New Haven, Conn.: Yale University Press, 1961).

3. Robert Dahl, *A Preface to Democratic Theory* (Chicago: University of Chicago Press, 1956), p. 137, emphasis added.

4. Earl Latham, "The Group Basis of Politics," *American Political Science Review* 46 (June 1952), 390.

5. E. E. Schattschneider, *The Semi-Sovereign People: A Realist's View of Democracy in America* (Hinsdale, Ill.: Dryden Press, 1960), p. 38.

6. See, for example, Sidney Verba and Norman Nie, *Participation in America* (New York: Harper and Row, 1972).

7. Schattschneider, *The Semi-Sovereign People,* p. 38.

8. See J. Leiper Freeman, *The Political Process* (New York: Random House, 1955); Douglass Cater, *Power in Washington* (New York: Random House, 1964); A. Grant McConnell, *Private Power and American Democracy* (New York: Alfred A. Knopf, 1966); Emmette S. Redford, *Democracy in the Administrative State* (New York: Oxford University Press, 1969).

9. Hugh Heclo, "Issue Networks and the Executive Establishment," in *The New American Political System,* ed. Anthony King (Washington, D.C.: American Enterprise Institute, 1979), p. 88.

10. Nelson Polsby, "How to Study Community Power: The Pluralist Alternative," *Journal of Politics* 22 (August 1960), 478–79.

11. Ira Sharkansky, *Whither the State? Politics and Public Enterprise in Three Countries* (Chatham, N.J.: Chatham House, 1979).

12. Heclo, "Issue Networks," p. 102.

13. Ibid.

14. Ibid.

15. Charles O. Jones, "American Politics and the Organization of Energy Decision Making," *Annual Review of Energy* 4 (1979), 105.

16. James Q. Wilson, *Political Organizations* (New York: Basic Books, 1973); Charles O. Jones, *Clean Air: The Politics and Policies of Pollution Control* (Pittsburgh, Pa.: University of Pittsburgh Press, 1975); Randall B. Ripley and Grace A. Franklin, *Congress, the Bureaucracy, and Public Policy,* 2d ed. (Homewood, Ill.: Dorsey Press, 1979); Anthony King, ed., *The New American Political System.*

17. E. E. Schattschneider, *Politics, Pressures, and the Tariff* (Englewood Cliffs, N.J.: Prentice-Hall, 1935), p. 288.

18. Ripley and Franklin, *Congress, the Bureaucracy, and Public Policy,* p. 20.

19. Ibid., pp. 22–23.

20. Schattschneider, *Politics, Pressures, and the Tariff,* p. 286.

21. Bruce I. Oppenheimer, *Oil and the Congressional Process* (Lexington, Mass.: Lexington Books, 1975), p. 118.

22. Schattschneider, *The Semi-Sovereign People,* pp. 4–5.

23. Oppenheimer, *Oil and the Congressional Process,* p. 63.

24. Schattschneider, *The Semi-Sovereign People,* p. 71, emphasis in original.

25. Ibid., p. 17.

26. Oppenheimer, *Oil and the Congressional Process,* p. 153.

27. Roger H. Davidson, "Subcommittee Government: New Channels for Policy Making," in *The New Congress,* ed. Thomas E. Mann and Norman J. Ornstein (Washington, D.C.: American Enterprise Institute, 1981), p. 100.

28. Amitai Etzioni, *The Active Society* (New York: Free Press, 1968), p. 293.

29. Hugh Heclo, "Review Article: Policy Analysis," *British Journal of Political Science* 2 (January 1972), 104.

30. Jones, *Clean Air,* p. 20.

31. Barbara J. Nelson and Thomas Lindenfeld, "Setting the Public Agenda: The Case of Child Abuse," in *The Policy Cycle,* ed. Judith V. May and Aaron Wildavsky (Beverly Hills, Calif.: Sage Publications, 1978), p. 18.

32. Jones, *Clean Air,* p. 136.

33. Schattschneider, *The Semi-Sovereign People,* p. 71.

34. "Dominant" interests are those of insiders—those part of the policymaking community relevant to the issue and who stand to benefit most directly from policy outcomes. They are, in the simplest sense, the winners in the game. "Subordinate" interests, by contrast, are those of outsiders—those not included in the game. They are the losers, even if they never took part directly in the fight.

35. Schattschneider, *The Semi-Sovereign People,* p. 71.

36. Ibid., pp. 34–35, emphasis in original.

Chapter Two. The Pesticides Paradigm

1. Charles O. Jones, *Introduction to the Study of Public Policy,* 2d ed. (North Scituate, Mass.: Duxbury Press, 1977), p. 26.

2. Roger W. Cobb and Charles D. Elder, *Participation in American Politics: The Dynamics of Agenda Building* (Boston: Allyn and Bacon, 1972), p. 85, emphasis added.

3. Charles Lindblom, *The Policymaking Process* (Englewood Cliffs, N.J.: Prentice-Hall, 1966), p. 13.

4. Peter Bachrach and Morton Baratz, *Power and Poverty: Theory and Practice* (New York: Oxford University Press, 1970), p. 15.

5. E. E. Schattschneider, *The Semi-Sovereign People: A Realist's View of Democracy in America* (Hinsdale, Ill.: The Dryden Press, 1960), p. 68, emphasis in original.

6. *New Republic,* April 29, 1946, p. 599.

7. George H. Gallup, *The Gallup Poll* (New York: Random House, 1972), 1:561–52.

8. Allen J. Matusow, *Farm Policies and Politics in the Truman Years* (Cambridge, Mass.: Harvard University Press, 1967), p. 111.

9. Ibid., pp. 112–13.

10. R. G. Bressler, "The Impact of Science on Agriculture," *Journal of Farm Economics* 40 (December 1958), 100.

11. Matusow, *Farm Policies and Politics,* p. 115.

12. Ibid., p. 41.

13. Ibid.

14. U.S. Department of Agriculture, *Yearbook of Agriculture 1948* (Washington D.C.: GPO, 1949), p. 125.

15. Schattschneider, *The Semi-Sovereign People,* p. 2.

16. John Blodgett, "Pesticides: Regulation of an Evolving Technology," in *The Legislation of Product Safety,* ed. Samuel S. Epstein and Richard D. Grady (Cambridge, Mass.: MIT Press, 1974), 3:204, emphasis added.

17. U.S. House of Representatives, Committee on Appropriations, Subcommittee on Agriculture, *Department of Agriculture FY1946,* 79th Cong., 1st sess., 1945, p. 512.

18. James Whorton, *Before Silent Spring: Pesticides and Public Health in Pre-DDT America* (Princeton, N.J.: Princeton University Press, 1974), p. 5.

19. Harrison Wellford, *Sowing the Wind* (New York: Grossman, 1972), p. 253.

20. James Simmons, *Saturday Evening Post,* January 6, 1945, as cited in Whorton, *Before Silent Spring,* p. 248.

21. John Perkins, *Insects, Experts, and the Insecticide Crisis* (New York: Plenum Press, 1982), p. 13.

22. John Perkins, "Reshaping Technology in Wartime: The Effect of Military Goals on Entomological Research and Insect-Control Practice," *Technology and Culture* 19 (April 1978), 183–84.

23. U.S. House of Representatives, Committee on Appropriations, Subcommittee on Agriculture, *Department of Agriculture FY1946,* p. 498.

24. Thomas Dunlap, *DDT: Scientists, Citizens, and Public Policy* (Princeton, N.J.: Princeton University Press, 1981), p. 65.

25. U.S. House of Representatives, Committee on Appropriations, Subcommittee on Agriculture, *Department of Agriculture FY1948,* 80th Cong., 1st sess., 1947, p. 810.

26. Dunlap, *DDT,* p. 35.

27. Ibid., p. 17.

28. Whorton, *Before Silent Spring,* p. xii.

29. Anthony King, "The American Polity in the Late 1970s: Building Coalitions in the Sand," in *The New American Political System,* ed. Anthony King (Washington, D.C.: American Enterprise Institute, 1979), p. 382.

30. Robert K. Carr, Marver H. Bernstein, et al., *American Government: Theory Politics, and Constitutional Foundation* (New York: Holt, Rinehart, and Winston, 1961), pp. 197–98.

31. Schattschneider, *The Semi-Sovereign People,* p. 30.

32. David Truman, *The Governmental Process* (New York: Alfred A. Knopf, 1951).

33. Jack Walker, "The Origins and Maintenance of Interest Groups in America," *American Political Science Review* 77 (June 1983), 403.

34. Wesley McCune, *The Farm Bloc* (New York: Greenwood Press, 1968), p. 262, emphasis in original.

35. Graham K. Wilson, *Interest Groups in America* (New York: Oxford University Press, 1981), p. 20.

36. Christiana M. Campbell, *The Farm Bureau and the New Deal* (Urbana, Ill.: University of Illinois Press, 1962), p. 5.

37. V. O. Key, *Politics, Parties, and Pressure Groups,* 5th ed. (New York: Thomas Crowell, 1964), p. 33.

38. Wilson, *Interest Groups in America,* p. 20.

39. Campbell, *The Farm Bureau and the New Deal,* p. 11.

40. McCune, *The Farm Bloc,* p. 132.

41. The National Grange, the eldest of the farm organizations, by this time was little more than a fraternal society. The Grange delved little into this issue, and is not discussed here.

42. Matusow, *Farm Policies and Politics,* p. 67.

43. Mancur Olson, *The Logic of Collective Action* (Cambridge, Mass.: Harvard University Press, 1964).

44. Walker, "The Origins and Maintenance of Interest Groups in America," p. 404.

45. *Department of Agriculture FY1948,* p. 404.

46. See, for example, the exchange between Representative Clarence Cannon (D, Mo.) and Secretary of Agriculture Clinton Andresen about the future of the farm population, U.S. House of Representatives, Committee on Appropriations, Subcommittee on Agriculture, *Department of Agriculture FY1948,* 80th Cong., 1st sess., 1947, p. 25.

47. See, for example, Clifford Hardin, *The Politics of Agriculture* (Glencoe, Ill.: Free Press, 1952).

48. George Galloway, *The Legislative Process in Congress* (New York: Thomas Crowell, 1953), p. 645.

49. Douglass Cater, *Power in Washington* (New York: Random House, 1964), p. 157.

50. Charles O. Jones, "Representation in Congress: The Case of the House Agriculture Committee," *American Political Science Review* 55 (June 1961), 358–67.

51. Cater, *Power in Washington,* p. 158.

52. Robert Holbert, *Tax Laws and Political Access: The Bias of Pluralism Revisited* (Beverly Hills, Calif.: Sage Publications, 1975), p. 8.

53. The provision on "substantiality" was in fact included as an amendment to the 1934 tax law by a senator seeking to curb the activities of the National Economy League. The sponsor did not wish, however, to harm "worthy" institutions, whatever they were. See Bruce R. Hopkins, *The Law and Tax-Exempt Organizations,* 3d ed. (New York: John Wiley, 1979), p. 177.

54. *Frothingham* v. *Mellon,* 262 U.S. 447 (1923); see Karen Orren, "Standing to Sue: Interest Group Conflict in the Federal Courts," *American Political Science Review* 70 (September 1976), 723–41.

55. James Q. Wilson, *American Government: Institutions and Policies,* 2d ed. (Englewood Cliffs, N.J.: Prentice-Hall, 1982), p. 418; see also chapter 16 in his *Political Organizations* (New York: Basic Books, 1973).

56. Ibid., p. 419, emphasis in original.

57. Jones, *Introduction to the Study of Public Policy,* p. 37.

58. Wilson, *Political Organizations,* p. 330.

Chapter Three. The Politics of Clientelism

1. Cited in Harrison Wellford, *Sowing the Wind* (New York: Grossman, 1972), p. 254.
2. Thomas Dunlap, *DDT: Scientists, Citizens, and Public Policy* (Princeton, N.J.: Princeton University Press, 1981), p. 61.
3. *Time,* June 30, 1947, p. 96.
4. *Science News Letter,* June 14, 1947, back cover.
5. Jane Stafford, "Insect War May Backfire," *Science News Letter,* August 4, 1944, p. 21.
6. "Total Insect War Urged," *Science News Letter,* January 6, 1945, p. 5.
7. John Terres, "Dynamite in DDT," *New Republic,* January 6, 1945, p. 415.
8. D. H. Killifer, "Is DDT Poisonous?" *Science News Letter,* January 2, 1946, p. 5.
9. W. H. Wiley, as quoted in James Whorton, *Before Silent Spring: Pesticides and Public Health in Pre-DDT America* (Princeton, N.J.: Princeton University Press, 1974), p. 98.
10. Ibid., pp. 98–99.
11. John Perkins, *Insects, Experts, and the Insecticide Crisis* (New York: Plenum Press, 1982), p. 3.
12. Whorton, *Before Silent Spring,* p. 114.
13. Quoted in ibid., p. 116.
14. Quoted in ibid., p. 136.
15. John Blodgett, "Pesticides: Regulation of an Evolving Technology," in *The Legislation of Product Safety,* ed. Samuel S. Epstein and Richard D. Grady (Cambridge, Mass.: MIT Press, 1974), 3:206.
16. Quoted in Dunlap, *DDT,* p. 50.
17. Quoted in Whorton, *Before Silent Spring,* p. 228.
18. Ibid., p. 229.
19. Ibid., p. 230.
20. Dunlap, *DDT,* p. 55.
21. Whorton, *Before Silent Spring,* p. 239.
22. Ibid., p. 241.
23. Dunlap, *DDT,* p. 63.
24. U.S. House of Representatives, Committee on Agriculture, *Federal Insecticide, Fungicide, and Rodenticide Act,* 79th Cong., 2d Sess., 1946, p. 2.
25. Ibid.
26. Ibid., pp. 1–2.
27. Ibid., p. 8.
28. Ibid., p. 4.
29. Ibid.
30. U.S. House of Representatives, Committee on Agriculture, *Federal Insecticide, Fungicide, and Rodenticide Act,* 80th Cong., 1st sess., 1947, p. 6.
31. Ibid., p. 10.
32. Ibid., p. 30.
33. Ibid., p. 31.
34. *Congressional Record,* May 12, 1947, p. 5050.

35. Ibid.

36. Ibid., p. 5051.

37. Ibid.

38. Ibid., emphasis added.

39. Blodgett, "Pesticides: Regulation of an Evolving Technology," 3:208.

40. Randall B. Ripley and Grace A. Franklin, *Congress, the Bureaucracy and Public Policy,* 3d ed. (Homewood, Ill.: Dorsey Press, 1984), pp. 103–07.

41. Steven Kelman, *Regulating America, Regulating Sweden: A Comparative Study of Occupational Safety and Health Policy* (Cambridge, Mass.: MIT Press, 1981), p. 113.

Chapter Four. The Apotheosis of Pesticides

1. John Kenneth Galbraith, *The Affluent Society* (Boston: Houghton-Mifflin, 1958), pp. 15–16.

2. "Eisenhower on the Presidency," interview with Walter Cronkite on CBS television, October 17, 1961, as quoted in Samuel P. Huntington, *American Politics: The Promise of Disharmony* (Cambridge, Mass.: Harvard University Press, 1981), p. 285n.

3. Huntington, *American Politics: The Promise of Disharmony,* p. 170.

4. Raymond Bauer et al., *American Business and Public Policy* (New York: Atherton, 1968), p. viii.

5. E. E. Schattschneider, *Politics, Pressures, and the Tariff* (Englewood Cliffs, N.J.: Prentice-Hall, 1935), p. 288.

6. Carl B. Huffaker, as quoted in John Perkins, *Insects, Experts, and the Pesticides Crisis* (New York: Plenum Press, 1982), p. 214.

7. *1982 Statistical Abstract of the United States* (Washington: GPO, 1982), p. 207.

8. U.S. House of Representatives, Committee on Appropriations, Subcommittee on Agriculture, *USDA Appropriations FY1954,* 83rd Cong., 1st sess., 1953, p. 1746; hereinafter referred to as the Whitten Subcommittee.

9. Clem Miller, *Member of the House* (New York: Scribners, 1962), p. 40.

10. Charles O. Jones, "From Committee Suzerainty to Subcommittee Battalions," in *Changing Congress: The Committee System,* ed. Norman J. Ornstein, American Academy of Political Science, *Annals* 411 (January 1974), 75–86.

11. Quoted in Roger H. Davidson and Walter Oleszek, *Congress and Its Members* (Washington, D.C.: Congressional Quarterly Press, 1981), p. 205.

12. John F. Bibby et al., *Vital Statistics on Congress* (Washington, D.C.: American Enterprise Institute, 1981), p. 63.

13. Ibid., p. 110.

14. Clarence Cannon, "Against the Item Veto," *Harvard Law Review,* April 21, 1960, as reprinted in the *Congressional Record,* May 6, 1960, p. A3916.

15. Richard F. Fenno, *Power of the Purse: Appropriations Politics in Congress* (Boston: Little, Brown, 1965), p. 418.

16. Ibid., p. 417.

17. Ibid., p. 37.

18. *Congressional Quarterly Almanac 1957* (Washington, D.C.: Congressional Quarterly Press, 1958), p. 93.

19. Quoted in Fenno, *Power of the Purse,* p. 135.

20. Ibid., p. 168.

21. Jamie L. Whitten, speech, as submitted in the Extension of Remarks, *Congressional Record,* September 18, 1961, p. A7391.

22. See, for example, Nick Kotz, *Let Them Eat Promises: The Politics of Hunger in America* (New York: Doubleday, 1971).

23. Whitten Subcommittee, *Department of Agriculture FY1950,* 81st Cong., 1st sess., 1949, p. 793.

24. Fenno, *Power of the Purse,* p. 311.

25. Quoted in ibid., p. 378.

26. Whitten Subcommittee, *Department of Agriculture FY1954,* 83rd Cong., 1st sess., 1953, p. 819.

27. Whitten Subcommittee, *Department of Agriculture FY1953,* 82nd Cong., 2d sess., 1952, p. 1000.

28. Ibid., p. 995.

29. Ibid., p. 1036.

30. *Congressional Record,* March 20, 1953, p. A1417.

31. Whitten Subcommittee, *Department of Agriculture FY1955,* 83rd Cong., 2d sess., 1954, p. 595.

32. Herbert Asher, "Committees and the Norms of Specialization," in *Changing Congress: The Committee System,* ed. Norman J. Ornstein, American Academy of Political Science, *Annals* 441 (January 1974), 72.

33. Fenno, *Power of the Purse,* p. 370.

34. Between 1948 and 1955, for example, FIFRA as the topic for discussion averaged but a half page per hearing transcript, compared to the hundreds of pages exhausted on various pest threats and programs.

35. Whitten Subcommittee, *Department of Agriculture FY1952,* 82nd Cong., 1st sess., 1951, p. 580.

36. Fenno, *Power of the Purse,* p. 413.

37. Rep. James V. Delaney, U.S. House of Representatives, Select Committee to Investigate the Use of Chemicals in Food Products, *Use of Chemicals in Food Products,* pt. 1, 81st Cong., 2d sess., 1950, p. 10; hereinafter referred to as the Delaney Committee.

38. *Consumer Reports,* March 1950, p. 135.

39. "Chemicals Introduced in the Production of Food," *American Journal of Public Health,* pt. 2 (May 1950), 122.

40. Ibid., p. 126.

41. *Consumer Reports,* July 1951, p. 329.

42. Quoted in Judy Bentley, "James V. Delaney," in *Citizens Look at Congress: The Ralph Nader Congress Project* (New York: Grossman, 1974), p. 5.

43. Delaney Committee, *Use of Chemicals in Food Products,* pt. 1, 1950, p. 2.

44. *New York Times,* December 3, 1950, p. 111.

45. Ibid.

46. Delaney Committee, *Use of Chemicals in Food Products,* pt. 1, p. 357.

47. Ibid.

48. Ibid., p. 364.

49. Ibid., p. 365.

50. Ibid.

51. Ibid., p. 508.

52. *New York Times,* March 9, 1952, p. 31.

53. *New York Times,* February 1, 1952, p. 23.

54. *Congressional Record,* April 30, 1953, p. A2286.

55. Ibid.

56. Thomas Dunlap, *DDT: Scientists, Citizens, and Public Policy* (Princeton, N.J.: Princeton University Press, 1981), p. 74.

57. *Congressional Record,* July 12, 1954, p. 10309.

58. *New York Times,* July 9, 1954, p. 21.

59. John Perkins, *Insects, Experts, and the Insecticide Crisis* (New York: Plenum Press, 1982), pp. 30–31, emphasis added.

60. Ibid., p. 31.

61. Dunlap, *DDT,* pp. 74–75.

Chapter Five. Creating a Public: The Eradication Campaigns

1. Charles O. Jones, *Introduction to the Study of Public Policy,* 2d ed. (North Scituate, Mass.: Dorsey Press, 1977), p. 39.

2. John Dewey, *The Public and Its Problems* (New York: Holt, Rinehart & Winston, 1927), pp. 15–16.

3. Roger W. Cobb and Charles D. Elder, *Participation in American Politics: The Dynamics of Agenda Building* (Boston: Allyn and Bacon, 1972), pp. 112–24.

4. U.S. House of Representatives, Committee on Appropriations, Subcommittee on Agriculture, *Department of Agriculture FY1958,* 85th Cong., 1st sess., 1957, p. 1053; hereinafter referred to as the Whitten Subcommittee.

5. *New York Times,* March 30, 1957, p. 57.

6. *New York Times,* April 24, 1957, p. 35.

7. *New York Times,* May 23, 1957, p. 32.

8. *New York Times,* May 25, 1957, p. 23.

9. *New York Times,* May 23, 1957, p. 32.

10. *New York Times,* June 2, 1957, p. 23.

11. Ibid.

12. *Audubon* 54, no. 3 (May–June 1952), 146.

13. *Audubon* 58, no. 3 (May–June 1956), 129.

14. *Audubon* 58, no. 4 (July–August 1956), 177.

15. Ibid., p. 178.

16. *Audubon* 56, no. 2 (March–April 1958), 68–69.

17. Ibid., p. 79.

18. As quoted by Charles D. Kelley, U.S. House of Representatives, Committee on Merchant Marine and Fisheries, Subcommittee on Fisheries and Wildlife, *Coordination of Pesticides Programs,* 86th Cong., 2d sess., 1960, p. 22; hereinafter referred to as the Dingell Subcommittee.

19. Testimony by Dr. Clarence Cottam, ibid., p. 29.

20. Committee on the Imported Fire Ant, *Report to the Administrator, Agricultural Research Service, U.S. Department of Agriculture,* National Academy of Sciences–National Research Council, 1967.

21. Kirby L. Hays, "The Present Status of the Imported Fire Ant in Argentina," *Journal of Economic Entomology* 51 (1958), 111–12.

22. Harrison Wellford, *Sowing the Wind* (New York: Grossman, 1972), p. 291.

23. Rachel Carson, *Silent Spring* (Greenwich, Conn.: Fawcett, 1962), p. 147.
24. Wellford, *Sowing the Wind,* p. 301.
25. Ibid., p. 37n.
26. John Devlin, "Fire Ant Alarms the South," *New York Times,* March 19, 1957, p. 40.
27. U.S. Senate, Committee on Agriculture and Forestry, *Fire Ant Eradication,* 85th Cong., 1st sess., 1957, p. 3.
28. Ibid., p. 2.
29. *Congressional Record,* March 29, 1957, p. 4753.
30. Whitten Subcommittee, *Department of Agriculture FY1958,* 85th Cong., 1st sess., 1957, p. 432, emphasis added.
31. *Congressional Record,* August 8, 1957, p. 13827.
32. Rachel Carson, *Silent Spring,* p. 147.
33. *New York Times,* December 23, 1957, p. 25.
34. Ibid.
35. As quoted in testimony before the Dingell Subcommittee, *Coordination of Pesticides Programs,* 86th Cong., 2d sess., 1960, p. 22.
36. Whitten Subcommittee, *Department of Agriculture FY1959,* 85th Cong., 2d sess., 1958, p. 584.
37. *Senior Scholastic,* April 12, 1957, p. 22; the baby referred to appears to be the same young man.
38. *Congressional Record,* April 17, 1958, p. A3477.
39. Ibid.
40. *New York Times,* December 25, 1957, p. 33.
41. *New York Times,* December 27, 1957, p. 11.
42. *New York Times,* January 8, 1958, p. 46.
43. *New York Times,* December 27, 1957, p. 11.
44. *Audubon* 56, no. 2 (March–April 1958), 78.
45. U.S. House of Representatives, Committee on Merchant Marine and Fisheries, Subcommittee on Fisheries and Wildlife, *Miscellaneous Fish and Wildlife Bills,* 85th Cong., 2d sess., 1958, pp. 7–8.
46. U.S. Senate, Committee on Interstate and Foreign Commerce, Subcommittee on Marine and Fisheries, *USDI Pesticides Research,* 85th Cong., 1st sess., 1958, p. 10.
47. Ibid., p. 12.
48. Ibid., p. 15.
49. Ibid., p. 16.
50. Ibid., p. 18.
51. Ibid., p. 35.
52. U.S. House of Representatives, Committee on Merchant Marine and Fisheries, *USDI Pesticides Research,* H. Rept. 85-2181, 85th Cong., 2d sess., 1958, p. 2.
53. Richard F. Fenno, *Power of the Purse: Appropriations Politics in Congress* (Boston: Little, Brown, 1965), p. 363.
54. U.S. Senate, Committee on Appropriations, Subcommittee on the Department of the Interior, *Department of the Interior,* 86th Cong., 2d sess., 1960, p. 444.
55. *New York Times,* December 3, 1957, p. 52.
56. *New York Times,* February 14, 1958, p. 33.
57. *New York Times,* February 22, 1958, p. 6.

58. *New York Times,* February 25, 1958, p. 45.

59. *New York Times,* June 24, 1958, p. 33.

60. Thomas Dunlap, *DDT: Scientists, Citizens, and Public Policy* (Princeton, N.J.: Princeton University Press, 1981), p. 89.

61. Notable in dissent was Justice William O. Douglas, who wanted the Court to tackle the broader question of the government's right to spray. Douglas, for the time anyway, was a minority of one.

62. John Blodgett, "Pesticides: Regulation of an Evolving Technology," in *The Legislation of Product Safety,* ed. Samuel S. Epstein and Richard D. Grady (Cambridge, Mass.: MIT Press, 1974), 3:70.

63. *New York Times,* November 11, 1959, p. 34.

64. Randall B. Ripley and Grace A. Franklin, *Congress, the Bureaucracy, and Public Policy,* 2d ed. (Homewood, Ill.: Dorsey Press, 1979), p. 149.

65. Whitten Subcommittee, *Department of Agriculture FY1961,* 86th Cong., 2d sess., 1960, p. 119.

66. *New York Times,* November 11, 1959, p. 20.

67. *Wall Street Journal,* November 20, 1959, p. 1.

68. Whitten Subcommittee, *Payments to Growers of Cranberries and Caponettes,* 86th Cong., 2d sess., 1960, p. 15.

69. Ibid.

70. Ervin L. Peterson, testimony before the Whitten Subcommittee, *Department of Agriculture FY1959,* 85th Cong., 2d sess., 1958, p. 431.

71. Dunlap, *DDT,* p. 91.

72. Whitten Subcommittee, *Department of Agriculture FY1960,* 86th Cong., 1st sess., 1959, p. 894.

73. Ibid., p. 896.

74. Whitten Subcommittee, *Department of Agriculture FY1961,* 86th Cong., 2d sess., 1960, p. 2851.

75. Ibid.

76. Dingell Subcommittee, *Coordination of Pesticides Programs,* 86th Cong., 2d sess., 1960, p. 5.

77. Ibid., p. 6; The FDA in October 1959 announced its intent to establish a zero tolerance for heptachlor on various commodities, which would force the USDA to switch fire ant control chemicals even as it claimed heptachlor's safety. Mirex, a new chemical with lower toxicity and less persistence, was substituted, but it too would come under attack in later years.

78. Whitten Subcommittee, *Department of Agriculture FY1961,* 86th Cong., 2d sess., 1960, p. 116.

79. Ibid.

80. *Congressional Record,* April 13, 1960, p. 7077.

81. Dingell Subcommittee, *Coordination of Pesticides Programs,* 86th Cong., 2d sess., 1960, p. 19.

82. Ibid., p. 29.

83. Ibid., p. 30.

84. Ibid., p. 35.

85. Ibid., p. 59.

86. Ibid., p. 61.

87. Ibid., p. 53.

88. Ibid., p. 41.

89. *National Journal* 4, no. 7 (February 19, 1972), 321; the Citizens Committee on Natural Resources was founded in 1954 as a non–tax exempt conservation lobby. This very small and chronically underfunded group in many ways prefigured the environmental organizations of the 1970s.

Chapter Six. Sea Change in the Sixties

1. Valerie Bunce, *"Do New Leaders Make a Difference?" Executive Succession and Public Policy Under Capitalism and Socialism* (Princeton, N.J.: Princeton University Press, 1981), p. 139.

2. Samuel P. Huntington, *American Politics: The Promise of Disharmony* (Cambridge, Mass.: Harvard University Press, 1981), p. 174.

3. Samuel P. Huntington, "The United States," in *The Crisis of Democracy,* ed. Michael Crozier et al. (New York: New York University Press, 1975), pp. 59–60.

4. Richard Brody, "The Puzzle of Participation in America," in *The New American Political System,* ed. Anthony King (Washington, D.C.: American Enterprise Institute, 1979), p. 316.

5. Jeffrey Berry, *Lobbying for the People* (Princeton, N.J.: Princeton University Press, 1977), p. 34.

6. Andrew McFarland, *Public Interest Lobbies: Decision Making on Energy* (Washington, D.C.: American Enterprise Institute, 1978), pp. 4–5.

7. Theodore White, *America In Search of Itself* (New York: Harper and Row, 1982), p. 168.

8. Huntington, *American Politics: The Promise of Disharmony,* p. 102.

9. *Congressional Record,* February 2, 1961, p. 1717.

10. *Congressional Quarterly Almanac 1961* (Washington: Congressional Quarterly Press, 1962), pp. 34–35.

11. *Congressional Quarterly Weekly Report,* September 16, 1966, p. 7.

12. James L. Sundquist, *Politics and Policy: The Eisenhower, Kennedy, and Johnson Years* (Washington, D.C.: Brookings Institution, 1968), p. 513; see also Roger H. Davidson et al., *Congress in Crisis: Politics and Congressional Reform* (Belmont, Calif.: Wadsworth, 1966).

13. Valerie Bunce, *Do New Leaders Make a Difference?* p. 258.

14. Sundquist, *Politics and Policy,* p. 346, emphasis added.

15. John F. Kennedy, Special Message to Congress on Natural Resources, February, 23, 1961, *Public Papers of the Presidents* (Washington, D.C.: GPO, 1961), p. 115.

16. John Kingdon, *Agendas, Alternatives, and Public Policies* (Boston: Little, Brown, 1984), p. 25.

17. James Reston, quoted in *The National Purpose,* ed. Henry R. Luce (New York: Holt, Rinehart & Winston, 1960), p. 112.

18. Sundquist, *Politics and Policy,* p. 345.

19. The FPCRB replaced the Interdepartmental Committee on Pest Control (ICPC), which had been little more than an *ad hoc* information clearinghouse.

20. John Blodgett, "Federal Pesticides Policy," master's thesis, Case Western Reserve University, 1969, p. 94.

21. Frank Graham, Jr., *Since Silent Spring* (Boston: Houghton Mifflin, 1970), p. 188.

22. For more detailed treatment of Carson's thesis and the ensuing controversy, see ibid., and Blodgett, "Federal Pesticides Policy."

23. Thomas Dunlap, *DDT: Scientists, Citizens, and Public Policy* (Princeton, N.J.: Princeton University Press, 1981), p. 103.

24. "Rachel Carson's Warning," *New York Times,* July 2, 1962, p. 28.

25. *Washington Post,* July 13, 1962, p. 17.

26. *Wall Street Journal,* August 3, 1962, p. 1.

27. Graham, *Since Silent Spring,* p. 49.

28. "USDA Comments on Rachel Carson's Articles in *The New Yorker,"* mimeographed press release, U.S. Department of Agriculture, August 1962, p. 1.

29. Ibid., p. 2.

30. John M. Lee, "Silent Spring is Now Noisy Summer," *New York Times,* July 22, 1962, sec. 3, p. 1.

31. Ibid., sec. 3, p. 11.

32. As cited in Graham, *Since Silent Spring,* p. 39.

33. William J. Darby, "Silence, Miss Carson," *Chemical and Engineering News,* October 1, 1962, p. 60.

34. Ibid., pp. 62–63.

35. Blodgett, "Federal Pesticides Policy," p. 76.

36. *New York Times,* September 22, 1962, p. 28.

37. Graham, *Since Silent Spring,* p. 49.

38. Ibid., emphasis added.

39. *New York Times,* September 22, 1962, p. 35.

40. *Audubon* 64, no. 6 (November 1962), 306.

41. Graham, *Since Silent Spring,* p. 72.

42. *Audubon* 65, no. 1 (January 1963), 24.

43. *Audubon* 66, no. 1 (January 1964), 38.

44. *New York Times,* April 3, 1963, p. 95.

45. CBS Reports, "The Silent Spring of Rachel Carson," transcript (New York: Columbia Broadcasting System, 1963), p. 23.

46. *New York Times,* April 4, 1963, p. 95.

47. Kingdon, *Agendas, Alternatives, and Public Policies,* p. 3.

48. See Anthony Downs, "Up and Down with Ecology: The Issue-Attention Cycle," *Public Interest* 28 (Summer 1972).

49. President's Science Advisory Council, *Use of Pesticides* (Washington, D.C.: The White House, May 15, 1963).

50. *New York Times,* November 10, 1963, p. 50.

51. *New York Times,* May 16, 1963, p. 1.

52. John Blodgett, "Pesticides: Regulation of an Evolving Technology," in *The Legislation of Product Safety,* ed. Samuel S. Epstein and Richard D. Grady (Cambridge, Mass.: MIT Press, 1974), 3:216.

53. Graham, *Since Silent Spring,* p. 77.

54. Dunlap, *DDT,* p. 115.

55. *Pest Control and Wildlife Relationships,* NAS-NRC Report 920A, B, and C (Washington: National Academy of Sciences, 1962–63).

56. Frank Engler, *Atlantic Naturalist,* October 1962, p. 23.
57. *Audubon* 64, no. 6 (December 1962), 358.
58. Graham, *Since Silent Spring,* p. 45.
59. Blodgett, "Pesticides: Regulation of an Evolving Technology," p. 224.
60. Thomas Kuhn, *The Structure of Scientific Revolution,* 2d ed. (Chicago: University of Chicago Press, 1970).
61. Blodgett, "Pesticides: Regulation of an Evolving Technology," p. 225.
62. The USDA long was unhappy about protest registration, which theoretically allowed chemical makers to market products without Department approval. Only 23 out of some 50,000 products were so registered between 1947 and 1963, but critics saw the loophole as affirmation that the federal tiger had no teeth. See U.S. Senate, Committee on Government Operations, Subcommittee on Reorganization and International Organizations, *Interagency Environmental Hazards Coordination,* 88th Cong., 1st sess., 1963, p. 15.
63. Ibid., p. 54.
64. Ibid., p. 76.
65. U.S. House of Representatives, Committee on Appropriations, Subcommittee on Agriculture, *USDA Appropriations FY1965,* 88th Cong., 2d sess., 1964, p. 437.
66. *Congressional Record,* June 18, 1964, p. 14248.
67. Graham, *Since Silent Spring,* p. 107.
68. *Congressional Record,* June 18, 1964, p. 14248.
69. U.S. Senate, Committee on Government Operations, Subcommittee on Reorganization and International Organizations, *Interagency Environmental Hazards Coordination,* pt. 6, 89th Cong., 2d sess., 1966, pp. 48–49.
70. Ibid., p. 34.
71. Ibid., p. 37.
72. U.S. House of Representatives, Committee on Appropriations, Subcommittee on Agriculture, *Department of Agriculture FY1965,* 88th Cong., 2d sess., 1964, p. 141.
73. *Congressional Record,* June 22, 1964, p. 14548.
74. *Congressional Record,* September 1, 1964, p. 21179.
75. Ibid., p. 21182.
76. *Congressional Quarterly Almanac 1968* (Washington, D.C.: Congressional Quarterly Press, 1969), p. 248.
77. *Audubon* 69, no. 5 (October 1967), 5.
78. For a fuller treatment of this case and the overall question of the tax code's impact on political action, see Robert L. Holbert, *Tax Laws and Political Access: The Bias of Pluralism Revisited* (Beverly Hills, Calif.: Sage Publications, 1975); this quote from p. 39.
79. *New York Times,* November 20, 1970, p. 40.
80. Dunlap, *DDT,* p. 143.
81. William A. Butler, speech before the Conference on the Twentieth Anniversary of *Silent Spring,* transcript, October 1962, p. 2.
82. See Graham, *Since Silent Spring;* Dunlap, *DDT.*
83. *Audubon* 69, no. 3 (May 1967), 30.
84. Dunlap, *DDT,* p. 146.
85. Ibid., p. 158.
86. See Dunlap, *DDT,* for particulars.

87. Ibid., pp. 197–98.

88. See, for example, Richard A. Liroff, *A National Policy For the Environment: NEPA and its Aftermath* (Bloomington, Ind.: Indiana University Press, 1976), and Frederick R. Anderson, *NEPA and the Courts* (Baltimore: Johns Hopkins University Press, 1973).

89. U.S. General Accounting Office, *Need to Improve Regulatory Enforcement Procedures Involving Pesticides,* September 10, 1968, p. 5.

90. U.S. House of Representatives, Committee on Government Operations, Subcommittee on Intergovernmental Relations, *Deficiencies in the Administration of FIFRA,* 91st Cong., 1st sess., 1969, p. 14.

91. William Rodgers, "The Persistent Problem of Persistent Pesticides: A Lesson in Environmental Law," in *Environmental Law Review 1971,* ed. H. Floyd Sherrod (Albany, N.Y.: Sage Hill Publishers, 1971), p. 344.

92. *Audubon* 71, no. 4 (July 1969), p. 183.

93. *National Journal,* November 1, 1969, p. 31.

94. U.S. Department of Health, Education and Welfare, *Report of the Secretary's Commission on Pesticides and Their Relationships to Environmental Health,* 1969, p. 5.

95. Ibid., p. 23.

96. Ibid., p. 7.

97. Blodgett, "Pesticides: Regulation of an Evolving Technology," p. 236.

98. *New York Times,* November 13, 1969, p. 1.

99. *Audubon* 72, no. 1 (January 1970), 116.

100. *New York Times,* November 14, 1969, p. 46.

101. Rodgers, "The Persistent Problem," p. 353.

Chapter Seven. Environmentalism, Elite Competition, and Policy Change

1. Hazel Erskine, "The Polls: Pollution and Its Costs," *Public Opinion Quarterly* (Spring 1972), p. 120.

2. V. O. Key, *Public Opinion and American Democracy* (New York: Knopf, 1961), pp. 285–86.

3. Gallup International, *The Gallup Poll Index,* report no. 60 (June 1970), p. 8.

4. Amitai Etzioni, *Demonstration Democracy* (New York: Gordon and Beach, 1970), p. 1.

5. John Kingdon, *Agendas, Alternatives, and Public Policies* (Boston: Little, Brown, 1984), p. 70.

6. Gaylord Nelson, interview with author, April 1983.

7. Benjamin Page and Robert Shapiro, "Effects of Public Opinion on Policy," presented at the annual meeting of the American Political Science Association, New York, September 1981, p. 29.

8. Ibid., p. 11.

9. Charles O. Jones, *Clean Air: The Policies and Politics of Pollution Control* (Pittsburgh, Pa.: University of Pittsburgh Press, 1975), p. 207.

10. Charles O. Jones, "Speculative Augmentation in Federal Air and Water Pollution Policy-Making," *Journal of Politics* 36, no. 2 (1972), 454.

11. See David Truman, *The Governmental Process* (New York: Knopf, 1951).

12. *National Journal,* August 8, 1970, p. 1718.

13. *National Journal*, July 24, 1971, p. 1557.

14. *National Journal*, February, 19, 1972, p. 321.

15. *Audubon* 72, no. 1 (January 1970), p. 118; ibid. 73, no. 1 (January 1971), p. 98; ibid. 75, no. 1 (January 1973), p. 115.

16. Page and Shapiro, "Effects of Public Opinion on Policy," p. 26.

17. Jones, *Clean Air*, p. 176.

18. David Vogel, "The Power of Business in America: A Re-appraisal," *British Journal of Political Science* 13 (January 1981), 19.

19. Ford Foundation, *1979 Annual Report* (New York: Ford Foundation, 1979), p. 15.

20. See also Richard A. Liroff, *A National Policy For the Environment: NEPA and its Aftermath* (Bloomington: Indiana University Press, 1976).

21. See Robert L. Holbert, *Tax Laws and Political Access: The Bias of Pluralism Revisited* (Beverly Hills, Calif.: Sage Publications, 1975).

22. Kingdon, *Agendas, Alternatives, and Public Policies*, p. 160.

23. See, for example, Roger H. Davidson and Walter S. Oleszek, "Adaptation and Change: Structural Innovation in The U.S. House of Representatives," *Legislative Studies Quarterly* 1 (January 1976), 37–65.

24. For a cogent discussion of postwar tenure trends, see Nelson W. Polsby, "The Institutionalization of the U.S. House of Representatives," *American Political Science Review* 62 (March 1968), 144–68.

25. See, for example, Davidson and Oleszek, "Adaptation and Change"; and U.S. House of Representatives, Select Committee on Committees, *Overview of Principal Legislative Developments of the 1970s*, H. Rept. 96–866, 96th Cong., 2d sess., 1980.

26. Liroff, *A National Policy for the Environment*, p. 33.

27. *Scenic Hudson Preservation Conference* v. *Federal Power Commission*, 354 F.2d 608 (2d Cir. 1965); *Association of Data Processing Organizations, Inc*. v. *Camp*, 397 U.S. 150 (1970) and *Barlow* v. *Collins*, 397 U.S. 159 (1970).

28. R. Shep Melnick, *Regulation and the Courts: The Case of the Clean Air Act* (Washington, D.C.: Brookings Institution, 1983), p. 10.

29. Ibid.

30. *Calvert Cliffs Coordinating Committee et al*. v. *U.S. Atomic Energy Commission et al.*, 449 F.2d 1109 (1971).

31. *National Journal*, September 18, 1971, p. 1972.

32. Melnick, *Regulation and the Courts*, p. 4.

33. E. E. Schattschneider, *The Semi-Sovereign People: A Realist's View of Democracy in America* (Hindsdale, Ill.: Dryden Press, 1960), p. 17.

34. Daniel A. Mazmanian and Jeanne Nienaber, *Can Organizations Change?* (Washington, D.C.: Brookings Institution, 1977), p. 2.

35. Hugh Heclo, *A Government of Strangers* (Washington, D.C.: Brookings Institution, 1977), p. 6.

36. See, for example, ibid.; and Richard Nathan, *The Plot That Failed: Nixon and the Administrative Presidency* (New York: John Wiley and Sons, 1975).

37. William Safire, "The Plot that Failed," *The New York Times*, May 1, 1975, p. 18.

38. U.S. House of Representatives, Committee on Agriculture, *Hearings on the Federal Environmental Pesticides Control Act of 1971*, 92nd Cong., 1st sess., 1971, p. 736.

39. Interview with author, April 1983.
40. Harrison Wellford, *Sowing the Wind* (New York: Grossman, 1972), p. 336.
41. *Hearings on the Federal Environmental Pesticides Control Act of 1971,* p. 201.
42. Ibid., p. 300.
43. Ibid., p. 2.
44. Ibid., p. 270.
45. Ibid., p. 246.
46. Ibid., p. 248.
47. U.S. House of Representatives, Committee on Agriculture, *FEPCA of 1971,* H. Rept. 92-511, 92nd Cong., 1st. sess., 1971.
48. Ibid., p. 15.
49. *Hearings on the Federal Environmental Pesticides Control Act of 1971,* p. 313.
50. Ibid., p. 75.
51. *National Journal,* November 20, 1971, p. 2322.
52. Ibid., p. 2323.
53. *Congressional Record,* November 8, 1971, p. 39976.
54. *FEPCA of 1971,* H. Rept. 92-511, p. 2.
55. *Congressional Record,* November 9, 1971, p. 40035.
56. Ibid., p. 40036.
57. Randall B. Ripley and Grace A. Franklin, *Congress, The Bureaucracy, and Public Policy,* 3d ed. (Homewood, Ill.: Dorsey Press, 1984), p. 135.
58. Richard F. Fenno, *Congressmen in Committees* (Boston: Little, Brown, 1973), pp. 190–91.
59. Charles O. Jones, *The United States Congress: People, Place, and Policy* (Homewood, Ill.: Dorsey Press, 1982), p. 184.
60. *National Journal,* March 25, 1972, p. 531.
61. *Congressional Quarterly Weekly Report,* May 13, 1972, p. 1063.
62. Ibid., p. 1065.
63. *National Journal,* September 2, 1972, p. 1401.
64. *Washington Post,* June 7, 1972, p. 3.
65. U.S. Senate, Committees on Agriculture and Commerce, *FEPCA of 1972,* S. Rept. 92-970, 92nd Cong., 2d sess., 1972.
66. *National Journal,* September 2, 1972, p. 1395.
67. Ibid., p. 1397.
68. *Congressional Quarterly Weekly Report,* October 14, 1973, p. 2638.
69. Ibid., p. 2637.
70. U.S. House of Representatives, H.Rept. 92-1540, 92nd Cong., 2d sess., 1972.
71. *Congressional Record,* October 5, 1972, p. 33921.
72. *Congressional Quarterly Weekly Report,* October 14, 1972, p. 2638.
73. Ibid., p. 2637.
74. *Congressional Record,* October 5, 1972, p. 33923.
75. Ibid., p. 33924.
76. Ibid.
77. John Blodgett, "Pesticides: Regulation of an Evolving Technology," in *The Legislation of Product Safety,* ed. Samuel S. Epstein and Richard D. Grady (Cambridge, Mass.: MIT Press, 1974), 3:265.

78. We will continue to refer to the law as FIFRA, both for the sake of convenience and because the 1972 law essentially was a major rewrite of the 1947 statute.

Chapter Eight. The Policy Pendulum: Pesticides Politics into the Eighties

1. E. E. Schattschneider, *The Semi-Sovereign People: A Realist's View of Democracy in America* (Hinsdale, Ill.: Dryden Press, 1960), p. 65.

2. Anthony Downs, "Up and Down with Ecology: The Issue-Attention Cycle," *Public Interest* 28 (Summer 1972), 39.

3. See, for example, Charles O. Jones, "American Politics and the Organization of Energy Decision Making," *Annual Review of Energy* 4 (1979), 99–110.

4. Downs, "Up and Down with Ecology," p. 40.

5. Eugene Bardach, *The Implementation Game* (Cambridge, Mass.: MIT Press, 1977), p. 3.

6. Jeffrey L. Pressman and Aaron Wildavsky, *Implementation,* 2d ed. (Berkeley and Los Angeles: University of California Press, 1979), p. xxi, emphasis added.

7. George C. Edwards, *Implementing Public Policy* (Washington, D.C.: Congressional Quarterly Press, 1980), p. 17, emphasis added.

8. Theodore Lowi, *The End of Liberalism,* 2d ed. (New York: W.W. Norton, 1979), p. 106.

9. Richard R. Nelson, *The Moon and the Ghetto: An Essay on Public Policy Analysis* (New York: W.W. Norton, 1977), p. 14.

10. Pressman and Wildavsky, *Implementation,* p. 113.

11. Interview with author, April 1983. Subsequent interviews are not footnoted.

12. Charles O. Jones, *Clean Air: The Politics and Policies of Pollution Control* (Pittsburgh, Pa.: University of Pittsburgh Press, 1975), p. 272.

13. Ibid.

14. Ibid.

15. William J. Keefe, *Congress and the American People* (Englewood Cliffs, N.J.: Prentice-Hall, 1980), p. 157.

16. *Chicago Tribune,* August 25, 1978, p. 4.

17. U.S. Senate, Committee on Environment and Public Works, *Environmental Protection Affairs of the 94th Congress,* 95th Cong., 1st sess., 1977, p. 261.

18. *Congressional Quarterly Almanac 1975* (Washington, D.C.: Congressional Quarterly Press, 1976), p. 203.

19. *Environmental Protection Affairs of the 94th Congress,* p. 286.

20. Thomas McGarity, "Registration of Pesticides in the Environmental Protection Agency," presented to the Conference on Pesticides, Harper's Ferry, Va., January 1983, p. 8.

21. *Environmental Defense Fund* v. *Costle,* 631 F. 2d 922, 932-37 (1980), cert. denied, 449 U.S. 1112 (1981).

22. See, for example, Robert A. Scott and Donald R. Shore, *Why Sociology Does Not Apply: A Study of the Use of Sociology in Public Policy* (New York: Elsevier, 1979); Laurence E. Lynn, ed., *Knowledge and Social Policy* (Washington, D.C.: National Academy of Sciences, 1978); and James S. Coleman, *Policy Research in the Social Sciences* (Norristown, N.J.: General Learning Press, 1972).

23. Life Sciences closed in 1976, and Allied Chemical was fined $13 million for various health and environmental violations. The corporation also set up an $8 million fund for Kepone research and to pay for employee medical expenses. EPA, in a RPAR "rush job," announced its intent to cancel the chemical, to which Allied agreed in 1977. See U.S. House of Representatives, Committee on Agriculture, *Kepone Contamination,* 94th Cong., 2d sess., 1976.

24. U.S. Senate, Committee on the Judiciary, Subcommittee on Administrative Practices and Procedures, *The Environmental Protection Agency and the Regulation of Pesticides,* 94th Cong., 2d sess., 1976, p. 1.

25. Ibid., p. 4.

26. Francine Schulberg, "The Proposed FIFRA Amendments of 1977," *Harvard Environmental Law Review* 2 (1977), 348.

27. U.S. House of Representatives, Committee on Agriculture, *Amendments to FIFRA,* H. Rept. 95-663, 95th Cong., 1st sess., 1977, p. 57.

28. Shirly Briggs, *Federal Regulation of Pesticides, 1910–1981* (Washington, D.C.: Rachel Carson Council, 1982), p. 11.

29. *Congressional Quarterly Weekly Report,* October 29, 1977, p. 2289.

30. *Congressional Quarterly Almanac 1980* (Washington, D.C.: Congressional Quarterly Press, 1981), p. 610.

Chapter Nine. The Endless Pesticides Campaign

1. Samuel P. Huntington, *American Politics: The Promise of Disharmony* (Cambridge, Mass.: Harvard University Press,, 1981), p. 216.

2. See, for example, Helen M. Ingram and Dean E. Mann, eds., *Why Policies Succeed or Fail* (Beverly Hills, Calif.: Sage Publications, 1980); Aaron Wildavsky, *Speaking Truth to Power: The Art and Craft of Policy Analysis* (Boston: Little, Brown, 1979); Jeffrey L. Pressman and Aaron Wildavsky, *Implementation,* 2d ed. (Berkeley and Los Angeles: University of California Press, 1979); and, Theodore Lowi, *The End of Liberalism,* 2d ed. (New York: W.W. Norton, 1979).

3. Council on Environmental Quality, *Report of the Council on Environmental Quality 1981* (Washington, D.C.: GPO, 1982), p. 17; the CEQ itself has been cut to almost nothing during the 1980s, and its role in the Reagan administration is negligible.

4. See, for example, Charles Lindblom, *Politics and Markets* (New York: Basic Books, 1977).

5. Quoted in *From Outrage to Action: The Story of the National Audubon Society* (New York: The National Audubon Society, 1982), p. 17.

6. As printed in *Congressional Quarterly Weekly Report,* July 19, 1980, p. 2056.

7. *Washington Post,* March 11, 1983, p. 1.

8. See, for example, Norman J. Vig and Michael E. Kraft, *Environmental Policy in the 1980s: Reagan's New Agenda* (Washington, D.C.: Congressional Quarterly Press, 1983).

9. Ruckelshaus subsequently resigned in early 1985.

10. John Todhunter, "Regulatory Changes at EPA," presented at the annual meeting of the Arkansas Agricultural Pesticides Association, Hot Springs, Arkansas, February 1982, p. 8, emphasis in original transcript.

11. Interview with author, April 1983; subsequent interview excerpts are not footnoted.

12. *Washington Post,* January 11, 1983, p. 13.

13. U.S. House of Representatives, Committee on Agriculture, Subcommittee on Department Operations, Research, and Foreign Agriculture, *EPA Pesticide Regulatory Program Study,* 97th Cong., 2d sess., 1982, p. 91.

14. Ibid., p. 115.

15. *Washington Post,* September 22, 1984, p. 12.

16. Mary Etta Cook and Roger H. Davidson, "Deferral Politics: Congressional Decision-Making on Environmental Issues in the 1980s," in *Public Policy and the Natural Environment,* ed. Helen M. Ingram and R. Kenneth Godwin (Greenwich, Conn: JAI Press, 1985), p. 47.

17. *Public Interest Profiles: National Audubon Society* (Washington, D.C.: Foundation for Public Affairs, 1983), p. F75.

18. *Congressional Quarterly Weekly Report,* February 20, 1982, p. 337.

19. *Congressional Quarterly Weekly Report,* June 15, 1982, p. 1109.

20. Ibid.

21. *Congressional Quarterly Weekly Report,* August 14, 1982, pp. 1983–84.

22. The EPA subsequently initiated cancellation proceedings against toxaphene, and formally banned the pesticide for most uses in November 1982, though exhaustion of existing stocks would be allowed. Environmentalists called the action electoral posturing, while EPA assistant administrator Todhunter accused Congress of muddying the regulatory waters. Industry officials took the ban in stride, arguing that toxaphene use was on the decline anyway. See the *Washington Post,* October 19, 1982, p. 17.

23. *Congressional Quarterly Weekly Report,* September 18, 1982, p. 2312.

24. Testimony before U.S. House of Representatives, Committee on Agriculture, Subcommittee on Department Organization, Research, and Foreign Research, *Amendments to FIFRA,* 98th Cong., 1st sess., 1983, mimeograph, p. 1.

25. *Congressional Quarterly Weekly Report,* May 7, 1983, p. 892.

26. *Washington Post,* May 4, 1983, p. 2.

27. *Washington Post,* April 4, 1984.

28. *Congressional Quarterly Weekly Report,* March 11, 1986, p. 620.

29. Ibid.

30. *New York Times,* March 20, 1986, p. 21.

31. *Washington Post,* May 24, 1985, p. 25.

32. Ibid.

33. Ibid.

34. Press release from the National Agricultural Chemicals Association, September 11, 1985, pp. 7–8.

35. *Congressional Record,* September 11, 1985, p. E3961.

36. *Chemical Marketing Reporter,* March 17, 1986, p. 11.

37. *Chemical Marketing Reporter,* April 7, 1986, p. 7.

38. *Congressional Quarterly Weekly Report,* May 17, 1986, p. 1128.

39. *Congressional Quarterly Weekly Report,* June 21, 1986, p. 1408.

40. Earl Latham, "The Group Basis of Politics," *American Political Science Review* 46 (June 1952), 390.

41. E. E. Schattschneider, *The Semi-Sovereign People* (Hinsdale, Ill.: Dryden Press, 1960), p. 38.

42. Kay Lehman Schlozman and John T. Tierney, *Organized Interests and American Democracy* (New York: Harper and Row, 1986), p. 307.

Chapter Ten. The Pesticides Perspective Revisited

1. *Washington Post,* April 17, 1984, p. 2.

2. To $215 billion. See *USA Today,* January 24, 1985, p. 1; see also the *1986 Statistical Abstract of the United States* (Washington, D.C.: GPO, 1986), p. 642.

3. *USA Today,* January 24, 1985, p. 1.

4. *Washington Post,* April 17, 1984, p. 2.

5. David Nyhan, "Reagan's Ignorant Policy and the Debacle Down on the Farm," *Boston Globe,* August 7, 1986, p. 19.

6. *Washington Post,* April 17, 1984, p. 2.

7. *Boston Globe,* March 23, 1986, p. 12.

8. *USA Today,* January 28, 1985, p. 1.

9. *1986 Statistical Abstract of the United States* (Washington, D.C.: GPO, 1986), p. 206.

10. *New York Times,* May 25, 1982, p. C1.

11. *Washington Post,* February 14, 1984, p. 8.

12. *Pittsburgh Press,* February 5, 1984, p. B1.

13. Ibid.

14. *Business Week,* February 20, 1984, p. 38.

15. *Pittsburgh Press,* February 14, 1984, p. B1.

16. John Dewey, *The Public and Its Problems* (New York: Holt, Rinehart & Winston, 1927), p. 126.

17. Hugh Heclo, "Issue Networks and the Executive Establishment," in *The New American Political System,* ed. Anthony King (Washington, D.C.: American Enterprise Institute, 1979), pp. 118–19.

18. Dorothy Nelkin and Michael Pollack, *The Atom Besieged: Antinuclear Movements in France and Germany* (Cambridge, Mass.: MIT Press, 1982), p. 100.

19. Burdett A. Loomis and Allan J. Cigler, "The Changing Nature of Interest Group Politics," in *Interest Group Politics,* ed. Allan J. Cigler and Burdett A. Loomis (Washington, D.C.: Congressional Quarterly Press, 1983), p. 1.

20. Roger H. Davidson, "Subcommittee Government: New Channels for Policy Making," in *The New Congress,* ed. Thomas E. Mann and Norman J. Ornstein (Washington, D.C.: American Enterprise Institute, 1981), p. 130.

21. Steven Kelman, *Regulating America, Regulating Sweden: A Comparative Study of Occupational Safety and Health Policy* (Cambridge, Mass.: MIT Press, 1981), p. 166.

22. Lawrence C. Dodd and Richard L. Schott, *Congress and the Administrative State* (New York: John Wiley, 1979), p. 176.

23. Roger Davidson, "Subcommittee Government," p. 130.

24. Quoted in Gregg Easterbrook, "What's Wrong with Congress?" *Atlantic* (December 1984), p. 72.

25. Dodd and Schott, *Congress and the Administrative State,* p. 239.

26. Kelman, *Regulating America, Regulating Sweden,* p. 111.

27. Dodd and Schott, *Congress and the Administrative State,* p. 173.

28. R. Shep Melnick, *Regulation and the Courts: The Case of the Clean Air Act* (Washington, D.C.: Brookings Institution, 1983), p. 373.

29. Ibid., pp. 373–74.

30. Hugh Heclo, *A Government of Strangers* (Washington, D.C.: Brookings Institution, 1977), pp. 224–25.

31. Joel D. Aberbach et al., *Bureaucrats and Politicians in Western Democracies* (Cambridge, Mass.: Harvard University Press, 1981), p. 251.

32. Heclo, "Issue Networks," p. 118.

33. Alfred Marcus, *Promise and Performance: Choosing and Implementing an Environmental Policy* (Westport, Conn.: Greenwood Press, 1980), p. 160.

34. E. E. Schattschneider, *The Semi-Sovereign People: A Realist's View of Democracy in America* (Hinsdale, Ill.: The Dryden Press, 1960), p. 3.

35. Ibid., p. 141.

36. John Kingdon, *Agendas, Alternatives, and Public Policies* (Boston: Little, Brown, 1984), p. 25.

37. Kenneth M. Dolbeare, *Democracy at Risk: The Politics of Economic Renewal* (New York: Chatham House, 1985), p. 3.

38. Schattschneider, *The Semi-Sovereign People,* p. 17.

Index

Pitt Series in Policy and Institutional Studies

Bert A. Rockman, Editor

Agency Merger and Bureaucratic Redesign
Karen M. Hult

The Aging: A Guide to Public Policy
Bennett M. Rich and Martha Baum

Clean Air: The Policies and Politics of Pollution Control
Charles O. Jones

Comparative Social Systems: Essays on Politics and Economics
Carmelo Mesa-Lago and Carl Beck, Editors

Congress Oversees the Bureaucracy: Studies in Legislative Supervision
Morris S. Ogul

Foreign Policy Motivation: A General Theory and a Case Study
Richard W. Cottam

Homeward Bound: Explaining Changes in Congressional Behavior
Glenn Parker

Japanese Prefectures and Policymaking
Steven R. Reed

Managing the Presidency: Carter, Reagan, and the Search for Executive Harmony
Colin Campbell, S.J.

Perceptions and Behavior in Soviet Foreign Policy
Richard K. Herrmann

Pesticides and Politics: The Life Cycle of a Public Issue
Christopher J. Bosso

Political Leadership: A Source Book
Barbara Kellerman

The Politics of Public Utility Regulation
William T. Gormley, Jr.

The Presidency and Public Policy Making
George C. Edwards III, Steven A. Shull, and Norman C. Thomas, Editors

Public Policy in Latin America: A Comparative Survey
John W. Sloan

Roads to Reason: Transportation, Administration, and Rationality in Colombia
Richard E. Hartwig

The Struggle for Social Security, 1900–1935
Roy Lubove